Fire Retardancy of Polymers

The Use of Intumescence

Fire Retardancy of Polymers
The Use of Intumescence

Edited by

M. Le Bras
LCAPCS, Ecole Nationale Supérieure de Chimie de Lille, France

G. Camino
Department of Chemistry, University of Torino, Italy

S. Bourbigot
LCAPCS, Ecole Nationale Supérieure de Chimie de Lille, France

R. Delobel
CREPIM, Bruay la Bussière, France

THE ROYAL
SOCIETY OF
CHEMISTRY
Information
Services

Papers presented at the 6th European Meeting on Fire Retardancy of Polymeric Materials (FRPM'97) organised jointly by the 'Laboratoire de Chimie Analytique et de Physico-Chimie des Solides' de l'E.N.S.C. de Lille and the 'Centre de Recherche et d'Etude des Procédés d'Ignifugation des Matériaux' at the University of Lille (France) on 24–26th September 1997.

Special Publication No. 224

ISBN 0-85404-738-7

A catalogue record for this book is available from the British Library

Published by The Royal Society of Chemistry,
Thomas Graham House, Science Park, Milton Road,
Cambridge CB4 4WF, UK

For further information see our web site at www.rsc.org

Printed by Athenaeum Press Ltd, Gateshead, Tyne and Wear, UK

Preface

An exponential development such as that of synthetic polymers in the last forty years has never previously happened in the history of mankind's use of materials. Indeed, it has been proposed that this Century be named the 'Plastic Age'.

Polymer chemists and engineers can now design and produce synthetic materials of an almost unlimited variety of properties which strongly affect our technological and socio-economic progress. However, in many fields such as electrical, electronic, transport, building, etc., the use of polymers is restricted by their flammability, whatever the importance of the advantages their use may bring. Polymers may indeed initiate or propagate fires because, being organic compounds, they decompose to volatile combustible products when they are exposed to heat.

Attempts to prevent polymer materials from burning are already reported in the earliest historical records available, that is Roman war chronicles in which wood treatment with inorganic compounds is described. Similar so called 'fire retardant' or 'flame retardant' treatments of natural polymer materials and textiles have been reported in the literature throughout the centuries.

The present diffusion of synthetic polymers has greatly increased the 'fire risk' and the 'fire hazard', i.e. respectively the probability of fire occurrence and its consequence either on humans or on structures. Thus, there is now an increasing demand for fire retardant systems capable of reducing both these risks whereas in earlier developments the reduction of fire risk was the major target. Moreover, the 'sustainable development' concept applied to this field implies that fire retardants should involve a low impact on health and environment during the entire life cycle including recycling and disposal.

For all these reasons , the versatile and effective class of halogen-based fire retardants is somewhat unsatisfactory. Indeed, while they perform a fire retardant action, they may increase optical density, corrosiveness and toxicity of smoke produced in fires, which have been shown to be major hazards in the experience acquired in fires occurring in the past decades. Furthermore, super-toxic products were detected when some aromatic halogenated fire retardants were exposed to heat. Thus, the halogen-based fire retardant can effectively reduce fire risk but may increase fire hazard as compared to the non-retarded material.

Among alternatives currently actively investigated, the 'intumescent' systems which undergo charring and foaming upon thermal degradation, forming a blown protective cellular char, are considered a particularly promising environmentally friendly approach. The first fire retardant system likely to show intumescent behaviour was reported in the literature by Gay Lussac for cellulosic fibres [Gay Lussac, J. L., *Annales de Chimie*, **18(2)** (1821) 211] and intumescent coatings for fire protection of wood or metal structures from the effect of fire have been in industrial use for about fifty years now. However, the regular use of additives to induce the intumescent behaviour in polymer materials is more recent. In particular, mechanistic studies on the process of intumescence, which are indispensable to develop effective fire retardants, have been carried out only in the last twenty five years.

These studies are particularly difficult because of the superposition of several chemical and physical processes concurring in the intumescent effect. A consistent progress in the understanding of the mechanisms involved was due to the relatively recent availability of new effective techniques for the characterisation of the molecular structure of intractable solids such as charred intumescent materials.

The evolution of this area of fire retardancy is well documented by the increasing number of papers concerning intumescence presented at the series of European Conferences on fire retardant mechanisms, materials and environmental-regulation aspects, regularly organised in Europe in the last fifteen years. It was then thought useful to collect these papers in this volume which is the first attempt ever to appear in the literature to supply a comprehensive updated view of this field. I hope that this effort may produce a useful tool for further research and development work in this field and I wish to thank the organisers of the Lille Conference and, in particular, Dr. Michel Le Bras for having made this possible.

Giovanni Camino

At the last edition of the European Meeting on Fire Retardancy of Polymeric Materials (FRPM'97) held in Lille, France, in 1997, a large number of papers dealt with the intumescence fire retardant approach (33 of the 96 contributions). They covered the most recent studies on intumescence mechanisms, new intumescent polymeric materials or textiles and comparative ecological reviews.

This book presents the reviews and the original work presented at FRPM'97, to which the Editors have added three invited reviews.

In all, twenty seven Research Teams (Academic Laboratories, Institutes or Industrial Groups) have contributed to this book and proposed different chemical, chemical engineering, or physical approaches of intumescence and some industrial applications. A comparatively short last section asks researchers the question: 'Is intumescence of polymeric materials an environmentally friendly process?'.

The location of papers in the sections of the book has not been easy. In practice, works dealing with mechanisms present new formulations and, of course, works on new materials propose a mechanism discussion as well.

Michel Le Bras

Acknowledgements

The Editors would like to thank:

Professor C. Cazé (*E.N.S.A.I.T., Roubaix, France*), Dr J. Gilman (*N.I.S.T., Gaithersburg, U.S.A.*), Professor A. R. Horrocks (*Bolton Institute of Higher Education, U.K.*), Dr B. K. Kandola (*Bolton Institute of Higher Education, U.K.*), Dr T. Kashiwagi (*N.I.S.T., Gaithersburg, U.S.A.*), Dr S. Levchik (*Byelorussian State University, Minsk, Belarus*), Dr B. Mortaigne (*D.G.A./C.R.E.A., Arcueil, France*), Professor G. Montaudo (*Dipartimento di Chimica, Catania, Italy*), Professor E. M. Pearce (*Polytechnic University of New York, U.S.A.*), Dr D. Price (*University of Salford, U.K.*), Professor J. Wang (*Beijing Institute of Technology, China*) and Professor E. D. Weil (*Polytechnic University of New York, U.S.A.*),

for their invaluable help with refereeing the papers included in this Volume.

Contents

New Intumescent Polymeric Materials

Flame Retarded Intumescent Textiles

Intumescence- an Environmentally Friendly Process?

Abbreviations

Polymers and products

ABS	acrylonitrile/butadiene/styrene
BPFC-DMS	9,9 bis-(4-hydroxyphenyl)fluorene/carbonate-poly(dimethylsiloxane) block polymer
CFR	carbamide-formaldehyde resin
CTBN	carboxy-terminated butadiene acrylonitrile
EBM	ethylene/butyl acrylate/ maleic anhydride terpolymer
ECPE	extended-chain polyethylene
EPDM	methylene propylene diene copolymer
ER	epoxy resin
ETFE	ethylene/tetrafluoroethylene
EVAx	ethylene/vinyl acetate (x wt. %)copolymer
HDPE	high density polyethylene
HIPS	high impact polystyrene
Kevlar	poly p-phenylene terephthalamide
LDPE	low density polyethylene
LRAM3.5	ethylene/butyl acrylate/ maleic anhydride terpolymer
MA, MEL or ML	melamine
MA-CY	melamine cyanurate
MA-PR	melamine pyrophosphate
Nylon 6	polyamide-6
PA-4.6	polyamide-4.6
PA-6	polyamide-6
PAN	poly(acrylonitrile)
PAS	poly(arylsulfone)
PBO	poly(p-phenylene-benzobisoxazole)
PC	polycarbonate
PCA	polycaproamide
PCBs	polychlorinated biphenyls
PCDDs	polychlorinated dibenzodioxins
PCDFs	polychlorinated dibenzofurans
PE	polyethylene
PEEK	poly(etheretherketone)

PES	poly(ethersulfone)
PET	poly(ethyleneterephthalate)
PEUFA	poly(ethyleneureaformaldehyde)
PFA	perfluoroalkoxy modified tetrafluoroethylene
PFR	phenol-formaldehyde resins
PFS	poly(phenylenesulfide)
PhFR	phenol-formaldehyde resins
PMA	poly(methylacrylate)
PMFIA	meta-phenylenediamine
PMMA	poly(methylmetacrylate)
PP	polypropylene
PP-g-MA	polypropylene-graft-maleic anhydride
PP-co-PE	ethylene/propylene copolymers
PPO	poly(phenyleneoxide)
PPS	poly(phenylenesulfide)
PS	polystyrene
PVA	poly(vinylalcohol)
PVB	poly(vinylbromide)
PVC	poly(vinylchloride)
PVC_2	poly(vinylidenechloride)
SAN	styrene/acrylonitrile copolymer
SBR	styrene/butadiene rubber
SMA	styrene/maleic anhydride copolymer
TAC	triallyl cyanurate
TCDD	2,3,7,8-tetrachlorodibenzo-p-dioxin
Twaron	poly p-phenylene terephthalamide
UHMWPE	ultra high molecular weight polyethylene

Additives

AGM	γ-aminopropylvinylsilane
AlMP	aluminium hydroxide methyl phosphonate
APB	diammonium pentaborate
APP	ammonium polyphosphate
APPh	ammonium pyrophosphate
AP750	ammonium polyphosphate-based formulation (Hostaflam AP750)
ATH	aluminium hydroxide
BCOH	β-cyclodextrine
BMAPs	binary metal-ammonium phosphates
BrPC	brominated polycarbonate
CA	citric acid
DAPA	diamide of methylphosphonic acid
DBDPO	decabromodiphenyl oxide
HBCD	hexabromocyclododecane
IST	N and P-based commercial products
MA,MEL	melamine
MFMPA	methylphosphonic acid
MgH	magnesium hydroxide
MOH	mannitol
OM	silane-containing oligomers
PCS	N and P-based commercial products
Petol	pentaerythritol
PH	phospham
PON	phosphorous oxynitride
P_x, PR	red phosphorus
PVA	poly(vinylalcohol)
PVTS	polyvinyltriethoxysilane
Pyrovatex	dialkyl phospho-propionamide
PY	diammonium diphosphate
RDP	resorcinol diphosphate
SOH	d-sorbitol
TAL	talc
TES	tetraetoxysilane
TFRI	dimethyl N-methylol phosphonopropionamide

THPC	tetrakis(hydroxymethyl)phosphonium chloride
TMM	trimethylol melamine
THPPO	trihydroxypropylphosphine oxide
TPO	triphenylphosphine oxide
TPP	triphenyl phosphate
XOH	Xylitol
XPM-1000	5,5,5',5',5",5"-hexamethyltris(1,3,2-dioxaphosphorinanemethan)amine 2,2',2"-trioxide
VTS	vinyltriethoxy silane
ZSM-5	zeolite ZSM-5
4A	zeolite NaA (4A)
5A	zeolite CaA (5A)
13X	zeolite 13X

Fire Retardancy of Polymeric Materials: Strategies

PHYSICAL AND CHEMICAL MECHANISMS OF FLAME RETARDING OF POLYMERS.

M. Lewin

Polymer Research Institute, POLYTECHNIC UNIVERSITY,
Brooklyn, New York
and HEBREW UNIVERSITY, Jerusalem, Israel

1. INTRODUCTION.

The rapid progress in the field of flame retardancy of polymers in the last 10 years expressed itself in the emergence of new additives, new application systems to an ever increasing diversity of products for which flame retardancy is a dominant requirement and in new standards and testing methods and instruments. Of particular importance became the environmental considerations which focussed attention on the need for new environmentally friendly flame retardant systems. These developments were accompanied by a pronounced effort to gain a better understanding of the underlying principles and mechanisms governing flammability and flame retardancy and to develop new mechanistic approaches for the emerging new flame retardancy systems.

The basic mechanisms of flame retardancy were recognised as early as 1947 when several primary principles were put forward[1]. These included the effect of the additive on the mode of the thermal degradation of the polymer in order to produce fuel-poor pyrolytic paths, external flame retardant coatings to exclude oxygen from the surface of the polymer, internal barrier formation to prevent evolution of combustible gases, inert gas evolution to dilute fuel formed in pyrolysis and dissipation of heat away from the flame front. Discovery of the flame inhibiting effect of volatile halogen derivatives subsequently led to the postulation of the radical trap - gas phase mechanism.[2] The gas phase and the condensed phase proposals have long been generally recognised as the primary, though not the only effective mechanism of flame retardancy. This situation is now being modified as new mechanisms of new flame retarding systems, especially those based on physical principles, evolve and as new insights into the performance of flame retardants is being gained.

This paper attempts to review some of the principles and mechanisms prevailing at present in the field of flame retardancy of polymers.

2. GENERAL CONSIDERATIONS.

A general scheme of the pyrolysis and combustion of polymers is presented in Figure 1[1a]. The polymeric substrate heated by an external heat source is pyrolysed with the generation of combustible fuel, ΔH_2. Only a part of this fuel, ΔH_1, is combusted in the flame by combining with atmospheric oxygen.. The other part, $\Delta H_3 = \Delta H_2 - \Delta H_1$, remains and can be combusted by drastic means, e.g., in the presence of a catalyst and by an excess of oxygen. A part of ΔH_1 is fed back to the substrate and causes its continued pyrolysis, perpetuating the combustion cycle. Another part is lost to the environment by dissipation. The energy needed to heat the polymer to the pyrolysis temperature and to decompose and gasify or volatilise the combustibles, and the amount and character of the gaseous products determine the flammability of the substrate. A flame retardant acting via a condensed phase mechanism alters the pyrolytic path of the substrate and reduces substantially the amount of gaseous combustibles (ΔH_2), usually by favouring the formation of carbonaceous char, CO_2 and water. In this case ΔH_2 decreases with increase in the amount of the flame retarding agent and the ratio $\Delta H_1/\Delta H_2$ remains essentially constant. In the gas phase mechanism ΔH_2 remains constant. ΔH_1 and the ratio $\Delta H_1/\Delta H_2$ decrease with increase in the amount of the flame retarding agent. The amount of heat returned to the polymer surface is therefore also diminished and the pyrolysis is retarded or halted as the temperature of the surface decreases. The flame retarding moiety has to be volatile and reach the flame in the gaseous form. Alternatively it has to decompose and furnish the active fraction of its molecule to the gaseous phase. The char remaining in the substrate will contain less of the active agent. The pyrolysis of the polymer should ideally proceed as if there would have been no flame retarding agent incorporated in it. The composition of the volatiles reaching the flame should therefore also not be influenced by the presence of the gas phase active agent[3].

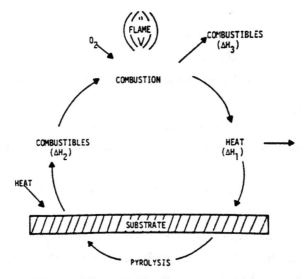

Figure 1. *Scheme of pyrolysis and combustion of a polymer.(from Reference 1a).*

3. GAS PHASE MECHANISM.

The gas-phase activity of the active flame retardant consists in its interference in the combustion train of the polymer. Polymers, like other fuels, produce upon pyrolysis species capable of reaction with air oxygen and produce the H_2-O_2 scheme which propagates the fuel combustion by the branching reaction[4]:

$$H + O_2 = OH + O \qquad (1)$$
$$O + H_2 = OH + H \qquad (2)$$

The main exothermic reaction which provides most of the energy maintaining the flame is:

$$OH + CO = CO_2 + H \qquad (3)$$

To slow down or stop the combustion, it is imperative to hinder the chain branching reactions (1) and (2). The inhibiting effect of halogen derivatives, usually chlorine and bromine, which are considered to operate via the gas phase mechanism, occurs by first releasing either a halogen atom, in case the flame retardant molecule does not contain hydrogen, or by releasing a hydrogen halide:

$$MX = M^{\cdot} + X \qquad (4)$$

$$MX = HX + M^{\cdot} \qquad (5)$$

where $M^{\cdot}$ is the residue of the flame retardant molecule. The halogen atom reacts with the fuel, producing hydrogen halide:

$$RH + X^{\cdot} = HX + R^{\cdot} \qquad (6)$$

The hydrogen halide is believed to be the actual flame inhibitor by affecting the chain branching:

$$H^{\cdot} + HX = H_2 + X^{\cdot} \qquad (7)$$

$$OH + HX = H_2O + X \qquad (8)$$

Reaction (7) was found to be twice as fast as (8) and the high value of the ratio H_2/OH in the flame front indicates that (7) is the main inhibiting reaction[5]. It is believed that the competition between reactions (7) and (1) determines the inhibiting effect. Reaction (1) produces 2 free radicals for each H atom consumed, whereas reaction (7) produces one halogen radical which recombines to the relatively stable halogen molecule.

3.1. Modes of Flame Retardant Activity of Halogen Derivatives.

Equation (7) represents an equilibrium with a forward reaction and a reverse reaction. The equilibrium constants for HBr and HCl[6] are:

$$K_{HCl} = 0.583.\exp(1097/RT) \qquad\qquad K_{HBr} = 0.374.\exp(16,760/RT)$$

The equilibrium constants decrease strongly with increase in temperature, which explains the decreasing effectivity of halogen derivatives in large hot fires[6]. Petrella [5] calculated that in the temperature range 500-1500 K the forward reaction predominates and K_{HBr} is much higher than K_{HCl} . Both are highly effective at the ignition temperature range of polymers. The flame retardant effectivity of the halogens was stated to be directly proportional to their atomic weights, i.e., $F:Cl:Br:I = 1.0:1.9:4.2:6.7$[7]. On a volumetric basis 13% of bromine was found to be as effective as 22% of chlorine when comparing the tetrahalophthalic anhydrides as flame retardants for polyesters[8,9]. A similar effect was found for PP, PS and PAN[3] and when comparing NH_4Cl to NH_4Br in cellulose[9]. The activity of the halogens is also strongly affected by the strength of the respective carbon- halogen bonds. The low bond strength of I-C and consequently the low stability of the iodine compounds, virtually exclude their use. The high stability of the fluorine derivatives and the high reactivity of the fluorine atoms in reactions (7) and (8) will prevent the radical quenching process in the flame. The lower bond strength and stability of the aliphatic compounds, their greater ease of dissociation as well as the lower temperature and earlier formation of the HBr molecules, are responsible for their higher effectivity as compared to the aromatic halogen compounds The higher stability of the last along with their higher volatility allow these compounds to evaporate before they could decompose and furnish the halogen to the flame.

The recently found high effectivities, as defined by the ratio of ΔO.I. / weight % of bromine, of 1.24 for NH_4Br encapsulated in PP as compared to 0.6 for aliphatic bromine compounds has been explained by the low dissociation energy of NH_4Br to HBr and NH_3 , which is much lower than the dissociation energy of the C-Br bond. The degree of dissociation is 38.7% at 320^0C so that sizeable amounts of HBr are readily available when PP begin to decompose[10].

3.2. The Physical Theory of Halogenated Flame Retardants

It should be realised that the radical trap activity is not the only activity of the halogenated flame retardants. The physical factors such as the density and mass of the halogen, its heat capacity, its dilution of the flame and thus decreasing the mass concentration of combustible gases in the flame, have a profound influence on the flame retarding activity of the agent. The proponents of the *physical theory* of the flame retardant activity of halogenated additives compare the halogen activity to that of inert gases, CO_2 and water[7]. The physical theory takes into consideration the basic parameters of the flame as well as the processes occurring in the solid phase leading to the production of the combustibles, and enables in certain cases to form an estimate of the amount of flame retardant agent needed to inhibit a flame. There appears to be no contradiction between the radical trap theory and the physical theory with regard to halogens. Both approaches complement each other. It is difficult to determine in a general way the relative contribution of each of the two modes of activity. This will usually depend on the structure and properties of the polymer and of the flame retardant as well as on the conditions and parameters of the flame and on the size of samples used.

4. THE CONDENSED PHASE MECHANISM.

As a rule there is in the condensed phase a chemical interaction between the flame retardant agent, which is usually added in substantial amounts, and the polymer. This interaction occurs at temperatures lower than those of the pyrolytic decomposition. Two principle modes of this interaction were suggested: dehydration and crosslinking. They have been established for a number of polymers including cellulosics and synthetics[11,12].

4.1. Dehydration.

The varying effectivity of phosphorus compounds in different polymers has been related to their susceptibility to dehydration and char formation, this explaining the decreasing activity with decreasing oxygen content of the polymer. Thus, whereas cellulosics are adequately flame retarded with ca 2% of P, 5-15% are needed for polyolefins[13]. The interaction of P derivatives with the polymers not containing hydroxyls is slow and has to be preceded by an oxidation. It has been suggested that 50-99% of the P derivatives are being lost by evaporation, possibly of P_2O_5 or other oxides formed in the pyrolysis of the P derivative[14]. This may be one of the reasons for the low yield.

Two alternative mechanisms have been proposed for the condensed phase dehydration of cellulosics with acids and acid-forming agents of phosphorus and sulphur derivatives. Both mechanisms lead to char formation[15]: (a) esterification and subsequent pyrolytic ester decomposition (see Scheme 1) and (b) carbonium ion catalysis (Scheme 2):

$$R_2CH\text{-}CH_2OH + ROH \text{ (acid)} \rightarrow R_2CH\text{-}CH_2OR + H_2O \rightarrow$$

$$R_2C{=}CH_2 + HOR \text{ (acid)} \qquad \text{(Scheme 1)}$$

$$R_2CH\text{-}CH_2OH \rightarrow R_2CH\text{-}CH_2OH_2^+ + H_2O \rightarrow$$

$$R_2C{=}CH_2 + H^+ \qquad \text{(Scheme 2)}$$

DTA and OI data indicated that P-compounds reduce flammability of cellulosics primarily by an esterification - ester decomposition mechanism (Scheme 1) which, being relatively slow, is affected by the fine structure of the polymer. Less-ordered regions (LOR) pyrolyze at a lower temperature than the crystalline regions and decompose before all of the phosphate ester can decompose, which decreases the flame retarding effectivity and necessitates a higher amount of P. Sulphated celluloses, obtained by sulphation with ammonium sulfamate, are dehydrated by carbonium ion disproportionment (Scheme 2) and show a strong acid activity which rapidly decrystallizes and hydrolyses the crystalline regions. The FR activity was accordingly found not to be greatly influenced by the fine structural parameters, and the same amount of sulphur was needed to flame retard celluloses of different crystallinities[15].

4.2. Crosslinking.

It was early recognised that crosslinking promotes char formation in pyrolysis of celluloses[16]. Crosslinking has been assumed to be operative in P-N synergism[17]. Crosslinking reduces in many cases, albeit not always, the flammability of polymers. Although it increases the OI of phenolics, it does not markedly alter the flammability of epoxides[18]. A drastic increase in char formation is observed when comparing crosslinked polystyrene (PS), obtained by co-polymerising it with vinylbenzyl chloride, to uncrosslinked PS. PS pyrolyzes predominantly to monomer and dimer units almost without char. Crosslinked PS yielded 47% of char[19]. Crosslinking and char formation were recently obtained by an oxidative addition of organometallics to polyester[20].

Crosslinking promotes the stabilization of the structure of cellulose by providing additional, covalent bonds between the chains, which are stronger than the hydrogen bonds, and which have to be broken before the stepwise degradation of the chain occurs on pyrolysis. However, low degrees of crosslinking can decrease the thermal stability by increasing the distance between the individual chains and consequently weakening and breaking the hydrogen bonds. Thus, although the OI of cotton increases marginally with increasing formaldehyde crosslinking, that of rayon markedly decreases (Figure 2)[21].

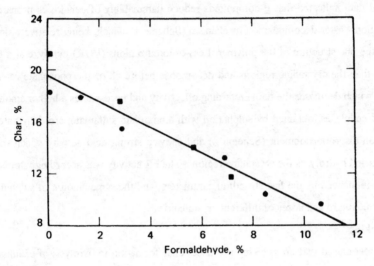

Figure 2. *Effect of Crosslinking on Char Yield; TGA Vacuum Pyrolysis Residue at 500⁰C of Rayon Fibers Crosslinked with Formaldehyde. (From ref. 21).*

The formation of char in celluloses is initiated by rapid *auto-crosslinking* due to the formation of etheric oxygen bridges between hydroxyl groups on adjacent chains. The auto-crosslinking is evidenced by a rapid initial weight loss, due to evolution of water, in the first stage of pyrolysis at 251⁰C, and is linearly related to the amount of char. Formaldehyde crosslinking of rayon interferes with the auto-crosslinking reaction, decreases the initial weight loss and reduces char formation, as shown in Figure 2[21]. It was suggested that crosslinking may increase the viscosity of the molten polymer in the combustion zone, thereby lowering the rate of transport of the combustible pyrolysis products to the flame[22].

4.3. Structural Parameters

In addition to bond strength and intermolecular forces, there are several other parameters, such as chain rigidity, resonance stability - aromaticity, crystallinity and orientation, which have a pronounced influence on pyrolysis and combustion. The linear correlations of Van Krevelen between O.I. and char and between the char forming tendency (CFT) and char residue, are well known[3]. The CFT (equation 9) is defined as the amount of residue at 850⁰C per structural unit, divided by 12, i.e. the amount of $\underline{C}$ equivalents per structural unit, where each group has its own CFT. These equations hold only for untreated polymers and for polymers containing condensed phase flame retardants.

They do not hold if halogen is present[3].

$$CR = 100 \times \Sigma(CFT)_i \times 12 / M \text{ units} \qquad (9)$$

Recent work on the relationship between chemical structures and pyrolysis and on the effects of introducing substituent functionalities into aromatic and heterocyclic structures on the modes of pyrolysis, was reviewed by Pearce[23,23a].

An interesting attempt to develop a generalised kinetic model of polymer pyrolysis was recently made by Lyon[24]. The model is based on some of the mechanisms important in the burning process, i.e., generation of combustible gases and char formation, but can be solved for the overall mass loss history of the specimen, for verification using laboratory thermogravimetric techniques. It is assumed that the pyrolysis proceeds in several steps.

1.The first, rate determining step, is the "dissociation" of the polymer, i.e., $P \Leftrightarrow I^*$, with k_p and k_{-p} the rate constants of the forward and reverse reactions . It is slow compared to the $I^* \Rightarrow GI$ (rate constant k_g) and $I^* \Rightarrow CI$ (rate constant k_c) reactions to form primary gas and char, respectively. It is an equilibrium..

2.I^* is an intermediate whose concentration is very low and its change in time is negligible (stationary state).

3. Only primary char is considered.

4. The thermo-oxidative environment in the pyrolysis zone of a burning solid is assumed to be anaerobic, and diffusion of oxygen and oxygen concentration in the pyrolysis zone are considered negligible and of no effect in the polymer combustion model.

Based on these assumptions a single first order law for the isothermal mass in burning polymer was worked out:

$$m/m_0 \equiv Y_c + (1 - Y_c) \exp(-k_p t) \qquad (10)$$

where $Yc \equiv k_c/k_g + k_c$ and m/m_0 is the ratio of char to initial mass and Y_c is the temperature dependant char yield.

Comparison of the model predictions with measured mass loss history of a phenolic triazine thermosetting resin in temperature scanning thermogravimetric experiments showed excellent agreement for the primary thermal decomposition step[23]. It appears that this model is still to be considered tentative and the assumptions underlying it should be further checked in the light of evidence on other polymers. This pertains primarily to the assumption that the presence of air during pyrolysis does not influence the kinetics of pyrolysis. In studies of isothermal pyrolysis of rayon it was found that in the presence of air the rate of weight loss at 251^0C was much higher than in vacuum pyrolysis. and that the rate decreased with the orientation of the fibers, i.e., it depended on the rate of diffusion of the air oxygen into the sample. In vacuum pyrolysis the rate *increased* with increase in orientation[25-27].

4.4. Fine Structural Parameters

In addition to orientation also crystallinity and degree of polymerisation (DP) have a strong influence on the energy required to melt and degrade polymers, on the rates of vacuum and air pyrolysis and on char formation and yield. The effect of the DP on the degradation temperature of rubber has been described[28]. Vacuum pyrolysis rates of purified celluloses were found to increase with increasing orientation and less-ordered regions (LOR) and inversely proportional to the square root of the DP[25,26]. The decrease in thermal stability with increasing orientation was ascribed to the straining of the hydrogen bonds. The extent of the autocrosslinking reaction, discussed earlier, was found to be directly proportional to the % of char. The char increases with the increase in LOR of the polymer.

The energy of activation of pyrolysis of cellulose was found to increase strongly with the increase in crystallinity, indicating different mechanisms operating for the crystalline and less-ordered regions[26].

4.5. Polymer Blends

Little is known on the effect of the fine structural parameters on the pyrolytic behavior of polymers other than cellulose. The inclusion of these parameters in mechanistic models might prove to be of considerable interest. One such area might be the pyrolysis and flame retardancy of blends, as evidenced in the case of cotton-wool blends.

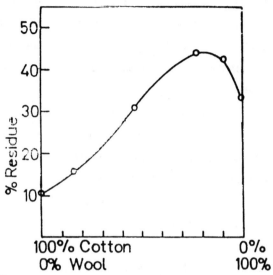

Figure 3. *Cotton - Wool Blends: Char Yield. (From ref. 27).*

The DSC endotherm of cotton at 350^0C, which is due to the decomposition of the levoglucosan monomer formed on pyrolysis, disappears with the addition of relatively small amounts of wool. Since levoglucosan is formed from the crystalline regions of the cellulose, its disappearance was attributed to the swelling decrystallization of the amino derivatives formed in the pyrolysis of the wool, which occurred at 225^0C, i.e., lower than the 300^0C at which the cotton pyrolysis begins. This is also manifested by a "synergism" in char production. The strong increase in char in these blends, which is beyond the char amounts predicted by the composition of the blend. This rise stems from the increase in the LOR due to the swelling. Consistent with the above is also the decrease in activation energy of the pyrolysis from 52.6 for cotton to 24.7 KCal/mole for the blend with 18% of wool. It is important to note that the above interaction between the ingredients of the blend is physicochemical in nature and depends on temperature. Pyrolysis-gas chromatography of a series of wool-cotton blends at

1000^0C for 30sec yielded all the peaks in the relative area ratios as expected from a simple additive calculation in the absence of any interaction. The degree of interaction of components in a blend is therefore to be considered as a kinetic process governed by temperature and time[28]. The increase in char does not result in improved flammability. Actually more additive is needed for the blend than for the individual components[28]. A similar situation exists in the case of cotton-polyester blends. In this case more flammable gases, such as ethylene, are evolved from the blend than from the individual components[28].

In blends of PVC with SMA a concave downward curve is obtained when plotting LOI against composition, indicating that a higher than calculated level of PVC is needed for improved flammability. This result can be interpreted as an antagonism which may be due to an interaction between the components[29]. An interesting mathematical model which computes the OI values of polymer blends from the OI values of the components and from the stoichiometry of the combustion of each, is shown in equation $(11)^{30}$.

$$1/Y = (1/Y_2 + m_1r_1/m_2r_2Y_1) (1 + m_1r_1/m_2r_2)^{-1} \qquad (11)$$

where Y_1 and Y_2 are the OI values of polymers 1 and 2; m_1 and m_2 the mass of the polymers and r_1 and r_2 are the calculated amounts oxygen to burn unit masses of of polymers 1 and 2. Such formulas may be very useful particularly for cases in which no interaction between the components occurs and if there is no barrier char present. Nevertheless the polymer blend area is rapidly developing and continued studies in this direction are highly interesting.

5. INTUMESCENCE , SYNERGISM AND CHAR

Flame retarding polymers by intumescence is essentially a special case of a condensed phase mechanism. The activity in this case occurs in the condensed phase and radical trap mechanism in the gaseous phase appears to not be involved. In intumescence the amount of fuel produced is also greatly diminished and char rather than combustible gases is formed. The intumescent char, however, has a special *active* role in the process. It constitutes a two-way barrier, both for the hindering of the passage of the combustible gases and molten polymer to the flame as well as the shielding of the polymer from the heat of the flame. In spite of the considerable number of intumescent systems developed in the last 15 years[31-33], they all seem to be based on the application of 3 basic ingredients: a "catalyst", a charring agent and a blowing agent.

The "catalyst" is usually a $\underline{P}$ derivative, in most cases ammonium polyphosphate (APP). The function of the APP is to *catalyse* the dehydration, which means that it is not supposed to be a main actual reactant in the system, but to catalyse reactions of other ingredients. The relatively large amounts of APP applied, i.e., in the range of 10-20% of the weight of the composition, are not compatible with the definition of a catalyst. It is therefore possible that the APP has an additional role in the system, and serves as an ingredient in the char structure. It is also possible that the low effectivity (EFF) of the APP is caused by volatilisation of phosphorus oxides formed in the pyrolysis. The FR EFF of APP[34-38] is low (0.31). The EFF of APP, however, increases greatly upon the addition of the co-additives. Depending on the chemical composition of these commercially available additives, which include the char formers and the blowing agents, which are considered as synergists, EFF values of up to 3.5 were obtained[35]. These values are much higher than the general value of 1.3 given by van Krevelen for $\underline{P}$ in PP, which does not include intumescent systems[3]. The synergistic effectivities[35-38] (SE) i.e. the ratio of the EFF of the FR agent together with the synergist to that of the agent alone, obtained in the intumescent systems, increase from values of 5.5 and 3.0 for APP + pentaerythritol (petol) and APP + melamine to values of 9.7 and 11.3, for other co-additives (Table 1).

5.1. *P-N Synergism*

Synergism , as currently used in FR terminology is a poorly defined term . Strictly speaking, it refers to the combined effect of two or more additives, which is greater than that predicted on the basis of the additivity of the effect of the components. The term SE used in this and previous publication of this laboratory is meant to serve as a general tool for the characterisation and comparing synergistic systems. In the following several synergistic systems will be briefly discussed and compared.

In Table 1 values of EFF and SE computed from literature data on a number of FR synergistic systems, are presented. An EFF of 4.0 and an SE of 1.75 are obtained for $\underline{P}$ in cotton treated with Pyrovatex (dialkyl phospho-propionamide) and TMM (trimethylol melamine) at a 2% P and 5% N level[39]. Although the ingredients of this system P , N and a polyhydric alcohol - cotton resemble an intumescent system the SE is much lower. A similar result is also obtained when treating cotton with ammonium sulfamate and urea[40].

The P-N synergism in the phosphorylation of cellulose is manifested by an increased rate of phosphorylation and yield of P, by an increased FR effect and improved physical properties. The magnitude of the effect varies from one N compound to another. Usually amine and amide derivatives are effective whereas nitriles are antagonistic, allegedly due to their volatility[39]. The synergism in the case of cellulose was ascribed (a) to a swelling effect on the polymer by the N derivatives[8]; (b) P-N bonds are more reactive in phosphorylation than P-O bond[40,41]. This consideration would also apply to the case of intumescence in PP during pyrolysis and combustion., since it is immaterial whether the effect occurs during processing or during pyrolysis. The net effect is the same: increase in FR effectivity.

The P-N bonds can also participate in the formation of the crosslinked networks, in which the P will be fixed and its volatilisation hindered[42].

Based on elemental analysis , infrared and XPS spectra ,Weil et al. confirmed recently the existence of P-N species along with phosphoric or polyphosphoric acids on the surface of char from burning an intumescent sample composed of EVA, melamine pyrophosphate and XPM-1000 [5,5,5',5',5'',5''-hexamethyltris(1,3,2-dioxaphosphorinanemethan)amine 2,2',2''-trioxide]. No evidence for the existence of the stable phosphorus oxynitride (PON) was found[43].

5.2. Halogen-Antimony Synergism.

Data on the SE of aromatic and aliphatic bromine derivatives with antimony trioxide, computed from reference (3) show SE values of 2.2 and 4.3 , respectively. Similarly, for aliphatic chlorine derivatives with antimony trioxide a SE value of 2.2 is computed for polystyrene (Table 1).

TABLE 1 *FR Effectivity (EFF) and Synergistic Effectivity (SE)of Intumescent and Other Systems.*

POLYMER	FR	SYNERG.	EFF	SE	REF.
PP	APP		0.31		35, 37
PP	APP	Petol	1.7	5.5	"
PP	APP	Melamine	0.92	3.0	"
PP	APP	Petol+Melam	2.4	7.7	"
PP	APP (EDAP)		2.1	6.8	"
PP	APP	Spinflam MF82	3.0	9.7	"
PP	APP	THEIC	2.8	9.0	"
PP	APP	Exolit IFR 23P	3.5	11.3	"
PP	APP	Petol+Cat. A	3.13	10.1	70
LRAM3.5 (EBM)	APP	Petol	1.76		71
"	APP	Zeolite 4A	3.16	10.2	"
Cotton	Pyrovatex		4.0		39
Cotton	Pyrovatex	TMM	7.0	1.75	"
PP	Arom. Br		0.45		3
PP	Arom. Br	Sb_2O_3	1.0	2.2	"
PP	Aliph. Br		0.6		"
PP	Aliph. Br	Sb_2O_3	2.6	4.3	"
PP	NH_4Br		1.24		10
ABS	Dechlorane+ Sb_2O_3		0.8		52
ABS	"	Br-epoxy	1.35	1.67	"
PC:PET/ 2:1	TPP		13.3		55
"	BrPC		1.7		"
"	BrPC	TPP		1.38	"
"	Br:P/7:3			1.57	"
PAN	APP		1.62		54
"	HBCD		1.21		"
"	HBCD+APP			1.55	"

The halogen-antimony synergism appears to depend both on condensed phase as well as vapor phase activities. It is believed that first some hydrogen halide is released from the halogen compound due to interaction with antimony trioxide or with the polymer. The HX reacts with the Sb_2O_3 producing either SbX_3 or $SbOX$[44,45]. Although some SbX_3 is found in the first stage of the pyrolysis,, the weight loss pattern requires the formation of less volatile Sb - containing moieties, obtained by progressive halogenation of Sb_2O_3[46]. SbOCl was found to decompose in several endothermic stages: at 245-280^{0}C to $Sb_4O_5X_2$; at 410-475^{0}C to Sb_3O_4X; above 685^{0}C solid Sb_2O_3 is reformed. During the transformations gaseous SbX_3 is evolved and released to the gas phase, whereas SbOX, which is a strong Lewis acid, operates in the condensed phase, facilitating the dissociation of the C-X bonds[47]. In the case of polyester the partial unzipping to the monomer is being replaced by charge transfer, bringing about the formation of higher molecular weight moieties, and crosslinking and consequently the formation of char.

The main effect of Sb_2O_3 is, however, in the gas phase. The antimony halides, after reaching the gas phase, react with atomic hydrogen producing HX, SbX, SbX_2, and Sb. Sb reacts with atomic oxygen, water and hydroxyl radicals , producing SbOH and SbO, which in turn scavenge H atoms. SbX_3 reacts with water, producing SbOH and HX. A fine dispersion of solid SbO and Sb is also produced in the flame and catalyses the H recombination. In addition it is believed that the antimony halides delay the escape of halogen from the flame, and thus increase its concentration, and at the same time it also dilutes the flame.

5.3. Bromine-Chlorine Synergism

The synergistic interaction between bromine and chlorine derivatives is discussed in several papers[48-52]. Cleave reported on such an effect in mixtures of chloroparafins with pentabromotoluene[48] in the presence of Sb_2O_3. Recently, Markezich reported on results obtained on ABS, HIPS and PP[53]. In most cases the maximum effect is found with a Br/Cl ratio of 1:1 and with 10-12% of the sum of chlorine and bromine. When using Dechlorane Plus and brominated epoxy resin (51% Br) with ABS in the presence of 5% Sb_2O_3, the FR EFF was calculated as 0.80. The SE obtained was 1.67. (Table 1). This synergism is in addition to the halogen-antimony synergism and is effective only in the presence of antimony trioxide. The synergistic effect increases with the amount of antimony and reaches a maximum at 6% trioxide.

Some light on the Br-Cl synergism was thrown in pyrolysis experiments carried out in the ion source cell of a mass spectrometer with mixtures of polyvinyl bromide (PVB) and polyvinyl chloride (PVC) or polyvinylidene chloride (PVC$_2$). Whereas high concentrations of HCl and HBr were found, the amounts of SbCl$_3$ and consequently the rate of interaction of HCl with the oxide were very small compared to SbBr$_3$, indicating that the Br-Cl synergism operates via the bromine-antimony route. This seems to be born out by the fact that little information is available on Br-Cl systems without antimony. Additionally, it is also conceivable that the radicals Br˙ and Cl˙ might recombine not only to Br$_2$ and Cl$_2$ but also to BrCl, which is polar and more reactive and will react with the H˙ radicals to produce HBr and another Cl˙ radical, thus increasing the effectivity. This may explain the increase in effectivity of the formulations containing only Br-based additives upon addition of chlorinated compounds.

5.4. *Bromine-Phosphorus Synergism*

There are several reports in the literature on P-Br synergism. Of particular interest is the case of polyacrylonitrile (PAN) treated with varying ratios of APP and hexabromocyclododecane (HBCD)[54]. As seen in Table 1, a SE value of 1.55 was calculated in this case. It was shown that the system acts via an intumescent mechanism. The bromine compound was proven by LOI and NOI tests not to operate in the gas phase but rather as a blowing agent to foam the char. This is a remarkable conclusion and it opens the way for reconsidering the mechanism of operation of bromine compounds as flame retardants in other polymers and bromine compounds.

Similar SE values are computed from data in Ref. 55 for a polycarbonate - polyethylene terephthalate (PC-PET) blend treated with varying ratios of triphenyl phosphate (TPP) and brominated PC[55]. A SE value of 1.38 is found. When a brominated phosphate with the ratio of bromine to phosphorus of 7:3 is added to the same blend, the SE value is 1.58. There are some indications though no clear evidence, that also in these cases bromine may serve, at least partly, as a blowing agent instead or together with the radical trapping activity in the gas phase. The Br-P SE values are considerably lower than the bromine-antimony values as well as the intumescent values, pointing to the possibility of a different mechanism.

It has recently been suggested that P compounds may replace antimony as a halogen synergist[56]. In the case of oxygen containing polymers, such as Nylon 6 and PET a strong synergism was shown and in the case of PET a decrease in the total amount of additives (Br-based and P-based) decreased by over 90% compared to Br-based and Sb additive. A similar synergistic activity of bromine and phosphorus was obtained for PBT, PP, PS, HIPS and ABS. A decrease of amount of total additive of 40% was obtained for PE[56]. The P-Br synergism was also demonstrated in a case when both atoms are part of the same molecule of a phosphonoalkyl, developed as a flame retardant for ABS. The results were based on comparative experiments with related compounds having only the bromine related structures or only P related structures[57]. A similar synergism was also lately obtained for uv-curable urethaneacrylate to which variable amounts of tribromo phenylacrylate and triphenyl phosphate were added. A Br:P ratio of 2.0 was found to yield the maximum synergism[58].

5.5. Gas-Phase Activity of Phosphorus Derivatives

Volatile, low boiling P compounds, such trialkyl phosphates and trialkyl and triaryl phosphine oxides, have been found to act as gas phase flame retardants[59,60]. Triaryl phosphates operate similarly in blends of polyphenylene oxide (PPO) with HIPS, in which the PPO act to form a protective char, i.e., a combined effect is obtained[61].

A study of phosphonium and phosphinoxide structures in THPC-urea polymers indicated a mixed vapor-condensed phase mechanism for PET. 3 polymeric compounds, having respectively (a) phosphonium groups: $-CH_2-P(Cl)(CH_2OH)_2-CH_2-NH-CO-NH-$

(b) phosphine oxide groups: $-CH_2-P(O)(CH_2OH)_2-CH_2-NH-CO-NH-$

and (C) phosphonium + phosphine oxide groups:

$-CH_2-C(H)(OH)-CH_2P^+(Cl^-)(CH_2OH)_2-CH_2-NH-CO-NH-CH_2-P(O)(CH_2OH)-CH_2-NH-CO-NH-$.

The relative activities of these materials in wool and PET-wool blends were in direct order to their volatilities, e.g., the phosphonium moiety was more effective in flame retarding wool and wool-PET blends than the phosphine oxide, whereas the mixed structure (c) showed intermediate activity. Accordingly a vapor-phase mechanism was inferred. All 3 polymers showed equivalent activity on PET, suggesting that both mechanisms apply to it[61,62].

Mass spectroscopy studies showed that triphenyl phosphate and triphenyl phosphine oxide break down in the flame to small radical species, such as P_2, PO, PO_2, and HPO, which can scavenge the H radicals in a similar way as the halogens[63,64].

It is of interest to note that the formation of $\underline{P}$ radicals was also suggested to occur during the pyrolysis and combustion of red phosphorus, which is believed to be a polymer. These radicals are assumed to react with oxygen in the polymer producing phosphoric ester structures[66]. The presence of the $\underline{P}$ radicals in red phosphorus-containing nylon was proven by EPR measurements. It has also been suggested that red phosphorus operates in the gas phase[65,66].

5.6. Relative importance of co-additives in intumescence

The synergistic effectivity of the PP-APP system is much higher than of any other synergism (Table 1). This basic difference emphasises the sensitivity of the intumescent system to the nature of the co-additives on one hand , and to the vast possibilities for further improvement of the system, on the other hand. The strong synergistic effect in the intumescent systems is not surprising when considering the highly complex nature of the chemical and physical interactions between the materials involved[67-68]. A series of processes is occurring during the pyrolysis and combustion: decomposition of APP with release of ammonia and water, phosphorylation of the pentaerythritol (petol) and of thermo-oxidised PP, dehydration, dephosphorylation, crosslinking, carbonisation, and formation of char structure. The blowing agent vaporises and decomposes to incombustible gases. At the same time a number of physical processes take place: diffusion and transport of combustible and incombustible gases through the polymer melt to the flame zone, transfer of the molten polymer and of the flame retarding moieties to the flame, diffusion and permeation through the char barrier. In addition complex heat transfer processes are taking place. The incipient char is being swollen and foamed to a cell structure.

All these processes occur in a very short time with various reaction rates. The ratios of the rates, which determine the sequence of the reactions and their timing, has a dominant effect on the properties of the final char and on the flaming behavior. The rates might , in principle, be influenced by suitable catalysts and possibly be engineered to higher effectivities.

Whereas the role of the polyhydric alcohol as a charring agent is clearly defined the function of the melamine appears to be more diverse: (a) endothermic action due to the volatilisation from the solid state and to decomposition in the flame. These effects, however, do not constitute a major contribution[69] as a part of the melamine is lost during the evaporation; (b) it may facilitate the phosphorylation, as discussed above; (c) it is believed to serve as a blowing agent in the intumescent formulation as a part of the volatiles emitted during pyrolysis, which include water, CO, CO_2, ammonia and hydrocarbons, all serving to foam the char.

Results of a series of experiments[70] designed to throw some light on the roles of melamine and petol in intumescent formulations with APP and PP, are summarised in Figure 4. The concentrations of PP and APP were kept constant at 75 and 16.6 weight %, respectively, while the amounts of petol and melamine in the remaining 8.4% of the formulation varied. Both petol and melamine are clearly seen to be synergists for APP, the effect being much more pronounced for petol. When applied together they are in fact co-synergists. However, compositions of petol:melamine 20:80 to 80:20 show virtually the same OI values of ca 30. This indicates that at least in this range petol and melamine are equivalent and interchangeable. Since petol shows a higher effectivity, the role of the melamine seems to be minor.

The upper curve in Figure 4 summarises an additional series of experiments, in which ca 1% of a metallic catalyst was added to the same formulations. In the presence of melamine only a very small effect of the catalyst is seen. However, in the absence of melamine a very pronounced effect is obtained[70]. It can be concluded that (a) melamine or other N-based blowing agents are not essential in the intumescent formulation; (b) other gases in the system perform the blowing function; (c) the phosphorylation occurs directly by APP possibly via phosphoramide structures, which is consistent with previous findings of their particularly high rate of phosphorylation compared to other phosphorylating agents[40]; (d) the effect of the catalyst is not connected with blowing activity.

There seem to be 2 possible effects of the catalyst: (a) facilitation or acceleration of the phosphorylation reaction and (b) stabilization of polyphosphoric acid by forming bridges or crosslinks between adjacent chains, thus hindering its dehydration to volatile oxides. This enables a fuller participation of the polyphosphoric acid in the phosphorylation as well as in the char structure[70].

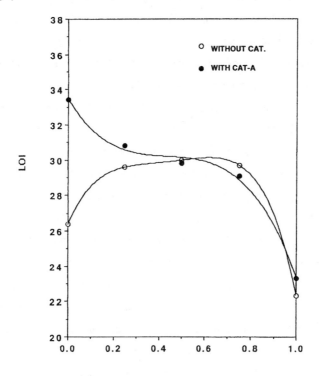

Melamine/Melamine+PETOL

Figure 4. *LOI vs. Melamine/M; Melamine + Petol; PP + APP; Upper Curve : Cat A - 1% ; No Melamine (From ref. 70).*

The stabilization of the APP requires a small amount of catalyst. Higher amounts, which would produce more crosslinks would render the polyphosphoric acid too stable and unreactive and would seriously interfere with its functions. Similar effects of talc and manganese dioxide on the intumescent flame retardancy of nylon, in which no melamine was used, were also recently reported by Levchik et al.[76].

Recently Bourbigot et al. reported on intumescent systems without applying blowing agents[71-75]. In one system terpolymers of ethylene- butyl acrylate and maleic anhydride were applied together with APP and petol and a zeolite-4A catalyst at 1.5% level, and gave a much better FR performance than the control without the catalyst. The OI increased with the increase in the amount of the oxygen containing components of the terpolymer. It is evident that also in these cases there was no need for an additional blowing agent. On the other hand it appears that the presence of the oxygen in the terpolymers assists in the formation of the char structure and assists in a better utilisation of the APP. It is of interest to note that the SE values calculated for the terpolymer ethylene-butyl acrylate-maleic anhydride (EBM; LRAM3.5) containing zeolite 4A[71] as well as for PP containing Cat A[70], are very close and are in the range of SE values for the most effective intumescent systems (Table 1). The catalytic function of the 4A zeolite appears to be specific and possibly linked to the structure of the char. The TGA residues are higher, indicating a stabilization of the char and a slower oxidation. Stabilization of the char to oxidation was shown earlier to be an important feature of intumescent systems and straight line correlation (r=0.99) were found between the residues after the last transitions (RAT) in TGA measurements in air and the OI, for six commercial intumescent systems. The temperature at which the last transitions occurred had no influence non the correlations and it appears that the last TGA transition signifies an important feature of intumescence[34-38]. In all cases no correlation was found between the OI values and the % char obtained in TGA at 500^0 C. On the other hand a linear correlation was found between the time of peak rate of heat release (TPRHR) as measured in the Cone calorimeter and the FR EFF for several intumescent systems[37]. Correlations were also indicated between OI values and several Cone calorimeter parameters: time of peak RHR, peak RHR, total heat release, mass loss rate and CO_2 at 300 sec[35,37]. Further work on the relationships between Cone calorimeter parameters and OI and other tests is needed and may contribute to a better understanding of flame retardancy.

5.7. Char Structure

The chars obtained in the intumescent systems are different from the chars studied by Van Krevelen[3], Lewin[25] and Pearce[23], discussed earlier, which were prepared by nitrogen or vacuum TGA pyrolysis at high temperatures and at a relatively slow rate. The intumescent combustion chars are prepared at lower temperatures and are not fully pyrolysed or oxidised, their rate of formation is high and involves thermo-oxidation. The mechanism of their formation is different. They are complex structures and serve as barriers to the passage of molten polymer and decomposition gases.

Several properties of char have been mentioned in the literature as desirable from the point of view of flame retardant performance: mechanical strength and integrity, coherence and adherence, openness of cells, impenetrability to gases and liquids. It has been stated that the above properties are more important than the volume swelling - degree of intumescence[77]. According to another view it is the kinetics of the charring processes which determine their behavior, and the exact timing of the expansion of the char is crucial; if it takes place too early after the dehydration, the char is ineffective and non-insulative. If the expansion occurs too late the char hardens before swelling takes place[31,32]. The expansion is caused by the migration of the gas bubbles, produced in the pyrolysis. The rate of the migration depends on the viscosity of the molten polymer mass in the combustion zone. This viscosity can possibly be regulated by controlled crosslinking and thus influence the char structure[22].

Gibov et al.[78] specify a number of essential char properties : thickness of char layer, high carbon content, low penetrability, high viscosity of pyrolyzing melt, homogeneous small pores. The addition of P derivatives to a phenol formaldehyde char surface layer on epoxy and polystyrene plates decreased the penetrability and increased the OI. Similar results were obtained for boric oxide. Polyphosphoric acid and boric oxide act as "liquid fillers of the pores" in the char and their effectivity depends on the viscosity and on the temperature of the preparation of the char. The presence of polyphosphoric acid on the surface of intumescent chars has also been established by Weil[43]. Gibov et al. also state, on the strength of X-ray and SEM observations, that the addition of P compounds changes the structure of the char, so that a more regular dispersion of the micropores and some crystalline regions are noticed[78].

A more detailed structure of P-containing chars is suggested by Le Bras and Bourbigot[75]. They visualise a multiphase structure with crystalline domains composed of macromolecular polyaromatic stacks, encapsulated in an amorphous phase matrix. The amorphous phase is composed of small polyaromatic molecules linked by P-O-C bonds to phosphate and alkyl species from the pyrolysed polymer and charring agent. Another interesting suggestion is that the addition of the zeolite 4A catalyst decreases the size of the amorphous domains resulting in a more compact and flexible structure, less susceptible to cracking and to the release of trapped combustible gases and free radicals. This suggestion was based on ^{1}H solid state NMR measurements. The order of magnitude of the crystalline and amorphous domains is in the range of 10-100 nm[79].

A recent study showed that it is possible to discern significant differences between intumescent chars also in the 5-50 micron scale by SEM scans. In most cases foam structures with well defined closed cells, could be observed on the surface as well as on the cross-section of chars, with diameters of 15-40 microns. The diameters varied in size and in the distribution of their sizes from formulation to formulation[35,37]. It is believed that systematic studies using imaging techniques might yield further important information on the structure of chars.

Non withstanding the significant progress in the study of char in recent years, in which diverse analytical methods were developed and applied, many of its basic and important properties and structural features can at present be estimated only in a descriptive and sometimes even vague manner. Reliable methods for determining volume, weight , density, mechanical properties, such as crush, resilience, hardness and toughness, coherence, permeability and other thermal and physical properties are still needed in order to further elucidate the structure and mechanism of action of char and in order to enable the development of effective modifications of char.

6. THE MECHANISM OF FLAME RETARDANCY BY PHYSICAL EFFECTS

Flame retardancy due to physical effects usually requires relatively large amounts of additives: 50-65% in the case of aluminium trihydrate (ATH) and magnesium hydroxide (MgH). The activity of these additives consists in: (a) dilution of the polymer in the condensed phase; (b) decreasing the amount of available fuel; (c) increasing the amount of thermal energy needed to raise the temperature of the composition to the pyrolysis level, due to the high heat capacity of the fillers; (d) enthalpy of decomposition - emission of water vapor; (e) dilution of gaseous phase by water vapor - decrease of amount of fuel and oxygen in the flame; (f) possible endothermic interactions between the water and decomposition products in the flame; (g) decrease of feedback energy to the pyrolyzing polymer; (h) insulative effect of the oxides remaining in the char[79-83]. Considerable attention is given to the particle size - surface area, as it is known to influence greatly the melt flow of the treated polymer, and to the crystal size of the additive. Hydrophobic coatings on the particles of the additives are applied to facilitate dispersion of the highly hydrophilic metallic oxides in the polymer. The possible effect of these coatings on the combustion behavior of the polymers is not clear. Another important property appears to be the "surface free energy" which determines the reactivity of the additive with acidic groups in polymers, which might produce crosslinks and reduce melt flow, as well as with acidic groups in air, i.e. carbon dioxide, or in acid rain, which might cause the 'chalking effect' and decrease flame retardancy[81-83].

Recently evidence has been presented, pointing to the possibility of an interaction between the endothermic additives and several polymers during pyrolysis and combustion, in the presence of some metallic catalysts[84-86] , such as nickel(II) oxide, manganese borate, 8-hydroxychinolino-copper and ferrocene. These interactions are believed to be the cause of a highly significant increase in char and OI. Other compounds, i.e., nickel(III) oxide and metal acetylacetonates gave negative results. Specific effects of transition metals, when co-precipitated with magnesium hydroxide as solid solutions, included : (a) facilitation and lowering the temperature of the dehydration of the additive; (b) catalysing dehydrogenation of the polymer; (c) promoting carbonization; (d) improving acid resistance[86]. The mechanism of the catalytic and antagonistic activities of metallic compounds has not yet been clarified. There are qualitative and quantitative differences between the effects of various ions and an effect of the ionic radius has been proposed. The elucidation of the mechanisms governing the activity of the metallic compounds in flame retardancy of polymers, with endothermic additives as well as with intumescent formulations, appears to be a highly intriguing challenge for polymer science today.

References

1.   R. W. Little, *'Flame Proofing Textile Fabrics'*, Reinhold, London, 1947.

1a. R. H. Barker and M. J. Drews, *'Final Report NBS-GCR-ETIP 76-22'*, NBS, Washington D.C., 1976.

2.   W. A. Rosser, H. Wise and J. Miller, in *'Proc. 7th Symposium on Combustion'*, Butterworth, 1959, p. 175.

3.   D. W. van Krevelen, *J. Appl. Polym. Sci. Appl. Polym. Symp.* 31, 1977, 269; *Chimia,* 28, 1974, 504; *Polymer,* 16, 1975,615.

4.   G. I. Minkoff and C. F. H. Tipper. *'Chemistry of Combustion Reactions'*, Butterworth, 1962.

5.   R. V. Petrella, in *'Flame Retardant Polymeric Materials'*, M. Lewin, S.M. Atlas and E. M. Pearce editors, Vol. 2, Plenum, 1978, p. 159.

6.   J. Funt and J. H. Magill, *J. Fire Flam..* 6, 1975, 28.

7.   E. R. Larsen, *'Kirk-Othmer Encyclopaedia of Chemical Technology'*, 3rd edition, Vol. 10, 1980, p. 373.

8   M. Lewin and S. B. Sello, in *'Flame Retardant Polymeric Materials'*, M. Lewin, S.M. Atlas and E. M. Pearce editors, Vol. 1, Plenum, 1975, 19.

9.   M. Lewin in *'Chemical Processing of Fibers and Fabrics'*, M. Lewin and S.B. Sello editors, Vol.2, Part B, Dekker, 1984, p. 1.

10. M. Lewin, H. Guttmann and N. Sarsour, in *'Fire and Polymers'*, G. L. Nelson ed., *ACS Symposium Series* 425, Washington DC, 1990, p. 130.

11. M. Lewin and A. Basch, in *'Flame Retardant Polymeric Materials'*, M. Lewin, S. M. Atlas and E. M. Pearce editors, Vol. 2, Plenum, 1978, p.1.

12. E. D. Weil in *'Flame Retardant Polymeric Materials'*, M. Lewin, S.M. Atlas and E. M. Pearce editors, Vol. 2, Plenum, 1978, p. 103.

13. J. W. Lyons, *"The Chemistry and Uses of Fire Retardants'*, Wiley-Intersci., 1970, p. 290.

14. S. K. Brauman, *J. Fire Retardant Chem.* 4, 1977, 18.

15. M. Lewin and A. Basch, *Textile Res. J.* 43, 1973, 693.

16. E. L. Back, *Pulp and Paper Mag. Canada, Tech. Sec.,* 68, 1967, 1.

17. J. E. Hendrix, G. L. Drake and R. H. Barker, *J. Appl. Polymer Sci.,* 16, 1972, 257.

18. J. Economy, in *'Flame Retardant Polymeric Materials'*, M. Lewin, S.M. Atlas and E.M. Pearce editors, Vol. 2, Plenum, 1978, 203.

19. Y. P. Khanna and E. M. Pearce, in *'Flame Retardant Polymeric Materials'*, M. Lewin, S.M. Atlas and E. M. Pearce editors, Vol. 2, Plenum, 1978, 43.

20. S. J. Sirdesai and C. A. Wilkie, *Polym. Preprints, ACS Polym. Chem. Ed.* 28, 1987, 149.

21. C. Roderig, A. Basch and M. Lewin, *J. Polym. Sci., Polym. Chem. Ed.,* 15, 1975,1921.

22. T. Kashiwagi, ' *Proc. 25th Symp. on Combustion'*, The Combustion Inst., 1994, 1423.

23. E. M. Pearce, in *'Rec. Adv. in FR. of Polym. Mat.'*, Vol. 1, M. Lewin and G. Kirshenbaum editors, BCC, 1990, p. 36

23a E. M. Pearce, Y. P. Khanna and D. Raucher, in *'Thermal Characterization of Polymeric Materials'*, E. Turi, ed., Academic Press, 1981, 793.

24. R. Lyon, in *'Rec. Adv. in FR. of Polym. Mat.'*, M. Lewin, ed., BCC, Vol. 8, in print, 1997.

25. M. Lewin and A. Basch, in *'Encyclopaedia of Polym. Sci. and Technol. Supplement 2'*, Wiley, 1977, p.340.

26. A. Basch and M. Lewin, *J. Polym. Sci. Polym. Chem. Ed.,* 11, 1973, 3071; 12,, 1974, 2063.

27. M. Lewin, A. Basch and B. Shaffer, *Cellulose Chem. Technol.,* 24, 1990, 477

28. B. Miller, J. R. Martin, C. H. Meiser and M. Gargiulo, *Textile Res. J.* 46, 1976, 530.

29. L. G. Bourland, *SPE ANTEC Proc.* 34, 1988, 631ff.

30. V. M. Lalayan, M. S. Skralivetskaya, V. A. Romanov and M. A. Khalturinskij, *Khim. Fiz.,* 8, 1989, 651.

31. D. J. Scharf, in *'Rec. Adv. in FR of Polym. Mat.'*, Vol. 2, M. Lewin and G. Kirshenbaum editors, BCC, 1991, 55.

32. D. J. Scharf, R. Nalepa, R. Heflin and T. Wus, *Fire Safety J,* 19 ,1992, 103.

33. G. Camino, L. Costa and L. Trossarelli, *Polym. Deg. Stab.,* 12, 1985, 213; *Polym. Degr. Stab.,* 6, 1984, 15; *Polym. Degr. Stab.,* 8, 1984, 243; *Polym. Deg. Stab.,* 12, 1985, 203.

34. M. Endo and M. Lewin, in *'Adv. in FR of Polym. Mat'.*, Vol. 4, M. Lewin ed., BCC, 1993, 171.

35. M. Lewin and M. Endo, in *'Adv. in FR of Polym. Mat'.*, Vol. 5, M. Lewin ed., BCC, 1994, 56.

36. M. Lewin, in *'Adv. in FR of Polym. Mat'.*, Vol. 6, M. Lewin ed., BCC, 1995, 41.

37. M. Lewin and M. Endo, in *'Fire and Polymers'* II, G.L. Nelson ed., *ACS Symp. Series* 599, 1995, 91.

38. M. Lewin and M. Endo, *PMSE ACS Preprints, 71*, 1994, 235.

39. J. Willard and A. E. Wondra, *Textile Res. J., 40*, 1970, 203.

40. M. Lewin, P. Isaacs, S. B. Sello and C. Stevens, *Textile Res. J. 47*, 1974, 700.

41. W. . Reeves, R. M. Perkins, R. M. Piccolo and G. L. Drake, *Textile Res. J., 40*, 1970,223.

42. J. T. Langley, M. J. Drews and R. T. Barker, *J. Appl. Polym. Sci., 25*, 1974, 243.

43. W. Zhu, E. D. Weil and S. Mukhopadhyay, *J. Appl. Polym. Sci., 62*, 1996, 2267.

44. J. W. Hastie, *J. Res. NBS, 77A*, 1972, 773; *Combust. Flame, 21*, 1973, 178; 401.

45. J. W. Hastie and McBee, *'NBS Final Report NBSIR'*, 1975, 75.

46. S. K. Brauman, *J. Fire Flam., 3*, 1976, 117; *J. Polym. Sci. Chem. Ed, 15*, 1977, 1607; *J. Polym. Sci. Chem. Ed, 17*, 1979, 1129.

47. L. Costa, L. Goberti, P. Paganetto, G. Camino and P. Squarzi, *'Proc. 3rd Meeting on FR Polymers'*, Torino, 1989, 19.

48. R. F. Cleave, *Plastics and Polymers,* 1970, 198.

49. I. Gordon, J. J. Duffy and N. W. Dachs, US Patent 4,000,114, 1976.

50. C. S. Ilardo and D. J. Scharf, US Patent 4,388,429, 1983.

51. D. D. O'Brien, US Patent 5,358,991, 1994.

52. R. L. Markezich and R. F. Mundhenke, in *'Rec. Adv. in FR of Polym. Mat.'*, M. Lewin ed., BCC, Vol. 6, 1995, p. 177.

53. A. Ballisteri, S. Roti, G. Montaudo, S. Papalardo and E. Scamporino, *Polymer, 20* , 1979, 783.

54. A. Ballisteri, G. Montaudo, C. Puglisi, E. Scamporino and D. Vitallini, *J. Appl. Polym. Sci., 28*, 1983,1743.

55. J. Green, in *'Rec. Adv. in FR of Polym. Mat'.*, M. Lewin ed., Vol. 4, BCC, 1993, 8.

56. R. Dombrowski and M. Huggard, in *'Proc. 4th Intern. Symp. Additives-95'*, Clearwater Beach, Fl, 1996, p. 1.

57. C. P. Yang and T. M. Lee, *J. Polym. Sci. Polym. Chem. Ed., 27*, 1989, 2239.

58. W. Guo, *J. Polym. Sci. Polym. Chem. Ed.*, 30, 1992, 819.

59. A. Granzow, *Accounts Chem. Res.*, 11, 1978, 177.

60. C. F. Cullis and M. M. Hirschler, *'The Combustion of Organic Polymers'*, Clarendon, 1981.

61. A. Basch, S. Nachumowitz, S. Hasenfrath and M. Lewin, *J. Polym. Sci. Polym. Chem. Ed.*, 17, 1979, 39.

62. A.Basch, B. Zwilichowski, B. Hirschman and M. Lewin, *J. Polym. Sci. Polym. Chem. Ed.*, 17, 1979, 27.

63. J. W. Hastie and D. W. Bonnel, *'NBS Report NBSIR* 80-2169', Washington, DC, 1980.

64. J. W. Hastie, *J. Res. NBS*, 77A, 1973, 733.

65. S. V. Levchik, G. F. Levchik, A. I. Balabanowich, G. Camino and L. Costa, *Polym. Deg. Stab.*, 54, 1996, 217.

66. A. Ballisteri, S. Foti, G. Montaudo, E. Scamporino, A. Arnesano and S. Calagari, *Macromol. Chem.*, 182, 1981, 1303.

67. G. Camino, L. Costa, L. Trossarelli, L. Costanzi and G. Landoni, *Polym. Deg. Stab.*, 8, 1984, 13.

68. G. Camino, L. Costa, L. Trossarelli, L. Costanzi and A. Pagliari, *Polym. Deg. Stab.*, 12, 1985, 213.

69. S. V. Levchik, in *'Rec. Adv. in FR of Polym. Mat'.*, M. Lewin, ed., BCC, Vol. 8, 1997, in print.

70. M. Lewin and M. Endo, Unpublished Results, 1992-1993, *Report to Showa Denko Company*, 1993.

71. S. Bourbigot, M. Le Bras, R. Delobel and P Bréant, *Polym. Deg. Stab.*, 54, 1996, 275.

72. S. Bourbigot, M. Le Bras and R. Delobel, in *'Rec. Adv. in FR of Polym. Mat.'*, M. Lewin ed., BCC, Vol. 7, 1996, p. 201.

73. C. Siat, S. Bourbigot and M. Le Bras, in *'Rec. Adv. in FR of Polym. Mat.'*, M. Lewin ed., BCC, Vol. 7, 1996, p.318.

74. S. Bourbigot, M. Le Bras and C. Siat, in *"Rec. Adv. in FR of Polym. Mat.'*, M. Lewin ed., BCC, Vol. 8, 1997, in print.

75. M. Le Bras and S. Bourbigot, in *"Rec. Adv. in FR of Polym. Mat.'*, M. Lewin ed., BCC, Vol. 8, 1997, in print.

76. S. V. Levchik, G. F. Levchik, G. Camino, L. Costa and A. I. Lesnikovich, *Fire Mater.*, 20, 1996, 183.

77. C. E. Anderson, J. Dziuk, W. A. Mallow and J. Buckmaster, *J. Fire Sci.*, 3, 1985, 161.

78. S. Bourbigot, M. Le Bras, R. Delobel, R. Decressain and J.-P. Amoureux, *J. Chem. Soc.*, *Faraday Trans.*, 92, 1996, 149.

79. P. R. Hornsby, *Fire and Mater.*, 18, 1994, 269.

80. P. R. Hornsby and C. L. Watson, *Polym. Deg. Stab.*, 30, 1990, 73.

81. F. Molesky, in *'Rec. Adv. In FR of Polym. Mat.'*, M. Lewin and G. Kirshenbaum ed., BCC, Vol. 1, 1990, p.92.

82. J. Levesque, in *'Rec. Adv. In FR of Polym. Mat.'*, M. Lewin and G. Kirshenbaum ed., BCC, Vol. 1, 1990, p.102.

83. O. Kalisky, R. J. Mureinik, A. Weismann and E. Reznik, in *'Rec. Adv. In FR of Polym. Mat.'*, M. Lewin and G. Kirshenbaum ed., BCC, Vol. 4, 1993, 140.

84. C.F. Cullis and M. M. Hirschler, *Euro. Polym. J.*, 20, 1984, 53.

85. K. Kanemitsuya, in *'Rec. Adv. In FR of Polym. Mat.'*, M. Lewin and G. Kirshenbaum ed., BCC, Vol. 2, 1991, 220.

86. S. Miyata, US Patent 5,401,442, 1995.

Intumescence: Mechanism Studies

CHAR-FORMING ADDITIVES IN FLAME RETARDANT SYSTEMS

E. D. Weil, W. Zhu, H. Kim, N. Patel,

Polymer Research Institute
Polytechnic University, Brooklyn, NY

L. Rossi di Montelera,

University of Turin, Italy

1. INTRODUCTION

Flame retarding poorly charrable polymers, especially with non-halogen systems, presents a challenge. Examples of difficult systems are the styrenics, the polyolefins and the aliphatic polyamides. These have a high heat of combustion and typically burn completely with very little char formation. Reaching a good flame retardant rating, such as UL 94 (ANSI/ASTM D635-77) V-0, usually requires such high loadings of additives that the result is either diminished heat distortion temperature (especially if the additive is soluble) or diminished impact strength (especially if the additive is an insoluble solid). Moreover, the additive in many cases adds cost, and additional problems of bloom, plate-out, mold fouling, deterioration of electrical properties, and reduced processability are often found. Several ways of escaping from these problems have been considered.

One approach is to use catalytic additives to initiate char formation. Our finding of the benefit of iron compounds in nylon-4,6 came from that approach[1], and our discovery of the char-promoting action of potassium carbonate in styrenic polymers containing diene rubbers[2,3] can also be explained by catalysis of oxidative crosslinking. A second approach is to include a flame retardant additive which itself provides substantial char; many examples are known, and we recently have studied a synergistic char-forming system which uses Monsanto's XPM-1000, a cyclic neopentyl phosphonate, as char source[4]. A third approach is to blend into the less char-forming polymer a better char-forming polymer; in the best circumstances, this polymer can also improve other properties as in the case of PPO-HIPS or PC-ABS blends.

We have found good results in combining these approaches, and in general we advocate using these tools in combination, a "systems approach" to flame retardancy[5]. In this paper, we will briefly summarise our work on each approach and we will also present some newer results.

2. METHODOLOGY

We prefer to utilise a top-down burning mode as exemplified by oxygen index (OI, ASTM D63/77 (1977)) and a bottom-up burning mode exemplified by UL 94. We usually find a poor correlation between these two types of small-scale burning tests. We consider that oxygen index is justified by its sensitivity to small changes in composition, its fair correlation to rate of heat release, and its continued use in flammability research and in the wire and cable field. UL 94 is justified for applications-oriented studies because of its widespread industrial usage, and its ability to visually reveal useful phenomenological information about the burning process.

3. REVIEW

3.1. Catalytic approaches

This has been a mode of attack by many research groups, to name a few, Wilkie at Marquette and the Lille Research Group, and will be adequately reported elsewhere.

Our own efforts with this approach are ongoing (whenever this approach seems appropriate in an industrially supported problem-solving project). Only some of the past results are reported.

3.2. Iron additives in a non-halogen system; catalytic action?

Our study of nylon-4,6 flame retardancy revealed a strong effect of iron compounds in a non-halogen flame retardant system containing polyphenylene oxide or other charrable polymer[1]. Here, the catalytic effect is easily postulated but not so clearly explained since iron compounds are able to catalyse dehydrogenation[6], oxidation[7], coupling and structural reorganisation of carbon.[8]. Empirically, in the nylon-4,6 study, we observed that the presence of ferric oxide led to smoother and more coherent char. We also found by X-ray diffraction that when ferric oxide was added to the nylon-4,6, the principal iron compound left in the char was Fe_3O_4. This finding might be construed as pointing to the iron compound acting as a oxidising reactant rather than as a catalyst, but on the other hand, Fe_3O_4 was found active, which points to both oxides being catalysts. In this iron-containing system, we cannot even exclude some flame zone effect perhaps from traces of iron carbonyl (flame emissivity was visibly greater). More work would be needed to elucidate the exact role of the iron additive.

3.3. Potassium carbonate in ABS or HIPS

While searching for charring catalysts for ABS or ABS-SMA blends, we found that potassium carbonate appeared to be a strong flame retardant by OI, although less so by UL 94[2,3]. We finally elucidated that this effect was caused by oxidative crosslinking of the rubber component of the ABS. Evidence was as follows: oven-ageing of the test bars under air but not under nitrogen strongly enhanced the effect; a zone of darkened solvent-insoluble material formed on the outside of the test bars under these conditions, and if removed, the underlying bar had enhanced flammability. The darkened zone material showed evidence of carboxylate bands by infrared; the nitrile band was undiminished and there was no evidence of carboxamide. Omitting the rubber component and testing potassium carbonate in SAN alone eliminated the flame retardant effect. Although we found no data in the literature on alkaline catalysis of oxidative crosslinking of diene rubbers, we did note that alkaline catalysis of auto-oxidation of various other organic substances is known[9].

The practical utility of this mode of flame retardancy is very doubtful, since it involves in effect a surface treatment (oxidative crosslinking) of an already-moulded plastic object - this does not fit the usual sequence of plastics manufacturing. Moreover, potassium carbonate is a water-soluble alkaline compound rather unsuitable as a plastics additive and we could not find any less-soluble less-alkaline relative of potassium carbonate which would elicit the same type of surface oxidative crosslinking.

3.4. Flame retardant additives which produce char and are self-catalyzing

It is possible to design the "carbonific" function and the catalytic function into a single molecule; sometimes also a "spumific" function can be included. This approach has been the subject of extensive work over many years, at American Cyanamid, Hoechst, Albright and Wilson, Borg-Warner, Montedison, and many university laboratories. A full review would be beyond the scope of this paper. Several studies are worth specific mention because they pioneered a substantial and ongoing development effort; an extensive study was made by Halpern on a series of pentaerythritol phosphates which are char-forming self-catalysing and intumescent[10]. This technology has been carried forward to commercial development by Great Lakes Chemical, one product being a bismelaminium salt of pentaerythritol spirobis-phosphate and another product being a bicyclic monophosphate alcohol from pentaerythritol.

Another further study was made by Wolf at Sandoz on neopentylene phosphates[11]; this led to at least one commercial compound, a bis(cyclic neopentyl) thiophosphoric anhydride used as an additive in viscose rayon. The patent literature is replete with many further cyclic phosphates and phosphonates as flame retardants; as examples, many such compounds have been synthesised and patented by researchers (Telschow *et al.*) at the Akzo Nobel laboratory[12] in Dobbs Ferry, New York.

3.5. Cyclic neopentyl phosphonate/melamine phosphate system; one additive as char-former and catalyst, another as spumific and catalyst

One recent product of this self-catalysing char-forming type came out of research at Monsanto in St. Louis, and is now in commercial development. This is XPM-1000, a cyclic neopentyl phosphonate introduced by Monsanto several years ago primarily for use in flexible urethane foams[13]. We noted that it has characteristics which appeared useful for electrical cable jacket flame retardancy, namely, thermal and hydrolytic stability and a potential for char formation as an additive in poorly-charring systems. We found, however, that by itself, it was not particularly effective in our ethylene vinyl acetate copolymer test system. Zhu postulated that it was too slow to form char, and therefore he tried it with additional additives expected to be catalytic for char formation. Melamine phosphates, in particular melamine pyrophosphate, were found to be best in this role.

We found an optimum flame retardant performance with combinations of XPM-1000 with melamine pyrophosphate, expressed by both oxygen index and UL-94 results. By substituting other char formers and other sources of phosphorus, we could establish that a main role of the XPM-1000 was to provide char, and a main role of the melamine pyrophosphate was to catalyse charring[4]. Another interesting finding made in this study was that by X-ray photoelectron spectroscopy, Mukhopadhyay could observe bands strongly indicative of P^V-NH-P^V or P^V-NH$_2$ structures in the char. We interpret this indication as a clue to one possible cause of P-N synergism, namely the formation of low-volatility polyphosphoramidic or polyphosphorimidic acids. We also note that the phosphorus content of the char is 74-88% titratable phosphate, by washing the char with cold water and then titrating the acidic washings by the classical phosphomolybdate method. Moreover, an infrared spectrum of the char showed the bands of phosphoric or polyphosphoric acid plus an additional small but sharp band at 1400 cm^{-1}, possibly ammonium.

3.6. Evidence for the importance of rate of char formation in the XPM-1000/melamine pyrophosphate system

In this system, we found that the flammability results as a function of additive level did not correlate to char yield. Interestingly, Zhu found a good correlation to rate of char. Measuring rate of char formation under actual burning conditions posed a challenging problem.

A simple but effective method was developed by Zhu[4]. He found that upon elevation of oxygen in the OI apparatus to 0.5-4% above the oxygen index value, so as to enable the sample to burn to the bottom of the bar, the rates of burn and the char yields tended to compensate so that their product, i.e. the *rate of char formation*, remained rather constant as the % oxygen varied. For this set of compositions at least, it was therefore possible to measure rate of char formation under reasonably "realistic" virtually steady-state burning conditions (admittedly at superatmospheric oxygen percentage). To permit the rate of char formation to be computed as the product of char yield, it is necessary to burn the entire specimen (it is mounted on a pin to facilitate a complete burn).

It is a reasonable point of view to consider rate of char formation to be just as fundamental as rate of burn or char yield. The only limitation of the method is that it depends on obtaining a char sufficiently coherent so that it can all be captured and weighed reliably, also the burning must be reasonably steady so that the measured burning time is reproducible.

A correlation to rate of char formation rather than char yield is a provocative finding. We find by surveying the literature, that this kinetic factor of rate of char formation has been much less considered, despite all the attention to char yield by Van Krevelen and others. We must admit to some misgivings about our mode of measurement of char forming rate, but alternatives such as thermogravimetric or DSC methods also seem dubious and too far removed from real fire conditions. We therefore present very tentatively our conclusion that rate of char formation is likely to be a critical factor in flame retardancy. We believe that further research on the kinetics of char formation under fire exposure conditions would be productive; the fast formation of a char layer on an object exposed to flame would obviously be highly protective.

3.7. Char formation in conjunction with a supporting material for effective barrier formation

We were funded by the Electric Power Research Institute (a U. S. utility organisation) to find new approaches to the flame-retarding of electrical power cable insulation. In our attempt to find a system suitable for use in primary insulation rather than only in cable jackets, we were limited to materials with good electrical properties.

In solving this problem in EPDM elastomer, we were able to make effective use of the formation of a barrier material comprising a major portion of an inexpensive calcined kaolin, normally used in insulation up to rather high loadings, plus PPO as a char-forming polymer (see below), plus melamine which does "double duty" both as an endothermic material and as a contributor to barrier formation[13a].

Table 1. *Effect of different PPO grades on flammability properties of EPDM formulations.*

PPO added	M.wt. (number average)[b]	OI (%)	UL 94 rating (at 1/8")	Char yield (%)	Char forming rate (mg/min)
HPP823	24,800	29.8	burning	25.3[d]	90[e]
HPP820	22,700	30.4	V-0	24.6	94
HPP821	13,300	33.7	V-0	22.8	103
PPOM1[c]	8,500	31.7	V-0	21.8	94
PPOM2[c]	3,200	31.9	V-0	22.9	96

[a] *All formulations contain 100 phr of EPDM, 100 phr of melamine, 60 phr of kaolin, 30 phr of PPO resin, and 3 phr of dicumyl peroxide used as curing agent.*

[b] *Provided by GE.*

[c] *Made by degradation of higher molecular weight PPO by method of H. S.-I. Chao and J. M. Whalen, Reactive Polym., 15, 9 (1991).*

[d] *Average CY from four runs with a standard deviation (S$_{n-1}$) of 0.4%..*

[e] *Average CR from four runs with a standard deviation (S$_{n-1}$) of 1.9 mg/min..*

3.8. Poly(2,6-dimethylphenylene oxide) (PPO) as a char forming additive for EPDM; an effective system with kaolin and melamine

The use of PPO as a char-forming polymer has a strong precedent in its use in NORYL where it provides char formation while also serving to benefit thermo-mechanical properties.

In our EPDM system, we recognised that the PPO would not be miscible, and would serve only as the char former, so we were determined to keep the level to a necessary minimum and to select the most effective grade of PPO for this purpose. We found that by using PPO together with a rather high loading of calcined kaolin, which is common practice in formulating EPDM insulation, and by including melamine for its endothermic and fuel-diluting effects, we could reach our flame retardant goal with tolerably low levels of PPO. The three additives, PPO, kaolin and melamine actually gave a synergistic result. All the barrier layers (char plus kaolin) from the effective compositions were dark, cement-like and coherent, whereas leaving out any of the three additives gave much poorer less-protective barrier layer and much poorer flame retardancy. An FTIR spectrum of the char shows bands consistent with PPO char and with melone (the thermoset condensation product of melamine).

More detailed study of the char-forming action of PPO; why is PPO particularly effective?

We were curious as to why PPO was superior to other char forming polymers which we tried as alternatives to PPO in the kaolin-melamine-char former-EPDM system. We had expected PPS, PEKK or a polyimide to perform at least as well as PPO but they did not. Zhu made the key observation that PPO melted and flowed prior to charring, whereas the PEKK and polyimide did not. PPS did melt before charring but did not flow as much as PPO.

This led to a tentative postulate: for a polymeric char-forming additive to contribute good protective barrier-type char, *it is desirable for it to melt and flow, at least enough to cohere, before charring*. This requirement probably could be deduced *a priori* since individual discrete particles of char would not be expected to provide a good barrier. Otherwise, merely dispersing carbon black into a polymer would make it flame retardant, which is contrary to common experience.

Another experiment supporting the melt-before-charring hypothesis

Another simple experiment was conducted to test the melting-before-charring hypothesis. A PPO resin which gives reasonably good flame retardant performance in our three-additive

EPDM formulation was heated in air at 200° for 5 h to give a yellowish oxidatively-crosslinked product which did not melt when burning in air. Consistent with the melt-before-charring idea, it gave poor flame retardant performance (LOI 25.7, burn in UL-94) in the EPDM system. The formulation failed to produce coherent char upon burning.

Why was one particular grade of PPO superior?

Much less obvious was the cause of a reproducible superiority in flame retardant contribution of PPO with a middle molecular weight PPO in the kaolin-melamine-PPO-EPDM system. We had obtained a range of PPOs from GE and synthesised a low M.wt. PPO by a known method.

A complete explanation of the flame retardancy differences was not possible because we do not have a complete knowledge of the structural and compositional details of these proprietary commercial products. We were unable to detect any feature of the molecular weight or the TGA behavior of the best PPO resin (HPP 821) which correlated with the best performance. However, a partial explanation for the superior performance of the best PPO proved to be challenging and enlightening.

As in the case of another flame retardant system discussed above, we expected to find a correlation of flame retardant performance to char yield. Once again, we found that this was not the case. The PPO which worked best was not the one giving the greatest char. Here again, the flame retardant performance related very well to char-formation rate as measured by Zhu's combustion method which was described above. In fact, the correlation of oxygen index to char forming rate was linear with a coefficient of correlation of 0.95.

Table 2. *Thermogravimetric characteristics of the various PPO resins under N_2 at 20°C/min.*

PPO resin	Temp. at 1 wt% decomposition (°C)	Temp. at max. DTG peak (°C)	TGA residue at 510°C (wt. %)	TGA residue at 700°C (wt. %)
HPP823	403	467	60.4	37.4
HPP820	416	476	57.2	35.7
HPP821	438	469	46.8	31.1
PPOM1	441	467	44.8	27.2
PPOM2	426	468	46.4	28.2

We now have found two instances, in rather diverse flame retardant systems, of a correlation of flame retardancy to char-forming rate rather than char yield. Not much has been published on the kinetics of char formation in burning systems, and we think this topic deserves further attention.

3.9. Novolacs as char-forming additives

The use of novolacs (non-thermoset oligomeric products of acid-catalysed phenol-aldehyde condensation) as char-forming components in flame-retardant polymer blends is not new[14,15] but appears to be a productive area for further work. Novolac resins range from materials containing volatile free phenols on one hand to thoroughly-devolatilized electronic encapsulating grade novolacs on the other hand. The colors can range from light tan to black. Their molecular weights range from about 200 up to about 30,000. They can be made from phenol, cresols or other alkylphenols. Formaldehyde is the usual aldehyde but glyoxal or other aldehydes are sometimes used. Many grades are available commercially, and they are easily made in the laboratory. The selection or even deliberate design of novolacs for optimum contribution to flame retardancy is certainly possible.

Some studies have been done already on the decomposition of novolacs in the absence of oxygen and in air. Using photothermal beam infrared spectroscopy, Morterra and Low examined the pyrolysis of phenol formaldehyde resins and found that formation of diphenyl ethers takes place, then aryl-aryl bonds are formed[16]. There are also a few studies which suggest structural correlations to char yield. An early study of char yields of phenol formaldehyde novolacs showed that the char yield was at a maximum at about 1.1:1 formaldehyde/phenol ratio[17]. On the other hand, more glass-like impervious carbons were formed if the formaldehyde/phenol ratio was higher[18].

A study initiated by L. Costa and G. Camino (University of Turin) and completed in the Polytechnic laboratory by L. Rossi di Montelera, further elucidated both the behavior of the novolacs themselves and their behavior as char formers in a thermoplastic (SAN). The first part of this work recently published[19], relates to the stability under air or nitrogen of the novolacs in the absence of other components. The degradation reaction was shown to involve oxidation steps regardless of whether the degradation was done under nitrogen or air. In the absence of air, the oxygen of the phenolic hydroxyl interacted with the methylene groups. In this study, phenol-formaldehyde novolacs were most stable, p-cresol-formaldehyde novolacs least stable.

More recently, triphenyl phosphate and novolacs of several types, including lightly crosslinked novolacs, were evaluated in SAN by use of TGA, infrared, and flammability measurements to attempt to elucidate structure/activity relationships and also to gain some mechanistic insight[20]. Briefly, what was found in this study was that triphenyl phosphate was synergised in flame retardant activity by novolacs from phenol and m-cresol, but not by a novolac from p-cresol. Light crosslinking of the novolacs interfered with their flame retardant contribution. Those novolacs that were effective in flame retardancy were found to have increased char yield and helped retain phosphorus in the condensed phase. Triphenyl phosphate, which normally is quite volatile at fire exposure temperatures, was partly volatilised and partly retained. By the time the char was formed, the retained phosphate had become partly hydrolysed to a phosphorus acid (most likely, a partial ester).

3.10. Combining char formers for enhanced flame retardancy results

This work reported at Salford in 1996 and previously published[5], is only briefly recapitulated here. We found in a study of flame retarding ABS that it was advantageous to use a combination of char forming additives such as PPO and novolac, along with a phosphate plus melamine. TGA showed that novolacs decomposed at a temperature somewhat below the main decomposition temperature of the ABS, whereas PPO decomposed at a temperature slightly above the main decomposition temperature of ABS.

We also noted that novolac addition tended to help in the first ignition of the UL-94 bars whereas PPO helped in the second ignition. Although the two steps were not sharply demarcated, we postulate that we are dealing with one additive (the novolac) which makes less-protective char quickly, and a second additive (PPO) which makes more-protective char more slowly. Empirically, we could demonstrate by statistical regression analysis on UL-94 burn time data that there was a synergism between the two char formers, i.e., a positive interaction term having statistical significance by the t-test.

Novolac selection by evaluation of commercially available products

We have discussed above some of the more systematic approaches to novolac selection. However, as a practical matter, we were faced with choosing between a large number of commercially available novolacs which vary in selection of starting phenol, aldehyde-phenol ratio, catalyst, reaction conditions, stripping, and other process variables often not revealed in detail by the resin manufacturers. In our study of phosphates plus novolacs as char formers in styrenic polymers, the best performance was given by commercial phenol-formaldehyde novolacs having molecular weights toward the high end of the range of available novolacs. Novolacs from o-cresol, m-cresol and higher alkylphenols were also found effective, but, consistent with the findings of Rossi, we got poor results with p-cresol novolacs.

3.14. A search for other useful char-forming blend components

We hope to explore further char-forming additives. It is tempting to synthesise polymers for this purpose but we are doubtful of the economics unless highly efficient systems can be found.

Perhaps most productive is to seek blends where the char forming polymer also contributes useful thermo-mechanical properties. The idea that it is desirable for the polymer to melt before charring may be a helpful guide. Char-forming polymers which seem reasonable to use by virtue of high char yields but which tend to char with insufficient flow might be modified to make them lower melting and thus better flame retardant adjuvants.

Table 3. *Flammability of Formulations Containing Various Char-forming Resins.*

Polymer added	OI (%)	UL 94 rating at 1/8"	Char Yield (%)	Char-forming Rate (mg/min)
PPO (HPP821)	33.4	V-0	25.2	126
PPS	25.7	burning	28.3	72
Polyimide	25.3	burning	31.1	72
PEKK	24.6	burning	27.3	53

a All formulations contain 100 phr of EPDM, 80 phr of melamine, 60 phr of kaolin, and 3 phr of DCP in addition to the 40 phr of the char-forming polymers indicated in the table.

The observed correlation of flame retardancy to char formation rate rather than char yield (Table 3) is provocative. We have no idea at this point how general the correlation may be, although we have now found it in two quite diverse flame retardant systems, i.e. the intumescent system of XPM-1000 and melamine pyrophosphate in EVAc and the present system of PPO, melamine, and kaolin in EPDM. Intuitively, it seems reasonable that char formation rate should be important. The sparsity of studies of this kinetic variable, in the face of the very extensive studies of char yield vs. flame retardancy, is surprising. One implication of this idea is that more effort should be put on char-forming catalysts.

4. CONCLUSIONS

We have reviewed and given new examples of char-forming systems which are based on catalysis (postulated), self-charring additives (some providing intumescence), and charrable polymers. Newer results are:

1. A system of a char-forming cyclic neopentyl phosphonate and a melamine phosphate serving as catalyst provides a synergistic intumescent effect. The maximum efficacy appears at the composition giving the fastest char formation, not the maximum char yield.

2. In this system, X-ray photoelectron spectrometry gave evidence for phosphorimidic or -amidic species which may provide a clue as to N-P synergism. A major fraction of the phosphorus in the char could be washed out with water and titrated as inorganic phosphate.

3. A system of polyphenylene oxide plus melamine plus calcined kaolin gives a synergistic flame retardant effect in peroxide-cured EPDM, apparently related to the formation of a cement-like char/melone/kaolin barrier. The optimum PPO appears to be the one giving the fastest char formation, not the one giving maximum char yield. Other char-forming polymers (PPS, polyimide, PEKK) were inferior to PPO in this formulation, which was attributable to their tendency to char before flowing, and to their lesser rate of char formation despite giving higher char yields.

4. In a styrenic polymer system, novolacs are useful char formers and some evidence was found showing and explaining a positive interaction of the novolac with other char formers such as PPO and with aromatic phosphates. Empirical testing and some systematic research provides guidelines for selection of phenolic char-forming additives.

References

1. E. D. Weil, N. Patel and R. Leeuwendal, U.S. Patent 5,071,894 (to Stamicarbon BV), (1991).
2. E. D. Weil and N. Patel, in *"Additives '95"*, Clearwater Beach, FL, Feb. 22-4, 1995.
3. E. D. Weil and N. Patel, in *"Conference on Recent Advances in Flame Retardancy of Polymeric Materials"*, Stamford, May 1995.
4. W. Zhu, E. D. Weil and S. Mukhopadhyay, *J. Appl. Polym. Sci.,* **62**(13), 2267 (1996).
5. E. D. Weil, W. Zhu, N. Patel and S. Mukhopadhyay, *Polymer Degradation and Stability,* **54**, 125-136 (1996).
6. R. E. Zielinski and D. T. Grow, *Carbon,* **30**, 295-299 (1992).

7. H. K. Kung and M. F. Kung, *Advances in Catalysis,* **33**, 159-238 (1985).

8. H. D. Marsh, D. Crawford and D. W. Taylor, *Carbon,* **21**(1), 81-7 (1983).

9. G. Sosnovsky and E. Zaret, *Organic Peroxides,* **1**, 517-560 (1970).

10. Y. Halpern, M. Mott and R. H. Niswander, *Ind. Eng. Chem., Prod. Res. & Dev.,* **23**, 233-238 (1984).

11. R. Wolf, *Ind. Eng. Chem., Prod. Res. & Dev.,* **20**(3), 413-420 (1981).

12. J. E. Telschow, U. S. Patents 5,362,898 (1994), 5,420,326 (1995), 5,536,863 (1996) (to Akzo Nobel N.V.); J. E. Telschow and E. D. Weil, U. S. Patents 5,235,085 (1993) and 5,237,085 (1993) (to Akzo N.V.); W. J. Parr, A. G. Mack and P. Y. Moy, U. S. Patent 4,801,625 (to Akzo America, Inc.) (1989).

13. F. E. Paulik and C. R. Weiss, U. S. Patent 5,276,066 (to Monsanto Co.) (1994).

13a. W. Zhu and E. D. Weil, in *"Fire & Polymers II, Materials and Tests for Hazard Prevention",* G. L. Nelson ed., *ACS Symposium Series* **599**, 199 (1995).

14. C. Taubitz *et al.,* U. S. Patent 4,618,633 (to BASF) (1986).

15. V. Muench *et al.,* U. S. Patent 4,632,946 (to BASF) (1986).

16. C. Morterra and M. J. D. Low, *Carbon,* **21**, 525-530 (1985).

17. S. J. Mitchell, R. S. Pickering and C. R. Thomas, *J. Appl. Polym. Sci.,* **14**, 175 (1970).

18. G. Bhatia, R. K. Aggarwal, M. Malik and O. P. Bahl, *J. Mater. Sci.,* **19**, 1022 (1984).

19. L. Costa, L. Rossi di Montelera, G. Camino, E. D. Weil and E. M. Pearce, *Polymer Degradation and Stability,* **86**, 23-5 (1997).

20. L. Costa, L. Rossi di Montelera, G. Camino, E. D. Weil and E. M. Pearce, *J. Appl. Polym. Sci.,* *submitted.*

Acknowledgements

The Polytechnic Authors wish to thank the Electric Power Research Institute and DSM for support and for permission to publish these results. We also thank Prof. Eli M. Pearce for help and support in our research. We are particularly grateful to Profs. L. Costa and G. Camino of University of Turin for encouraging Dr. Rossi to come to our laboratory to complete her interesting thesis research on the novolacs, and for permission to include here the preliminary results.

MECHANISTIC STUDY ON INTUMESCENCE

G. Camino and M. P. Luda

Dipartimento di Chimica IFM dell' Università,
Via P. Giuria 7, 10125 Torino, Italy

1. INTRODUCTION

Early fire retardant systems were required to minimise fire risk that is the probability of fire occurrence. Whereas now a continuous trend is observed towards the development of polymeric materials with also reduced fire hazard that is with low impact on people and property involved in fire. This is enforced by mandatory regulations or trade specifications due to strategical evaluations on polymer materials use. A wide field of applications of polymeric materials is involved, ranging from advanced technological uses such as in aircraft, to common uses such as in upholstered furniture. Moreover, the "sustainable development" concepts of modern industrial chemistry are now being applied to fire retardants which require to minimise their health and environmental impact throughout their life cycle including: synthesis, processing, use with possible occurrence of fire, recycling and disposal.

These trends have led to the reduction of the utilisation of halogen-based fire retardant systems, owing to relevant negative side effects which they may produce during the entire life cycle. For example, increase of formation of obscuring, corrosive and toxic smoke occurs when they perform their fire retardant action. Besides, super toxic compounds were detected when some widely used brominated aromatic fire retardants were exposed to heat during manufacturing or use or in fires. Indeed, international bodies, such as United Nations and World Health Organisation, concerned with environmental health impact of chemicals, have undertaken an extensive and deep survey of halogen based fire retardants[1]. Thus, basic research was promoted to support the development of environmental friendly systems which interrupt the self-sustained combustion cycle of the polymer at the earliest step that is the thermal degradation in condensed phase which originates the combustible volatiles thus avoiding environment contamination by volatile products of combustion. A special attention has been devoted to the development of systems which on heating decompose with formation of a large amount of thermally stable residue ("char"). This "char" should be able of acting as a thermal shield for heat transmission from the flame to the polymer and as a physical barrier hindering diffusion of volatiles towards the flame and of oxygen towards the polymer.

"Intumescent" chars resulting from a combination of charring and foaming of the surface of the burning polymers may be able to fulfil these tasks producing an effectively insulating multicellular structure. The fire retardance approach based on intumescent systems is indeed being widely developed because it is characterised by a desirable environmental behaviour.

The results of recent studies indicate that viable strategies to induce or enhance "charring", involve chemical processes in the condensed phase which result in the creation of unsaturations in the oligomeric degradation products of the polymeric material. Further thermal or catalysed crosslinking of these high temperature multifunctional oligomers creates the thermally stable residue. If gases are simultaneously made to evolve between gelation and vitrification of the material, a blown "intumescent" char may be formed which can play both the shielding and barrier role with high effectiveness. Thus, chemical and physical processes (charring reactions and blowing effect respectively) are both involved in intumescence phenomenon.

The relatively low effectiveness of the intumescent fire retardant systems commercially available as compared to halogen systems, is due to the fact that the relationship between the occurrence of the intumescence process and fire protecting properties of the resulting foamed char is not yet understood. This makes the fire retardant effectiveness of intumescent systems difficult to predict *a priori*. Within the frame of a long term basic research project concerned with the systematic study of this relationship, the results are reported here which were obtained in the case of polypropylene, PP fire retarded with the combined intumescent additive ammonium polyphosphate, APP-Poly(ethyleneurea-formaldehyde), PEUFA. In particular, since it is known that the fire retardant effectiveness of APP based systems may depend on APP crystalline form[2], we have used a number of samples of different source and crystalline structure.

2.EXPERIMENTAL

2.1. Materials

Five APP from different manufacturers and of different crystalline structure were studied, which are reported in table 1.

Table 1. *APP studied in the present work.*

sample	trade name	manufacturer	crystalline form
1	exolit422	Hoechst	II (100%)
2	amgard MC	Albright & Wilson	I (100%)
3	batch 11	Albright & Wilson	I (5%) II (75%) III (20%)
4	batch15	Albright & Wilson	I (100%)
5	cros 480	Cros	I (100%)

In addition five samples of APP from Monsanto containing different amounts of form II (60, 68, 82, 98 and 100% respectively) were used for IR calibration.

PEUFA, [poly(1-methylen-2-imidazolidinone):

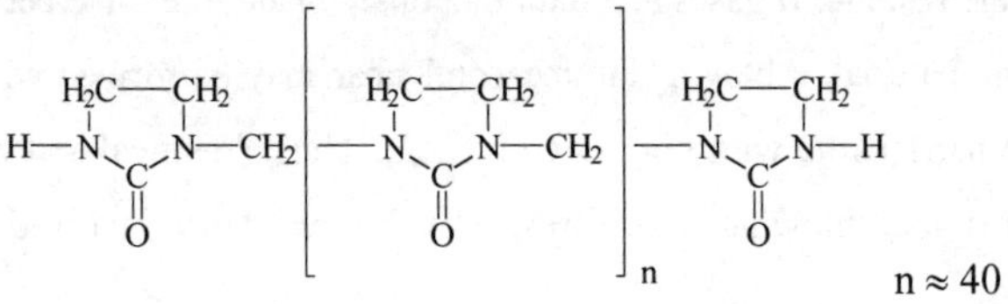

$$n \approx 40$$

and polypropylene (PP; by Himont, Moplen FLF 20) of MFI 10-15 dg/min. (ASTM D 1238/L) were supplied by Himont.

Binary mixtures PEUFA/APP were prepared by manual grinding in a mortar whereas ternary mixtures PP- PEUFA/APP were prepared in a Brabender mixer (AEV 330) at 55 rpm at 190°C under nitrogen flow.

2.1. Techniques

Thermogravimetry (TG) was carried out on a DuPont 951 Thermobalance driven by a TA 2000 Control System at 10°C/min under nitrogen flow of 60 cm^3/min, unless otherwise indicated.

The rate of ammonia evolution was measured by means of an ammonia gas sensing electrode (GSE) (Phoenix Electrode Company) immersed in a basic water solution (NaOH 0.1N) in which the nitrogen flow sweeping the gaseous products, evolved from the heating sample in the termobalance, was bubbled. Calibration was performed with known amount of NH_4Cl in 10M NaOH solution.

Conductivity measurements of evolved ammonia were also carried out in order to confirm GSE response: the exhaust from TG (up to 300°C) was bubbled in a predetermined volume of H_2SO_4 of known title (300 cm^3, 6.0 E-3N) and the decrease of conductivity was monitored in continuos. The decrease of conductivity due to 1.0 mg of NH_3 was evaluated by using NH_4OH. Agreement between GSE and conductivity methods was quite good (± 5%).

Foaming measurements were carried out in static air at a heating rate of 20°C/min in a home made equipment described elsewhere[3] using disks of 22 mm of diameter of the binary mixtures APP/PEUFA prepared by cold sintering under pressure.

Infrared spectra were run using a Perkin Elmer FTIR 1710. Films on KBr were cast from viscous samples and polymer solutions whereas pellets were prepared by sintering of crystalline powder samples with KBr.

Diffraction patterns of APP of table 1 were supplied by the manufacturer and checked with a Philips 1050/81 equipment in Himont (Ferrara) using $K\alpha$ radiation from Cu

3. RESULTS AND DISCUSSION

The understanding of the mechanism of intumescence, that is the interactions between the components of the intumescent additive APP and PEUFA during combustion, requires the detailed knowledge of the thermal behaviour of each component heated separately.

3.1. Infrared characterisation and thermal behaviour of APP

Five crystalline forms of APP are known [4]. Crystals of form I can be prepared by heating equimolar mixture of ammonium orthophosphate and urea under blanket of anhydrous ammonia at 280°C. Their unit cell dimensions were not determined because of lack of crystals large enough for single-crystal-X ray study and impossible to be enlarged by recrystallisation or tempering without conversion to other forms. However, their diffraction pattern is well defined and is used for identification and quantification. Form II crystals, of orthorhombic symmetry, are easily obtained from form I tempering at 200-375°C. Form III was observed as an intermediate in the conversion of form I to form II. Form IV is monoclinic and form V orthorhombic.

In scheme 1 the possible polymorphic transitions accomplished by tempering at selected temperature and then quenching are indicated according to reference 4.

Tempering Temperature, °C	Transition		
100-200	V	→	mixture I and II
200-375	I	→	II
250-300	V	→	II
300	I	→	III (intermediate) → II
300-370	IV	→	II
330-420	I	→	V
385	II	→	V
450-470	I or II	→	IV + glass

Scheme 1. *Polymorphic transitions of ammonium polyphosphate.*

The more common forms available in commercial APP are I and II. Identification of the crystallographic forms is usually done by the diffraction pattern of the powdered polyphosphate.

However, we found that a much simpler method can be used based on IR since APP crystals of form I show three specific IR absorptions at 760, 682 and 602 cm^{-1} which are absent in the spectrum of form II. Whereas the band at 800 cm^{-1} (δ P-O-P) is independent on the crystalline form (Figure 1) IR spectra of standard APP containing known percentage of form I and II (measured by diffraction patterns) have been run and the ratio between the absorbance of the band at 682 and 800cm^{-1} has been plotted towards the percentage of form I. A straight line is obtained which can be used as a calibration curve for an easy and quick determination of the crystallinity of unknown APP.

The thermal behaviour of APP of different crystalline forms has been studied. Cristallinity may involve considerable differences in the degree of polymerisation; for example, form I of APP contains chain with average sequences of PO$_3^-$ tetrahedra much shorter (50-200 units) than form II.

Typical TG in nitrogen of form I and II APP are shown in Figure 2. Two main degradation steps (Temperature of maximum rate of degradation T$_{max}$ 370 and 640°C) for form II APP and three for Form I APP (T$_{max}$ 335, 620 and 835°C) are recognisable. In addition, both samples lose weight slowly at a constant rate between 400 and 500°C.

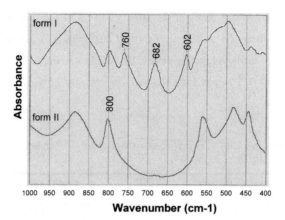

Figure 1. *IR spectra of APP of form I (sample 2, top spectrum) and II (sample 1, bottom spectrum).*

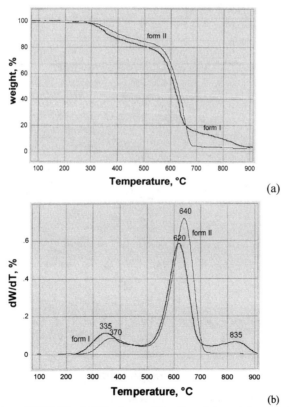

Figure 2. *TG (a) and DTG (b) of APP of crystalline form I (——) [sample 2] or II (- - - -) [sample 1]. Nitrogen flow 60cm³/min; heating rate 10°C/min; Platinum sample holder.*

The thermal behaviour seems to be strictly related to the original crystalline form; form I appears to be thermally weaker than form II beginning to loose weight earlier (onset 250 instead of 280°C); furthermore, the remarkable weight loss in the last step (835°C, 17 wt. %) is typical of form I APP whilst form II slowly looses only 3.5% of the starting weight in this temperature range.

Table 2. *Total weight loss and NH₃ and H₂O evolved from APP samples during the first step of degradation.*

sample	crystalline form	NH_3, wt. %	weight loss 1st step wt. %	difference
1	II	9.5	13.5	4
2	I	4	16	12
3	I:II:III 5:75:20	5.6	13.1	7.5
4	I	9.1	16.1	7
5	I	5.3*	19	13.7

: in two steps

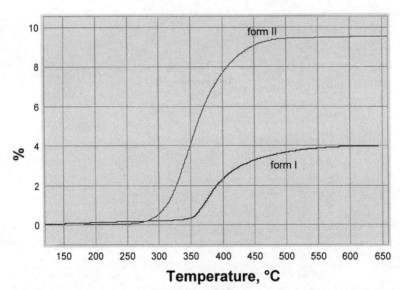

Figure 3. *Ammonia evolution from APP of crystalline form I (- - - -) [sample 2] or II (——) [sample 1]. Specific electrode detector, sample atmosphere: Nitrogen flow 60cm³/min; heating rate 10°C/min.*

The same final residual weight is however reached at 890°C, 2%. Water and ammonia are the products evolved in first step of thermal degradation of APP[5] (335°C, 370°C). In table 2 the amount of total ammonia evolved within the first degradation step (to 500°C) for all samples investigated is reported together with the weight loss.

The amount of ammonia evolved ranges from 4 to 9.5% and seems to depend more on type of sample than on crystalline form of APP (Table 1). However NH_3 is detected by the gas sensing electrode at a lower temperature from form II than I, as shown in Figure 3 although detectable weight loss in TG shows a reverse ranking. This might be due to impurities in form I (e.g. urea) which are absent in form II because form II is obtained on previous heating of form I. Ammonia evolved is, in all cases, lower than stoichiometric (17%) possibly because its evolution is hindered by pH decrease due to formation of polyphosphoric acid units:

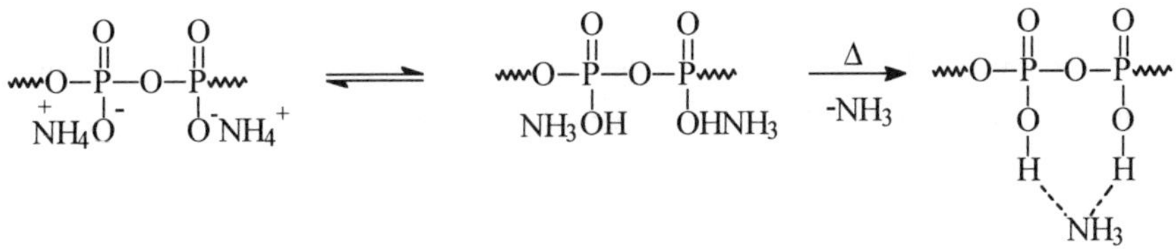

(1)

The amount of water eliminated can be assumed to correspond to the difference between weight loss and NH_3 evolved in the same experiment (Table 2) since there is no evidence of evolution of other products on this step. Water can be eliminated by condensation of free acid OH groups giving the anhydride crosslinks of ultraphosphate:

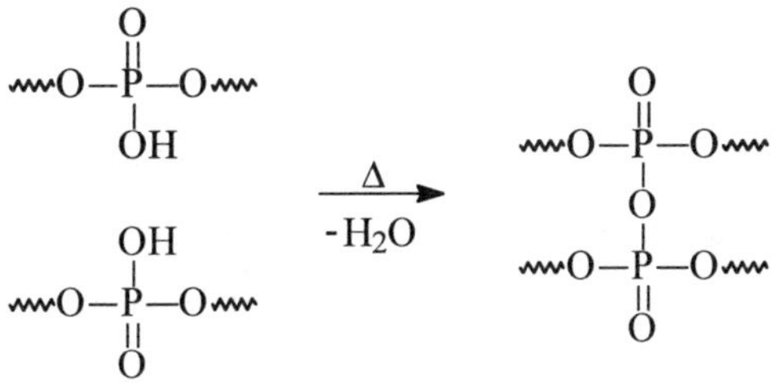

(2)

Polyphosphate   ultraphosphate

or by dehydration of ammonium salt to phosphoramide:

$$\underset{\underset{O^-NH_4^+}{|}}{\overset{\overset{O}{\|}}{\text{www}O-P-O\text{www}}} \quad \xrightarrow[-H_2O]{\Delta} \quad \underset{\underset{NH_2}{|}}{\overset{\overset{O}{\|}}{\text{www}O-P-O\text{www}}} \qquad (3)$$

Condensation of phosphoramidic groups could eliminate ammonia forming phosphorimidic crosslinks:

$$2 \quad \underset{\underset{NH_2}{|}}{\overset{\overset{O}{\|}}{\text{www}O-P-O\text{www}}} \quad \xrightarrow[-NH_3]{\Delta} \quad \begin{array}{c} \overset{\overset{O}{\|}}{\text{www}O-P-O\text{www}} \\ | \\ NH \\ | \\ \underset{\underset{O}{\|}}{\text{www}O-P-O\text{www}} \end{array} \qquad (4)$$

whereas condensation between phosphoramidic groups and acid groups leads to ammonia elimination and anhydride crosslinks or water elimination and phosphorimidic crosslinks:

$$\underset{\underset{NH_2}{|}}{\overset{\overset{O}{\|}}{\text{www}O-P-O\text{www}}} + \underset{\underset{OH}{|}}{\overset{\overset{O}{\|}}{\text{www}O-P-O\text{www}}} \xrightarrow[-NH_3]{\Delta} \begin{array}{c} \overset{\overset{O}{\|}}{\text{www}O-P-O\text{www}} \\ | \\ O \\ | \\ \underset{\underset{O}{\|}}{\text{www}O-P-O\text{www}} \end{array} \qquad (5)$$

$$\underset{\underset{NH_2}{|}}{\overset{\overset{O}{\|}}{\text{www}O-P-O\text{www}}} + \underset{\underset{OH}{|}}{\overset{\overset{O}{\|}}{\text{www}O-P-O\text{www}}} \xrightarrow[-H_2O]{\Delta} \begin{array}{c} \overset{\overset{O}{\|}}{\text{www}O-P-O\text{www}} \\ | \\ NH \\ | \\ \underset{\underset{O}{\|}}{\text{www}O-P-O\text{www}} \end{array} \qquad (6)$$

Formation of phosphoramidic and phosphorimidic groups can both also account for partial retention of nitrogen in the condensed phase.

The main step of thermal degradation (620°C, form I; 640°C, form II) regards the breakdown of the ultraphosphate structure with elimination of high boiling chain fragments and formations of P_2O_5 which can exist in four crystalline forms of different stability. The residue left by P_2O_5 volatilisation (670°C, form I; 680°C, form II) is likely to be a thermally stable -P-N- compound which is generally formed when P and N containing structures survive above 400°C. However this process leaves a larger residue in form I than in form II, which slowly disappear at higher temperature.

3.2. Thermal behaviour of PEUFA

We have already thoroughly investigated the thermal degradation of PEUFA, which acts as the char source in the intumescent system[6]. The nature of the volatile products together with the chemical evolution of PEUFA during the degradation process, indicates that the volatilisation process of PEUFA initiates by statistical chain scission. The resulting macroradicals undergo competing reactions leading to unsaturated oligomers which can either evaporate or crosslink to a thermally stable compact char (20% at 600°C, 18% at 800°C).

3.3. Thermal behaviour of mixtures PEUFA/APP

Most of the PEUFA in mixtures with APP decomposes at a lower temperature than expected because of the action of APP or of its low temperature degradation products[5]. The thermal stability of the mixtures follows the same order of the crystalline forms of APP (compare figures 2 and 4).

Two steps of weight loss are present in the temperature range < 400°C (T_{max} 300-310°C, 30% weight loss and, T_{max} 380°C 12÷15% weight loss). The IR spectra of the residues of degradation of the mixture PEUFA/APP 1/1 w/w collected at the end of each of these steps, carried out at the onset temperature (260° and 318°C respectively) to constant weight are shown in Figure 5. Similar results were obtained with all APP samples of Table 1.

On heating the mixtures to completion of the first step of weight loss, a physical mixture of water-soluble inorganic phosphorous moieties, such as phosphoric and polyphosphoric species, and water-insoluble charred material is obtained. The IR of the total residue of the first step (Figure 5, upper spectrum) shows the presence of P-O-C bonds ($990cm^{-1}$)[7] which are partially hydrolysed by hot water treatment and consequently reduced in the insoluble residue (Figure 5, middle spectrum). The phosphoric ester formation is a common feature found to occur on heating polymers containing reactive functional groups when they are heated in the presence of ammonium polyphosphate, as for example aliphatic polyamides[8].

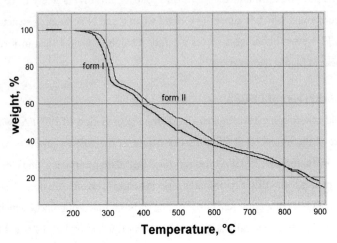

Figure 4. *TG of mixture PEUFA/APP 1/1 w/w; APP crystalline form I (- - - -) [sample 2] or II (——) [sample 1]. Sample atmosphere: Nitrogen flow 60cm³/min; heating rate 10°C/min; Platinum sample holder.*

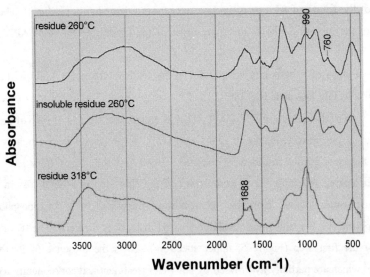

Figure 5. *IR spectra of the residues of degradation of the mixture PEUFA/APP 1/1 w/w collected after the end of each step; APP in form II (sample 1).*

PEUFA chains are likely to be attacked by polyphosphoric acid moieties, as soon as they are formed by heating ammonium polyphosphate, which lowers the temperature of PEUFA degradation of 60-80°C.

(7)

Indeed, protonated PEUFA units may undergo ring opening by proton transfer from position a or b in reaction 8. Polyphosphoric acid units are reformed while structures containing double bonds are created (I and III, route 8a and 8b respectively) and chain scission occurs (route 8-b). Alternatively, route 8c may be followed which leads to carbonium cation phosphate IV.

IV could either evolve to a phosphate ester bond seen in IR of Figure 5 or it can give more stable secondary carbonium ions (V and VII) by hydride transfer, as described in reaction 9a and 9b, and chain scission with formation of N-methylimidazolidinonic end groups containing structures (VI). Further proton transfer gives back phosphoric acid units and unsaturated structures VIII and IX.

Thus PPA would catalyse the thermal degradation of PEUFA giving products of degradation similar to those obtained at higher temperature on heating PEUFA alone[6].

However, since unsaturations are created at lower temperature in the presence of APP, crosslinking to char in the second step of degradation would be more likely in PEUFA/APP mixtures which competes with chain fragmentation and chain fragments volatilisation in pure

PEUFA. Furthermore, crosslinking is likely to be catalysed by polyphosphoric acid. This would explain larger thermally stable residue formed in PEUFA/APP as compared to that calculated assuming an additive behaviour[5].

$$(8)$$

$$(9)$$

In the second step of thermal degradation, besides crosslinking, thermal degradation of PEUFA units not yet reacted with APP can occur evolving volatiles products which blow the building char (intumescence). Indeed, the intumescent process from degrading PEUFA/APP mixtures starts above 300°C in the second degradation stage of degradation as shown in figure 6 in which different crystalline forms of APP have been used. The foaming behaviour depends on whether form I or II APP is used. In particular, the volume of the system suddenly increases 40 times at 310°C with form I whereas it increases somewhat less (30 times) and in a larger range of temperature (from 325 to 335°C) with form II.

The IR of the residue of the second step of degradation reveals the presence of highly condensed polyphosphoric structures; together with absorptions at 1688 and 760 cm^{-1} indicating that imidazolidinonic rings are still present in the intumescent char (figure 5, bottom spectrum).Therefore, together with condensation of the charring material by double bond polymerisation, crosslinking through polyphosphate chains should also occur mostly by P-N-C bonds (phosphoroimide) (Reaction 5).

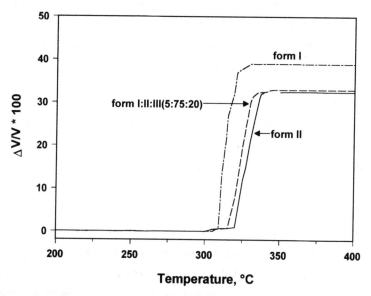

Figure 6. *Foaming behaviour of mixtures PEUFA/APP 1/1 w/w using APP of different crystalline form: form I (-----) [sample 2];form II (———) [sample 1] and form I + II + III [sample 3]; V= volume of the specimen.*

$$\text{(10)}$$

(Chemical reaction scheme 10: PEUFA end-group containing N–H reacts with polyphosphoric acid (OH–P(=O)–O–P(=O)–OH) to give the condensed product with a P–O–P linkage plus H$_2$O.)

Successive condensations lead to a thermally stable residue with flexible segments (the polyphosphate segments) and rigid segments (the PEUFA segments) either original or crosslinked by double bonds polymerisation. This residue can be blown by the simultaneously evolved gases from degradation of residual PEUFA and water of condensation. This could explain why PEUFA heated alone gives a compact char which cannot be blown whereas a multicellular structure is seen by SEM in the char obtained from PEUFA/APP mixtures[5]. The crosslinked PEUFA segments (hard segments) supply the thermal stability and mechanical resistance to the intumescent char whereas polyphosphate segments (soft segments) provide the flexibility for blowing under evolving gas pressure.

3.4. Effect of the polymer matrix

Introduction of an intumescent additive in the polymer matrix might, in principle, affect the intumescent process either from the chemical or the physical point of view. For example mutual modification of the decomposition process or change of the melt viscosity could take place which might affect the intumescent behaviour in the presence of the polymer. Alternatively the polymer could act as an inert diluent in the intumescent process.

Thermogravimetry shows that in a mixture of polypropylene (PP) and 30% of the intumescent additive APP/PEUFA 1/1 w/w the low temperature weight loss step due to the decomposition of the intumescent additive in the polymer matrix occurs at a somewhat higher temperature than calculated assuming an additive behaviour between PP and the mixture APP/PEUFA (319 instead of 301 °C). Whereas the degradation of the polymer as well as the amount of residue either from the degradation of the additive (462°C) or at 600°C are that expected.

4. CONCLUSION:

In conclusion, in APP/PEUFA mixtures, chemical reactions occur between APP and PEUFA on heating which are responsible for the intumescent behaviour. APP catalysing the thermal degradation of PEUFA, creates the structure precursor of the char at relatively low temperature where competition of crosslinking of unsaturated degraded PEUFA structures to char is favoured against volatilisation. Moreover, polyphosphoric acid can condense with PEUFA chain fragments forming phosphorimidic bonds. The yield and thermal resistance of the char from PEUFA may thus improve in presence of APP. Furthermore, the structure of the char includes soft polyphosphate segments which allow for the blowing action of evolving gaseous degradation product. Morphology and volume of the foamed char, which affects the fire retardant effectiveness of the additive, depend on the mixture composition in terms of APP/PEUFA ratio. The thermal behaviour of the additive may be somewhat modified by introduction in a polymer matrix as shown for PP.

The type of crystalline structure of APP somewhat affects the rate of thermal degradation processes occurring on heating these mixtures, resulting in a modification of the occurrence of intumescence which is a rate controlled phenomenon. This is likely to be due to the fact that these reaction occur in condensed phase at a temperature below complete melting of APP.

References

1. G. Camino, M. P. Luda, L. Costa, *"European Meeting on Chemical Industry and Environmental"* (Volume I), E. Casal ed., (1993). 221
2. G. Bertelli, Himont; personal communication.
3. Bertelli G.; Camino G.; Marchetti E.; Costa L.; Casorati E., Locatelli R., *Polym. Deg. and Stab., 25,* 1989, 277.
4. C. Y. Shen, N.E. Stahlheber, D.R. Dyroff, J. Am. Chem. Soc. **91**, (1969) 63.
5. G.Camino, M. P. Luda, L. Costa , ACS Symposium series, **599**, (1995) 76.
6. G. Camino, M. P. Luda, L. Costa, M. Guaita, Makromol. Chem. Phys., **197**, (1996) 41.
7. S.V.Levchik, L. Costa, G. Camino, Polym. Deg. Stab. **36**, (1992) 229.
8. S.V.Levchik, G. Camino, L. Costa, G. F. Levchik, Fire and material. **19**, (1995) 1.

FIRE RETARDED INTUMESCENT THERMOPLASTICS FORMULATIONS, SYNERGY AND SYNERGISTIC AGENTS – A REVIEW

M. Le Bras and S. Bourbigot,

Laboratoire de Chimie Analytique et de Physico-Chimie des Solides, ENSCL
BP 108, F-59652 Villeneuve d'Ascq Cedex, France.
Michel.le-bras@ensc-lille.fr

1. INTRODUCTION

Intumescent technology has recently found a place in polymer science as a method of providing flame retardancy to polymer formulations, especially polypropylene (PP)-based formulations[1]. Intumescent systems interrupt the self-sustained combustion of the polymer at its earliest stage, i.e. the thermal degradation with evolution of the gaseous fuels.

The intumescence process result from a combination of charring and foaming of the surface of the burning polymer (observed between 280°C and 430°C under air using the PP/ammonium polyphosphate/pentaerythritol (PP/APP/PER) model system[2]). The resulting foamed cellular charred layer which density decreases with temperature[3] protects the underlying material from the action of the heat flux or of the flame.

1.1. Chemistry of intumescence

Generally, intumescent formulations contain three active additives : an acid source (precursor for catalytic acidic species), a carbonific (or polyhydric) compounds and a spumific (blowing) agent. In a first stage (T < 280°C), the reaction of the acidic species with the carbonisation agent takes place with formation of esters mixtures.

The carbonisation process takes then place at about 280°C (via both Friedel-Craft reactions and a free radical process[4]). In a second step, the spumific agent decomposes to yield gaseous products which cause the char to swell (280 ≤ T ≤ 350°C). This intumescent material decomposes then at highest temperatures and looses its foamed character at about 430°C (temperatures ranges are characteristic of the extensively studied PP/APP/PER system).

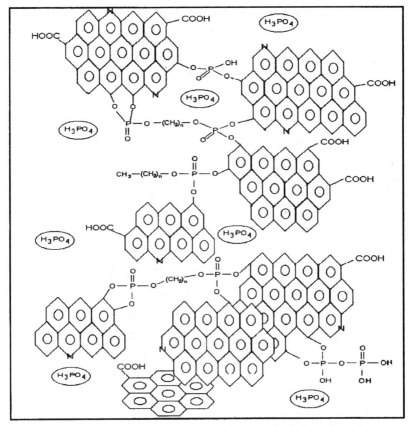

Figure 1. *Intumescent coating resulting from a polyethylenic formulation (additive: APP/PER) heat treated at 350°C.*

The carbonaceous material formed from the additives, plays two different chemical parts in the fire retardancy process:

-it contains free radical species which react with the gaseous free radical products formed during the degradation of the polymer. These species may also play a part in termination steps in the free radical reaction scheme of the pyrolysis of the polymer and of the degradation of the protective material in the condensed phase,

-it is a support for acidic catalytic species which react with the oxidised products formed during the thermo-oxidative degradation of the material.

The material resulting from the degradation of an intumescent formulation is an heterogeneous material. It is composed of « trapped » gaseous products in a phosphocarbonaceous cellular material, i.e., the condensed phase.

This condensed phase is a mixture of solid phases and of liquid phases (acidic tars) possessing the dynamic properties of interest which allows the trapping of the gaseous and liquid products resulting from the degradation of the polymer. The carbonaceous fraction of the condensed phase consists in polyaromatic species which are organised in stacks characteristic of a pre-graphitization stage (Figure 1).

The phosphocarbonaceous material is then a several phases material composed of crystalline macromolecular polyaromatic stacks bridged by polymer links and phosphate (poly-, di- or orthophosphate) groups, crystalline additive particles and an amorphous phase which encapsulates the crystalline domains. This amorphous phase is composed of small polyaromatic molecules, easily hydrolysed phosphate species, alkyl chains formed via the degradation of the additive system and fragments of the polymer chain. It governs the protective behaviour of the coating: this phase has to be voluminous enough to perfectly coat the crystalline domains and has to show an adequate rigidity/viscosity which yields the dynamic properties of interest (it avoids dripping and accommodate stress induced by solid particles and gas pressure).

1.2. Protection via intumescence

The proposed protection mechanism is based on the charred layer acting as a physical barrier which slows down heat and mass transfer between the gas and the condensed phases. The limiting effect for fuel evolving is proved by the presence of the polymer chains in the intumescent material. More, the layer inhibits the evolution of volatile fuels via an «encapsulation» process related to its dynamic properties. Finally, it limits the diffusion of oxygen to the polymer bulk. The limitation of the heat transfer (thermal insulation) is illustrated by a recent study of Siat et al. presented in this Book.

The stability of the intumescent material limits consequently the formation of fuels and leads to self-extinction in standard conditions. Oxygen consumption calorimetry in a cone calorimeter (according to NBS-IR 82:2604) confirms the low rate of the degradation related to the presence of a surface intumescent material (typical example presented in figure 2[5]).

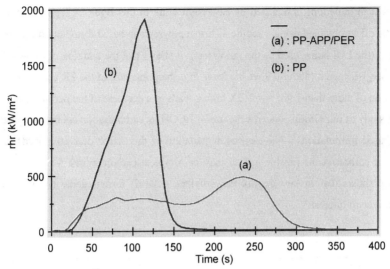

Figure 2. *Rates of heat release (rhr) of PP and PP-APP/PER under an heat flux of 50 kW/m².*

2. SYNERGY AND SYNERGISTIC AGENTS

2-1. Composition of the additive system

The intumescent additive systems developed in our Laboratory are mixtures of at least three ingredients:

-a precursor of a carbonisation catalyst such as APP, diammonium diphosphate (PY) or diammonium pentaborate (APB),

- a carbonisation agent such as PER, xylitol (XOH), mannitol (MOH), d-sorbitol (SOH), β-cyclodextrine (BCOH) or polyamide-6 (PA-6),

- a synergistic agent such as zeolites, clay materials, BCOH and an ethylene-vinyl acetate (8 %) copolymer (EVA-8) or other functionalised polymers.

The copolymer generally plays two parts in the formulations: first, it increases the Limiting Oxygen Index (LOI, ASTM D2863/77) and, secondly, it is a compatibiliser of the additives in the polymer (as an example, EVA-8 maximises the interfacial bonding and so prevents rejection of the mineral additive throughout a polymer matrix [6]).

Addition of an active agent in an additive system may lead either to an additional effect[7], an antagonistic effect[8] or a synergistic effect[7, 8, 9]. The part played by zeolites in intumescent formulations is reported by Bourbigot et al. in this Book. A typical example of synergistic effect obtained using a zeolite 13X in a polystyrene-based formulation is presented in figure 3 (the LOI being used as the rating test). It shows that the addition of a low amount of the synergistic agent (about 0.9 wt. %) leads to a sharp increase of the FR performance and that addition of more than 1 wt. % of 13X higher leads to a decrease of the performance.

A recent study of our Group concerns the use of BCOH as carbonisation agent in intumescent polyols-based formulations[7]. The thermal degradation of this starch derivative leads to high amounts of carbonaceous residue and so may be a typical carbon source for intumescence. The study shows that in low density polyethylene (LDPE) formulations, BCOH is not a carbon source of interest.

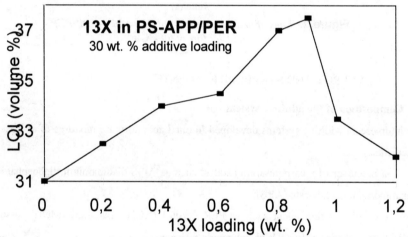

Figure 3. *LOI values of PS-APP/PER/13X formulations vs. the amount of zeolite 13X.*

Nevertheless, it may be added either as a synergistic agent or as an antagonistic agent in LDPE-based formulation in association with other polyols produced by the agrochemical industry[7] (FR performances of two typical intumescent systems are presented in Figure 4). These studies show that an additive may play very different parts depending on its concentration in the polymeric material and its association with the other different additives or fillers.

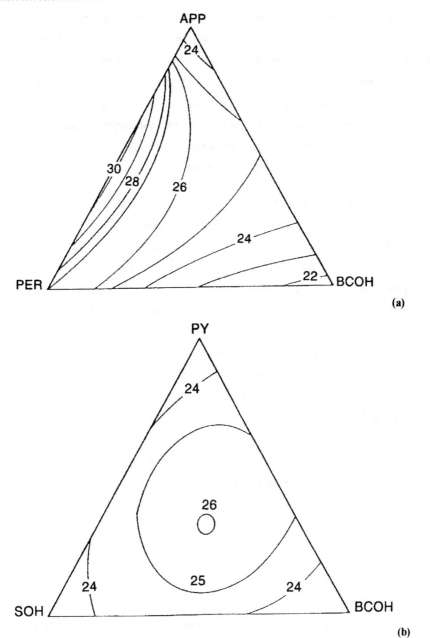

Figure 4. *LOI values of LDPE-APP/PER/BCOH (a) and LDPE-PY/SOH/BCOH (b) formulations in the Scheffé {3-3} lattice used for experiments with mixtures (from Ref.7).*

2.2. Chemical and physical parameters affecting the FR performance

The part played by the synergistic agents does not remain well known. A direct relation between the LOI and the carbonisation mode and/or the amount of stable carbonaceous materials formed from FR formulations had been previously reported[10, 11]. A recent study[4] which considers several new intumescent formulations or carbonising additive systems shows that this relation does not exist whatever is the considered carbonaceous material (intumescent, unexpanded or "high temperature" material) and proposes that the FR performance depends on the « quality of the carbon ».

2.2.1 Acidic character of the char.

A comparison of the thermal behaviour of additive systems[2] containing different phosphate species shows that comparatively low LOI values may be related to the formation of phosphorus oxides which evolve in the conditions of a fire and, so, to the decrease of the phosphate species amount in the char.

More, a previous study of the thermal behaviour of PP/APP/PER and the PP/APP/PER/4A formulations[9] shows that addition of the synergistic agent 4A in APP/PER modify the thermo-chemical behaviour of the phosphates species: phosphorus oxides (bands at about -45 and -55 ppm in MAS-DD NMR spectra) are not observed in the spectrum of heat treated APP/PER/4A. Moreover, addition of low amounts of zeolites limits the formation of condensed polyphosphate species and increases the content of acidic orthophosphate and diphosphate species in the intumescent material.

This trend observed studying several other additive systems leads to propose that a synergistic agent allows the upholding of acidic species in the intumescent material in a large temperature range. It is now well known that the intumescent carbonaceous material is formed by the thermal degradation of esters and Friedel-Craft reactions. The acidic species are catalysts for the formation of « carbon » via the formation of esters (reaction with the oxidised products of the degradation of the polymer) which then degrade.

2.2.2 Free radical in the char.

The « carbon » forms also via a free radical process. Electron spin resonance spectra of intumescent additive mixtures and of intumescent formulations after a thermal treatment at a temperature > 200°C, always present a signal (generally a lorentzian or a gaussian signal or their combination) at $g \approx 2$ related to the presence of π radical[12] in the protective coating formed. These esr study shows that a concentration in free radicals at least equal to 10^{21} spins/kg in the intumescent and in the unexpanded residue is always required to obtain LOI $\geq$ 28%. Moreover, a spin concentration at least equal to 10^{21} spins/kg in the « high temperature » residue is always required to obtain LOI $\geq$ 30%. In addition, this study shows that systems which do not allow a UL-94 classification (ANSI/ASTM D635-77), always present a low concentration ($< 10^{21}$ spins/kg) of free radical species.

The residues are not only composed of carbonaceous species but may be « composite » material containing free acid species and mineral phases. So, the evaluated paramagnetic character of intumescent materials and of residues is not that of the single « carbon » phase. It has been previously shown that a relation exists between the oxygen index values sequence (as an example: LOI $_{polymer-APB/PER}$ < LOI $_{polymer-APP/PER}$ < LOI $_{polymer-APP/PER/4A}$) and the corresponding C*/C sequence (C*/C: ratio of the free radicals concentration (spin/kg) on the atomic concentration of carbon (atom/kg) in the carbonaceous materials resulting of heat treatments of the additive mixtures). In this typical example, the synergistic agent 4A increases sharply the concentration of paramagnetic species in the intumescent material and then increases the reactivity of the carbon.

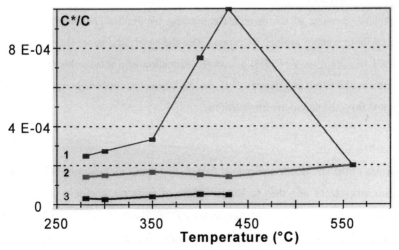

Figure 5. *Relative free radical concentrations C* in the APP – PER-4A (curve 1), the APP - PER (curve 2) and APB - PER (curve 3) systems vs. the temperature.*

The enhanced FR character is then explained by a reaction between gaseous free radical products arising from the degradation of the polymer (scission of the polymer chain or oxidative process) and the « activated » carbon species in the intumescent coating or in the high temperature residue. Such reactions explain the presence of polymer links in the protective coating which forms in fire conditions and thus leads to a decrease of the evolution of the flammable gaseous mixture from the system to the flame. Moreover, the reaction is, in fact, a termination-recombination process which may decrease the rate of the pyrolytic degradation of the carbon-based material and of the polymeric matrix.

2.2.3 Molecular dynamic

In an heterogeneous material, it is possible to observe the effect of spin diffusion with a Goldman-Shen pulse sequence[17] using ^{1}H solid state NMR[18]. Previous studies of our Group considering spin-spin relaxation times, have shown that an intumescent carbonaceous material is composed at least by two distinct structural phases undergoing respectively slow and fast decaying. On the basis of the theoretical considerations of Cheung & Gerstein[19] the mean width, b, of the slow relaxation domains may then be computed. These domains are proposed as amorphous domains in the material[15]. The results of the comparative study of the APP/PER and of the APP/PER/4A additives systems in an ethylene - butyl acrylate- maleic anhydride terpolymer (Lotader) are reported in Table1.

The amorphous domains of the materials containing the zeolite are always comparatively smaller than those without zeolite. We propose that the synergistic agent (LOI $_{Lotader-APP/PER/4A}$ = 39% for LOI $_{Lotader/APP/PER}$ = 30%) allows the formation of a most coherent structure and then orientates the structural property of the carbonaceous materials resulting from the additives and from the polymeric formulations.

2.2.4 crystalline phases.

The decrease of the size of the amorphous domain may leads to numerous collision between the polyaromatic stacks and then to the formation of large pre-graphitized domains. Such domains are observed by X-ray diffraction in the APP/PER system heat-treated at T $\geq$ 430°C (crystalline graphite: d_{002} = 0.367, size of the crystalline domain $\geq$ 55nm) or in the ABP/PER systems heat treated between 200 and 430°C (crystalline phase of either H_3BO_3 or B_2O_3).

Table1. *Estimation of the size of the amorphous domains of the formulations vs. the temperature of the thermal treatment (from Bourbigot et al.[14, 16]).*

systems	Temperature °C	spin diffusion coefficient. $10^{-20}m^2s^{-1}$	b $10^{-9}m$
APP/PER	280	1.5	3.1
	350	1.7	5
	430	1.4	5
APP/PER/4A	280	1.7	2.8
	350	2.1	2.8
	430	1.9	3.2
Lotader/APP/PER	280	1.7	3.7
	350	2.5	5.7
	430	4.0	7.1
	560	4.8	8.6
Lotader/APP/PER/4A	280	1.4	3.4
	350	2.8	4.8
	430	3.0	5.0
	560	3.7	6.1

Such crystalline species may also be obtained in intumescent materials when one of the additive decomposes before reacting with the others. A typical example is the APP/BCOH system heat-treated in which BCOH does not react with the phosphate species but degrades first to form crystalline products (phosphate and carbonaceous species) stable in the temperature range (about 350°C) where the intumescent coating forms (Figure 6).

The crystalline « rigid » domains in the intumescent materials are initiators for the formation of cracks in the coating. As a consequence, «encapsulated» gases (products of the degradation of the polymer and of the degradation of the coating) evolve and may contribute as « fuels » to the flame. The presented synergistic effect of BCOH in the LDPE-PY/SOH system may be explained by the formation in the high temperature range (T ≥ 400 °C) of turbostratic carbon without any graphite-like phase. This carbon increases the thermal stability of the intumescent coating and, as a consequence, the FR property of the system.

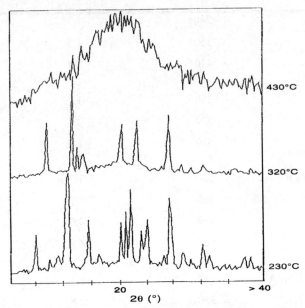

Figure 6. *XRD spectra of APB/PER vs. the temperature of the thermal treatment.*

3. CONCLUSION

Synergy in intumescent formulations may be explained by the physico-chemical behaviour of the materials. The phenomenological study of the action of synergistic agents, considering several intumescent systems, shows that synergistic effects in FR intumescent formulations result from several additional or antagonistic effects assigned to chemical and/or physical characteristics of the char. The FR property :

1- increases when the amount of acidic species in the char increases,

2- increases when the amount of « active carbon » in the intumescent materials increases. « Active carbon » correspond to free radical polyaromatic species or free radical trapped in polyaromatic stacks,

3- increases when a coherent structure forms in the mobile (NMR sense) phase of the intumescent material. This structural change affects the dynamic property of interest of the material,

4- decreases when a part of the turbostratic carbon which constitutes the protective char transforms (structural change and/or coalescence) into graphite,

5- decreases when the size of the crystalline phases (additives or carbon) in the protective material increases,

We have previously proposed that combination of these effects allows a good representation of the synergistic effect[4].

References

1   Camino, G., Costa, L. and Trossarelli, L., *Polym. Deg. & Stab.*, 1984, **7**, 25-31.
2   Delobel, R., Le Bras, M., Ouassou, N. and Alistiqsa, F., *J. Fire Sci.*, 1990, **8**(3-4), 85-108.
3   Bourbigot, S., Morice, L., and Leroy, J.-M., this Book.
4   Le Bras, M., Bourbigot, S., Delporte, C., Siat, C., and Le Tallec, Y., *Fire & Materials*, 1996, **20**, 191-203.
5   Bourbigot, S., Le Bras, M. and Delobel, R., *J. Fire Sci.*, 1995, **13**(1-2), 3-22.
6   Siat, C., Bourbigot, S., and Le Bras, M., in « 7th *BCC Conference - Recent Advances in Flame Retardancy of Polymeric Materials* », *ed. Lewin, M., BCC* Stamford, (1997) pp. 318-326
7   Le Bras, M., Bourbigot, S., Le Tallec, Y., and Laureyns, J., *Polym. Deg. & Stab.*, 1997, **56**, 11-21.
8   Le Bras, M., and Bourbigot, S., *Fire & Materials*, 1996, **20**, 39-49.
9   Bourbigot, S., Le Bras, M., Bréant, P., Trémillon, J.-M., and Delobel, R., *Fire & Materials*, 1996, **20**, 145-154.
10 Montaudo, G., Scamporino, E., and Vitalini, D., *J. Polym. Sci., Polym. Chem.*, 1983, **21**, 3361.
11 van Krevelen, D. W., *Polymer*, 1975, **168**, 615-620.
12 Lewis, I. C., and Singer, L. S., *J. Phys. Chem.*, 1981, **85**, 354-360.
13 Golman, M., and Shen, L., *Phys. Rev.*, 1966, **144**, 321.
14 Bourbigot, S., Le Bras, M., Trémillon, J.-M., and Delobel, R., *J. Chem. Soc., Faraday Trans.*, 1996, **92**(18), 3435-3444.
15 Cheung, T. T. P., and Gerstein, B. C., *J. Appl. Phys.*, 1981, **52**, 517.
16 Bourbigot, S., Le Bras, M., Delobel, R., Decressain, R., and Amoureux, J.-P., *J. Chem. Soc., Faraday Trans.*, 1996, **92**(1), 149-158.

MECHANISM OF ACTION OF HALOGEN-FREE FIRE RETARDANTS AND DEVELOPMENTAL APPROACHES TO DESIGN OF NEW FIRE RETARDANTS WITH REDUCED ENVIRONMENTAL AND HEALTH CONCERNS

L. Costa,

Dipartimento di Chimica Inorganica, Chimica Fisica e Chimica dei Materiali,
Universita di Torino, via P. Giuria 7, Torino 10125, Italy

J.-M. Catala,

Institute Charles Sadron,
Rue Boussingault 6, F-6708 Strasbourg, France

K. M. Gibov,

Research Institute for Chemical Sciences,
Ulikhanov 106, 480100 Alma Aty, Kazakhstan

A. V. Gribanov,

Research Institute for High Molecular Compounds,
Bolshoi Pr 31., 119004 Saint.Peterburg, Russia

S.V.Levchik

Institute for Physical Chemical Problems, Byelorussian State University,
Leningradskaya 14, 220080 Minsk, Belarus

and N. A. Khalturinskij

Research Institute for Synthetic Polymeric Materials,
Profsoyuznaya Street 70, 117393 Moscow, Russia

1. INTRODUCTION

The present tendency observed in some European countries is to decrease the use of halogenated flame retardants for polymeric materials because of the suspicion of environmental and toxicological impact. As the previous experience showed, the alternative to halogen-containing flame retardants might be additives which fire retard polymers through the so-called intumescent mechanism. The general mechanism of fire retardant action of these additives is multi-step and complex.

The additives (organic or inorganic) having, as a rule, phosphorus and nitrogen atoms in their structure, promote carbonization of the polymer on heating and therefore decrease the amount of combustible volatile products. Simultaneously, the amounts of smoke and toxic gases decrease because of the general decrease of volatiles. The formed carbonaceous char plays the role of a barrier which protects the polymer from heat feedback from the flame and hinders both oxygen access to the polymer surface and diffusion of combustible gaseous products of degradation to the flame. The best protective effect of the char is reached if an intumescent layer with proper physical and mechanical properties is formed. Therefore, the fire can be stopped or at least its propagation slowed down due to this complex action of the intumescent type fire retardants.

This project was aimed at supplying scientific background for development of new halogen-free fire retardants for various plastics and thermosets. To reach the goal a consortium of six research teams was created and a multidisciplinary approach included synthesis, processing, combustion testing, thermal decomposition study, characterisation of the products, mechanistic studies and modelling was used.

2. RESEARCH ACTIVITIES

A new method of synthesis of binary metal-ammonium polyphosphates $Me(NH_4)_n(PO_3)_m$, cyclotriphosphates $Me(NH_4)_nP_3O_9$, cyclotetraphosphates $Me(NH_4)_nP_4O_{12}$, linear triphosphates $Me(NH_4)_nP_3O_{10}$ and linear tetraphosphates $Me(NH_4)_nP_4O_{13}$ was developed[1,2]. These phosphates are more thermally stable than ammonium polyphosphate (APP), but similar to APP they evolve polyphosphoric acid at thermal decomposition and show fire retardant activity in aliphatic nylons[3]. As the literature shows,[4,5] inorganic phosphates are often less efficient than organic phosphorus-containing products. Therefore in this project various organic phosphonates were synthesised and studied as fire retardant additives. Some examples of thermally and hydrolytically stable phosphonates are given in the scheme 1[6].

Combustion and thermal decomposition experiments showed that the phosphonates produce an intumescent char. The shielding effect of the char can be improved by co-addition of melamine to DMF-1 or pentaerythritol to DFM-2,3. As measured by LOI DMF-1 shows higher efficiency in polystyrene (PS) than in polyethylene (PE) or polypropylene (PP), whereas DFM-2,3 and MFMPA are more effective in PP than in polyethylene PE or PS.

Scheme 1. *Examples of thermally and hydrolytically stable phosphonates.*

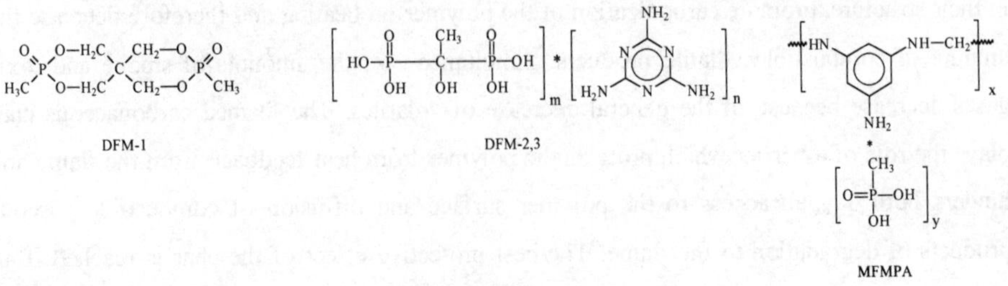

The shielding effect of the intumescent char formed on the surface of burning or pyrolysed polymers was estimated by a thermocouple placed inside the polymer specimen or beneath it. The polymer was heated either by a cone heater using the cone calorimeter setup[7] or by a laser beam[6]. It was shown[8,9] that inorganic fillers added to the formulations with APP improve thermal insulation properties of the intumescent char even if the volume of foaming decreases. The correlation exists between fire retardant efficiency of phosphonates and the shielding effect provided by the char.

A mathematical model of the char formation on the surface of the polymer subjected to the linear heating was developed[10]. Taking into consideration the competitive reactions of gasification and crosslinking[11], this model predicts the rate of char growing. The model was applied to phenol-formaldehyde resin fire retarded by 20% of B_2O_3. It gave good correspondence of experimental and predicted kinetic curves of increasing viscosity of the melt and evolution of the gases.

3. SELECTED RESULTS

3.1 Fire retardant action of phospham

Phospham is a cyclomatrix inorganic polymer of high thermal stability. The structure of phospham is not established and seems to be dependent on the method of preparation[12], however the chemical analysis usually shows the following formula $(PN_2H)_x$. Its fire retardant efficiency in aliphatic polyamides (PA-4.6) was studied by E. D. Weil and N. G. Patel.[13]

In our project phospham was prepared from hexachlorotricyclophosphazene by ammoniation and further cross-linking through thermal elimination of ammonia[14]. It is a white or white-grey powder which loses ca. 12% of its weight at heating to 600°C in inert atmosphere.

Phospham is an efficient fire retardant additive to PA-6 as it is shown by LOI measurements (Table 1.). The highest increment of LOI is observed at 10 wt. % of phospham, whereas at further loading LOI grows more slowly. The char is observed on the top of burning specimens with phospham. As Table 1 shows PA-6 is involved in the charring in the presence of phospham. For example, the formulation with 10 wt. % of phospham gives 17.5 % of char.

Table 1. *Oxygen index and char yield measured in combustion of PA-6 fire retarded by phospham.*

Phospham, wt. %	LOI, Vol. %	Char yield, wt. %
-	25.2	-
10	29.2	17.5
20	31.6	23.5
30	34.8	32.2

Figure 1 shows that there is no correspondence between experimental and calculated thermogravimetry curves of the formulation PA-6 + 20 wt. % phospham. In spite of high thermal stability of pure phospham it destabilises PA-6 in the formulation. Thermogravimetry curves clearly show that phospham reacts with PA-6 accelerating thermal decomposition of the polymer and producing a char the yield of which is higher than that predicted by the calculated curve.

To find out the chemical mechanism of interaction of phospham with PA-6, solid residues were collected at different steps thermal decomposition as well as chars after combustion and characterized by infrared, ^{13}C and ^{31}P solid state NMR or by EPR. Figure 2 shows infrared of solid residues obtained in inert atmosphere at different temperatures and in combustion.

In the initial spectrum absorptions at 1150 and 905 cm[-1] are likely to belong to phospham[12], whereas all other bands are characteristic absorptions of PA-6[15]. Big changes of the infrared pattern are observed at 340°C where only ca. 20% of weight loss occur. PA-6 mostly decomposes to produce primary amides which indicates preferable scission of alkyl

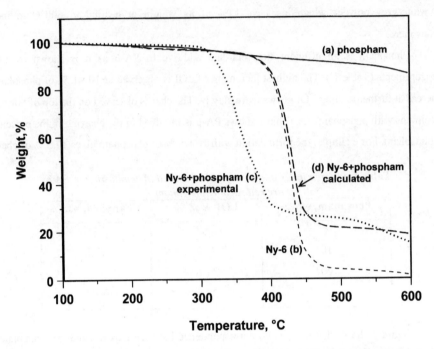

Figure 1. *Thermogravimetry curves of phospham (a), PA-6 (b) and experimental (c) or calculated (d) curves of PA-6 + 20% wt. % phospham. Nitrogen flow 60 cm³/min, heating rate 10 °C/min.*

amide bonds -NH-CH$_2$-[16]. Four strong bands (1360, 1160, 1010 and 740 cm^{-1}) appear in the region of phosphorus-containing functional groups. This indicate that instead of P-N and P=N stretchings of phospham, P-N-C and P=N-C stretchings having symmetric and asymmetric modes appear[17]. It means that phospham reacts with PA-6.

At further heating to 370°C the bands at 1360 and 740 cm^{-1} vanish. It means that at the end of the main step of weight loss phosphonitriles and phosphoamines degrade. At 370°C characteristic absorptions of aliphatic groups disappear, whereas at 450°C the N-H stretching band vanishes. The band at 1630 cm^{-1} is likely to belong to aromatic char[18]. Infrared residue of the char in combustion is similar to the residue from thermogravimetry obtained at 450°C.

^{31}P solid state NMR shows that the main absorption peak of phospham shifts from +2.3 to -14.9 ppm upon heating to 500°C PA-6 added by phospham. In agreement with the infrared data this proves interaction of phospham with nylon[19]. ^{13}C NMR shows that the carbonaceous part of the solid residue obtained in the presence of phospham consists mostly of aromatic species, whereas without phospham aliphatic -C=C- and -C=N- functionalities are mostly detected.

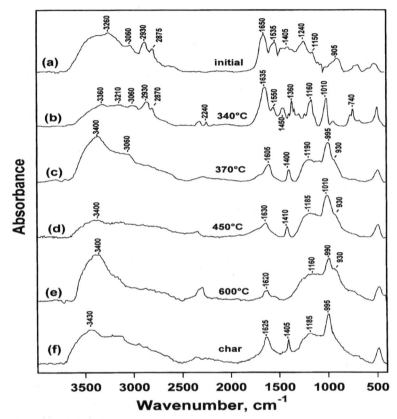

Figure 2. *Infrared spectra of initial PA-6 + 10% phospham (a), solid residues collected in thermogravimetry in inert atmosphere (b-e) and char in combustion (f). Pellets in KBr.*

This fact is also confirmed by EPR (Figure 3). From the literature it is suggested[20,21] that stable free radicals might be responsible for the carbonaceous char yield. The higher concentration of the stable free radical in solid residue the higher char yield is expected. Solid residues obtained from PA-6 fire retarded by phospham give much higher concentration of stable free radicals than residues from pure PA-6 or phospham.

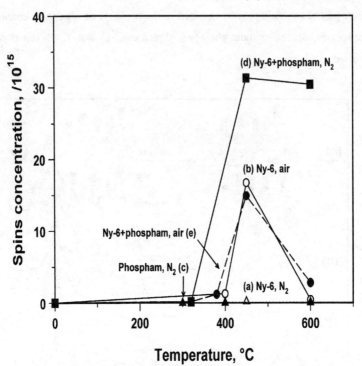

Figure 3. *Stable free radicals concentration obtained in solid residues from PA-6, phospham or PA-6 + phospham.*

3.2. Phosphorus modified Mg(OH)₂ and Al(OH)₃.

To increase the fire retardant efficiency of magnesium dihydroxide or aluminum trihydroxide (alumina trihydrate) these hydroxides were modified by dimethyl phosphite. The reaction was carried out by heating the hydroxides with dimethyl phosphite (molecular ratio 1:1) for 9 hours at 150°C in a methyl chloride solution. The obtained magnesium hydroxide methyl phosphonate (MgMP) or aluminum hydroxide methylphosphonate (AlMP) are more thermally stable than the unmodified hydroxides.

Fire retardant performance of polyolefines with added MgMP or AlMP is shown in Table 2. It is seen that AlMP is more effective than Al(OH)₃, whereas MgMP is less effective than Mg(OH)₂.

As it was shown in the literature[22] Mg(OH)₂ or Al(OH)₃ do not provoke charring of PE or PP. Our thermogravimetry experiments show that MgMP is also not efficient for increasing of solid residue, whereas AlMP promotes char formation from both PE and PP (Figure 4). For example, the formulation PE + 40 wt. % Al(OH)$_3$ gives ca. 22% of solid residue at 700°C, whereas the formulation PE + 40 wt. % AlMP gives 27% of solid residue. The increasing of solid residue is more pronounced in PP, where AlMP gives 40% of solid residue compared to 22% produced by Al(OH)$_3$.

Table 2. *Oxygen index for polyethylene or polypropylene fire retarded by magnesium or aluminum hydrates or by the modified hydrates.*

Polymer	Additive		LOI
	Name	wt. %	Vol. %
PE	-	-	17.5
	MgMP	20	19.0
	MgMP	40	21.0
	Mg(OH)$_2$	40	22.5
	AlMP	20	19.5
	AlMP	40	24.0
	Al(OH)$_3$	40	21.5
PP	-	-	17.5
	MgMP	20	18.5
	MgMP	40	20.0
	Mg(OH)$_2$	40	23.5
	AlMP	20	20.0
	AlMP	40	22.5
	Al(OH)$_3$	40	22.0

The shielding effect of the char was measured by using the following set-up: Pellets of the fire retarded formulations of 1 mm thickness and 10 mm of diameter with the thin thermocouple in the center were prepared by hot pressing. The pellets were exposed to the irradiation of CO$_2$ pulsed laser whereas the offset of the thermocouple was recorded. The better shielding effect shows AlMP in PP than in PE which is in agreement with thermogravimetry but not with combustion tests.

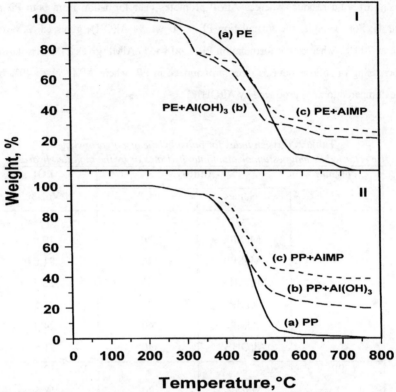

Figure 4. *Thermogravimetry curves of PE, PE + 40 wt. % Al(OH)₃ or PE + 40 wt. % AlMP (I) and of PP, PP + 40 wt. % Al(OH)₃ or PP + 40 wt. % AlMP (II) (static air atmosphere, heating rate 10 °C/min).*

3.3 Modelling of intumescence

The stability of any liquid foam depends on two main factors: (1) surface energy which is a thermodynamic factor and (2) viscosity of liquid which is a kinetic factor. If melted polymer foams the second factor is usually dominating. During the heating of the polymer its viscosity might increase because of cross-linking. With increasing temperature the viscosity of the polymer melt firstly decreases due to increasing of macromolecules segments mobility, but then may increase due to cross-linking of the polymer chains. The melt can trap gas which evolves simultaneously with polymer degradation only if the melt viscosity is higher than the critical level η_m. Foaming is observed in the region $\eta_m \rightarrow \infty$ till gelation or solidification occur.

According to Macosco *et al.*[23,24] the viscosity of thermosets can be approximated:

$$\eta = \eta_0 exp(-E_a/RT)\, L^\alpha$$

where η_0 and α ($\alpha \approx 3, 4$) are some empirical constants, E_a is the activation energy of viscous flow, L is an average longest length of the linear chain in the branching molecule, R is gas constant and T is temperature. L mostly depends on the concentration of active functionalities and kinetic parameters of reactions responsible for cross-linking. To estimate L the classical branching theory of Flory[25] is used.

The temperature interval where viscosity suitable for intumescence is observed ($\eta_m \to \infty$) depends on the kinetics of cross-linking and on the heating rate. If the heating rate is relatively high, this interval might shift to the temperature above 500°C where carbonization occurs. In this case intumescence is poor. This explains why the upper layer of intumescent coatings is usually very little foamed. If cross-linking proceeds too rapidly (for example, at a high concentration of active functionalities), then the temperature or the time interval corresponding to the ($\eta_m \to \infty$) is short and intumescence is poor.

If gas for foaming is produced in the same reaction where cross-linking occurs (e.g. polycondensation) the amount of gas which can be trapped by the polymer melt does not depend on heating rate. If kinetics of gas evolution and cross-linking are different the heating rate becomes an important factor.

4. CONCLUSIONS

It is shown that phospham is an efficient fire retardant additive to polyamide 6 because it promotes charring of the polymer. In spite of high thermal stability phospham reacts with the polyamide producing phosphonitriles and phosphoamines which decompose at higher temperature and give char. Free radical mechanism is suggested to contribute to the charring.

Alumina trihydrate modified by dimethyl phosphite shows higher fire retardant efficiency than untreated alumina trihydrate. This is not the case with magnesium dihydroxide, which loses fire retardant activity after the treatment by dimethyl phosphite. The modified alumina trihydrate increases char yield from polypropylene and polyethylene. Insulating effect of the char increases with increasing char yield but these parameters do not correlate with the oxygen index.

It is suggested a mathematical model of the formation of an intumescent char. It is shown that the intumescence extend depends on the kinetics of cross-linking and gases production as well as the rate of the polymer surface heating.

It is demonstrated an efficiency of the multidisciplinary approach applied to the complex tasks of development of new fire retardant systems.

References

1. A. F. Selevich, G. F. Levchik, A. I. Lesnikovich and S. V. Levchik, Byelorussian Patent 00260-01 (to Byelorussian University), 1993.
2. A. F. Selevich, S. V. Levchik, A. S. Lyakhov, G. F. Levchik, A. I. Lesnikovich and J.-M. Catala, *J. Solid State Chem.*, 1996, **125**, 43.
3. G. F. Levchik, S. V. Levchik, A. F. Selevich, L. Costa and V. A. Lutsko, in this Volume;
4. J. W. Lyons, *The Chemistry of Uses of Fire Retardants*, Wiley, New York, 1970.
5. B. B. Kopylov, S. N. Novikov and L. A. Oksent'evich, *Polymeric Materials with Low Flammability*, Khimiya (in Russian), Moscow, 1986.
6. I. Reshetnikov, A. Antonov, T. Rudakova, G. Aleksjuk and N. Khalturinskij, *Polym. Degrad. Stability*, 1996, **54**, 137.
7. S. V. Levchik, G. Camino, L. Costa and G. F. Levchik, *Fire Mater.*, 1995, **19**, 1.
8. S. V. Levchik, G. F. Levchik, G. Camino, L. Costa and A. I. Lesnikovich, *Fire Mater.*, 1996, **20**, 183.
9. G. F. Levchik, S. V. Levchik and A. I. Lesnikovich, *Polym. Degrad. Stability*, 1996, **54**, 361.
10. V. Sh. Mamleev and K. M. Gibov, *J. Appl. Polym. Sci.* (submitted for publication).
11. K. M. Gibov and V. Sh. Mamleev, *J. Appl. Polym. Sci.* (submitted for publication).
12. T. N. Miller and A. A. Vitola, *"Inorganic Phosphorus-Nitrogen Compounds"*, Zinatne (in Russian), Riga, 1986.
13. E. D. Weil and N. G. Patel, *Fire Mater.*, 1994, **18**, 1.
14. R. A. Shaw and T. Ogawa, *J. Polym. Sci., Part A,* 1965, **3**, 3343.
15. D. Lin-Vien, N. B. Colthup, W. G. Faterley and J. G. Grasselli, *"The Handbook of Infrared and Raman Characteristic Frequencies of Organic Molecules"*, Academic Press, Boston, 1991.
16. S. V. Levchik, L. Costa and G. Camino, *Polym. Deg. Stab.,* 1992, **36**, 229.
17. N. B. Colthup, L. H. Daly and S. E. Wiberley, *"Introduction to Infrared and Raman Spectroscopy"*, Academic Press, Boston, 1990

18. J. Zawadskii, in *"Chemistry and Physics of Carbon. A Series of Advances"*, ed. P.A. Thrower, Marcel Dekker, New York, 1990, Vol.21, p.147.

19. D. G. Gorenstein, in *"Phosphorus-31 NMR. Principles and Application"*, ed. D. G. Gorenstein, Academic Press, Orlando, 1984, Chapter 1.

20. I. C. Lewis and L. S. Singer, in *"Chemistry and Physics of Carbon. A Series of Advances"*, eds. P. L. Walker Jr. and P.A. Thrower, Marcel Dekker, New York, 1986, Volume 17, p.1.

21. M. Le Bras, S. Bourbigot, C. Siat, C. Delporte and Y. Le Tallec, *Fire Mater.* 1996, **12**, 191.

22. J. Rychly, A. Bucsi, K. Csomorova, L. Rychla and D. Simakova, *Makromol. Chem., Macromol. Symp.,* 1993, **74**, 193.

23. E. M. Valles and C. W. Macosco, *Macromolecules*, 1979, **12**, 521.

24. S. A. Bidstrup and C. W. Macosco, *J. Polym. Sci., Ser. B*, 1990, **28**, 691.

25. P. J. Flory, *"Principles of Polymer Chemistry"*, Cornell University Press, Ithaca, 1953, chapter 9.

Acknowledgements

This research was sponsored by the International Association for the Promotion of the Cooperation with Scientists from the Independent States of the Former Soviet Union under the INTAS project 93-1846.

INTUMESCENT CHARS

I. S. Reshetnikov, M. Yu. Yablokova

Institute of Synthetic Polymeric Materials,
70, Profsoyuznaya street., Moscow, 117393, Russia

N. A. Khalturinskij

Semenov Institute of Chemical Physics,
4, Kosygina street, Moscow, 117977, Russia

1. INTRODUCTION

Intumescent fire retardant compositions[1] have found great application due to some valuable properties, such as low weight of the coating and high protection properties. The main peculiarity of these systems is the ability to produce under the action of external heat flow a protective surface layer - foamed char. The thickness of this layer is, as a usual, tens times thicker than the initial coating, what provide effective protection of the substrate. To distinguish this char from other similar systems, foamed char, produced during intumescence, is frequently called as intumescent char.

Chemical processes, take place during intumescence, already have been described and overviewed in detail[2-5]. Besides, some attempts to estimate mathematical model of intumescence also have been carried out [6 and Refs. therein]. However a very little attention has been paid to the investigation of the char structure and properties and up to now comprehensive analysis of these aspects is absent. Existing isolated works devoted to different char characteristics will be noted in appropriate places in the present work.

The aim of this work is twofold. From the one hand, an attempt was made to give a brief overview of previous results. From the other hand, investigation of the char structure and properties is accompanied by original experimental results.

2. CHEMICAL STRUCTURE

Chemical structure of the intumescent chars, produced during burning, is, as a matter of fact, the most investigated aspect among char parameters. The main attention was drawn to the ammonium polyphosphate (APP) based char-forming (Brauman[7]) and intumescent compositions (Gnedin et al.[8], Bourbigot, Le Bras, Delobel et al.[9-12], Costa and Camino[13-14], and others[15-18]).

Studying pyrolysis of char forming composition polyethylene terephthalate with APP, Brauman[7] suggested that APP does not only create a protective layer but also takes a part in chemical reactions in condensed phase.

Investigations of intumescent compositions were focused mainly on the APP with pentaerythritol (PER) and melamine (ML) in polypropylene (PP) and others polymeric matrixes. Studying intumescent additive APP - PER (1.5:1 wt./wt.) in various matrixes: PP, polyethylene (PE) and polystyrene (PS)), Gnedin et al.[8] using IR spectroscopy have shown that char chemical structure practically does not depend on the matrix type and that the contents of C and P atoms in the char residue corresponds to these of pure intumescent additive. Moreover, Delobel et al.[9], studying PP formulations with APP-PER additive using ^{31}P MAS-NMR spectroscopy, have shown that formation of higher polyphosphate chains occurs during intumescence and that a sharp relative increase in the amount of orthophosphates species is founded in the char residue. The same behaviour is observed for PP formulations with diammonium pyrophosphate (APPh) and PER additive but, instead of orthophosphates, pyrophosphate species are detected. Recently, Bourbigot et al.[10] investigated using XPS structure of the bulk and surface of the char, produced during APP-PER system thermal treatment. The Authors verified that P/C ratio increases at high temperatures (500°C) on the char surface, when in bulk this ratio decreases. Furthermore, they showed that O/C ratio follows a same trend. They proposed a migration of phosphonates to the surface, followed by oxidation. From *O1s* spectra, it was shown that -O-/=O= ratio sharply increases at high temperatures and that, from *C 1s* spectra, oxidised carbon / aromatic carbon ratio is approximately 0.7 at 550°C.

For a ethylene-butyl acrylate-maleic anhydride terpolymer formulation with APP-PER additive Bourbigot et al.[11] even suggest possible structures of the char residue for different treatment temperatures. Analysing data of IR, ^{13}C NMR and DD-MAS NMR ^{31}P measurements they showed that char after a thermal treatment at 560°C consists in an aromatic structure cross-linked by -O-(POOH)-O- or -CH$_2$- links with isolated H$_3$PO$_4$ fragments.

Also, investigating some self-intumescent phenol-formaldehyde type resins (PFR) compositions, it was shown by Camino, Costa et al. [14] that at high temperatures (500°C), bands structure of the residue becomes complex and that the main structure of the polymer still present changes towards a poly-aromatic structure.

Mashlyakovsky et al.[15] have studied pentaphtalic alkyd-based intumescent with APP, PER, ML and melamine-formaldehyde resin (MFR) mixture as intumescent additive systems. The study of the chemical composition showed that the introduction of APP leads to an increase of linked carbon and nitrogen in the residue. The Authors proposed that APP is a catalyst for the thermo-oxidative destruction of the alkyd in a first stage and delays this destruction in the second one because of formation of cross-linked structures.

Marosi et al.[16-18], studying PP- and PE-based composites loaded by B-Si containing compounds, have showed (using XPS), that B-Si compounds migrate to the char surface layer during pyrolysis. The observed increase of the linked oxygen contents in the surface layer, when N, C and P contents decrease, is explained by the presence of Si(R)-O-Si(R)-O fragments in the oxidised char surface.

Testing the part played by APP in the formation of char structure, IR-analysis of intumescent chars (produced during pyrolysis of CFR with APP-sorbitol intumescent additive at 500°C) has been carried out. Characteristic absorptions at 980 cm^{-1} and 496 cm^{-1}, corresponding to the valence and deformation oscillations of PO_3- groups respectively are used to measured the content of phosphate species in the char (Figure 1). This phosphate content remains practically unchanged up to approximately 0.6 of total char cap height and then sharply drops. Explanation of such a behaviour will be given in the section about physical properties of the chars in this text.

Several factors affect the char chemical structure: the chemical structure of the matrix (for the reactoplasts) and components of intumescent additive, the conditions of pyrolysis (such as temperature and/or oxygen content) and, thirdly, different fillers (e.g. pigments) which traces can be found in the char. The two first factors were discussed above.

Fillers and special additives action on the intumescence process has to be considered. As a rule, phosphorus compounds are introduced to increase char yield. It is known that practically all the phosphorus stays in the condensed phase in the form of polyphosphoric acids and ultra-phosphates[13]. Some attempts to introduce special additives, in particular zeolites - for synergetic effect - indicates that there is practically no remaining zeolites traces in the residue, at least at high temperatures[12]. Decreasing of the pyrolysis temperature leads to the increase of polymer matrix chains fraction and of oxygen contents (catalytic reactions of thermo-oxidation) in the residue. Particular mechanism depending on the peculiarities of composition components will be discussed further.

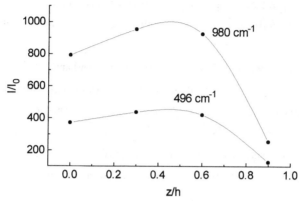

Figure 1. *Intensities of the characteristic IR absorptions of phosphate species vs. position in the intumescent cap.*

3. PHYSICAL STRUCTURE

Physical structure of chars has not been so widely investigated, despite it affects all processes taking place during burning: movement of liquid degradation products through the char cap (Gibov et al.[19-20]), thermo-protective properties of the cap (Anderson at al.[21] and others[22-24]). Moreover, only a few works devoted to the structural features of the char exist which present only a discussion on the influence of structure on the flammability (Gnedin et al.[25-26], Brauman[27-30]) or contains only averaged experimental results without detailed investigation[30]. It should be noted here that Gnedin et al.[26] have made an attempt to propose the relation between char structural parameters (char cap height (foaming ratio) n_k, and char yield m_k) and flammability (limiting oxygen index - *LOI*):

$$LOI = 1.6*n_k + 406.3*m_k + 23.49*n_k m_k - 309* m_k^2 n_k - 0.1*n_k^2 + 0.58*n_k^2 m_k - 974.49*m_k^2$$

In addition, some works were devoted to modelling of the formation of porous char structure during burning of intumescent compositions (Butler et al.[31]). One of the main aspect of intumescent char physical structure is its uniformity[4] and its porosity[32]. Brauman[28] propose two chars classes: dense and weakly porous chars. The type of char depends on the polymer matrix. So, for example, char formed during burning of polycarbonate (PC) is dense. Dense char provides effective protection, retarding processes of heat and mass transfer. A previous investigation of the action of metal salts and oxides in various polymers[29] has shown that introduction of Zn(II), Fe(III) and Mo(VI) salts leads to the formation of the char with a porous structure at the surface and a dense structure in the bulk. Introduction of Sn, Ni and Ba salts, in turn, leads to formation of the char with a foamed structure throughout its whole volume.

We studied the action of sorbitol - APP additive in a carbamide - formaldehyde resin (CFR). Electron scanning microscopy allows the study of the internal structure (Figure 2a) and the surface structure (Figure 2b) of the char. The investigated char has an uniform structure not only in particular point of the cap but for the overall cap. More, it is observed that surface char layer is enriched by pores with very small diameters. This observation may be related to the previously proposed difference of the chemical structures in surface and bulk.

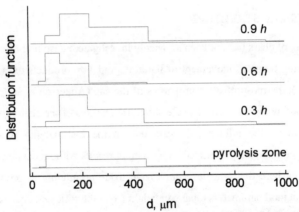

Figure 3. *Distribution of average pore diameters vs. position in the direction of intumescence.*

Distribution of the apparent average pore diameters in different positions in the direction of intumescence (Figure 3) may be computed from the scanning electron microscopy pictures of char cap cuts. All histograms have a maximum in a very narrow diameter range and that diameters present a distribution function with nearly a Gaussian shape.

We may now comment layer the previously reported change in porosity observed in the 0.6·h position in the cap height. The apparent sharp decrease of the pores diameter may be explained by the structure of the intumescent additive system It is known that intumescence in such systems occurs in a low temperatures range[33], approximately 210-250°C. APP is partially consumed during chemical reactions.

Residual APP (or solid phosphate species resulting from its thermal degradation) particles stay in the char. Ultimately, APP degradation leads to the formation of phosphorus oxides. When temperature becomes higher than the phosphorus oxides evolution temperature (about 560°C), apparent particles diameters decrease. After the complete solid residue degradation (0.9·h layer, from variation in the APP contents presented in Figure 1), pores diameters return to the "normal" value. We propose that formation of the foamed char structure is ended in the pyrolysis zone and that further transformation do not practically occur

Structure of the char depends also on its chemical transformation during heating. Two different behaviours have been previously reported. In the first case, char keeps its structure during degradation, as observed using PP formulations with standard intumescent additive (Bourbigot et al.[33]). On a second way, char can produce thermally stable graphite-like structures, for example: carbon-containing compounds capable to the mesa-phase transitions (Kipling et al.[34]). Authors suggested that mesa-phase transition at temperatures higher than 400°C corresponds to the orientation of large polycyclic molecules with formation of large arranged regions which may lead to the formation of high-oriented graphite. A Raman spectroscopy study allows to observe the formation of the two types of the structural organisation of the carbon[33].

Char resulting from a carbamide-formaldehyde (CFR)-based material with the APP-Sorbitol additive system presents a very high arrangement (scanning electron micrograph in Figure 4), even with formation of flaky structures.

Other different factors affect char structure. The first one is the properties of the polymer melt. So, Brauman[28] reported that char morphology is affected by the viscosity of the melt, the possible cross-linking in the melt from degrading polymer and finally, the depth of the molten layer. Gibov et al.[35] assumed that intumescence occurs in a narrow range of the polymer melts viscosity and that porosity and pores diameter of char depend on this melt viscosity. It may also be supposed that surface tension of polymer melt and plasticization of additives can also affect foam structure, but this last assumption has not yet been investigated.

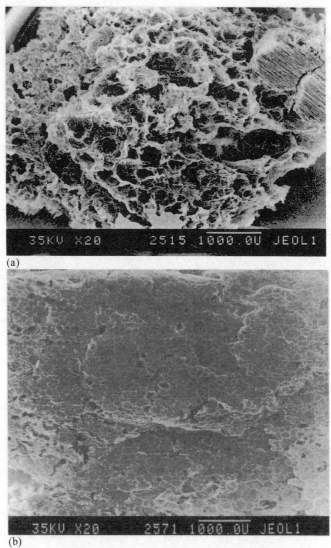

(a)

(b)

Figure 2. *Electron scanning microscopy picture of the structure of the char: a - in bulk, b – on surface.*

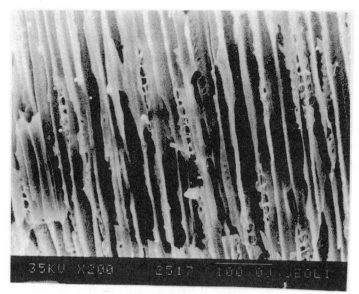

Figure 4. *Graphite structures in the char*
Product of the pyrolysis of CFR with APP-Sorbitol additive.

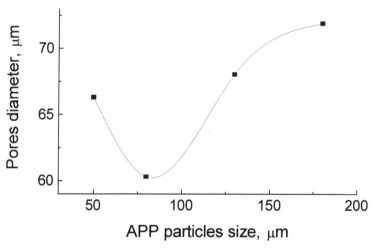

Figure 5. *Average pores diameters vs. APP particle size.*

The second may the texture of intumescent additive's components. It has been already shown[36] that aggregation of polyol particles in the intumescent additive affects the foamed structure uniformity. Besides, if one of the consistent parts of intumescent additive is a powder, which does not mould at the characteristic temperature of intumescence, it can be expected that particle size also affects pores diameters because intumescence takes place in the solid-melt interface. To check this point, intumescent systems APP-sorbitol with various APP particles size have been studied. Each system leads to char with an uniform structure, but average pores diameter depends on the APP particles size (Figure 5). This observation confirms our preliminary conclusion, i.e. intumescence reactions takes place in the melt - solid particle interface. It is an example of physical structure investigations which may help to the understanding of the intumescence chemical processes.

4. THERMAL STABILITY

Now, the main application of intumescent compositions, particularly intumescent coatings, is the fire protection of constructions and building. In fires, efficiency of heat protection depends not only on the coke yield, the char cap height and the char structure, but also on the thermal stability of the protective char layer.To study this last point, standard tests of flammability (such as limiting oxygen index (LOI) and UL-94 rating) are not suitable, because time of heat flux action is not taken into account. Investigation may be usefully carried out considering thermo-protective properties (TPP)[24], mass loss in isothermal conditions[37-38] or even high temperature pyrolysis method[39].

Comparison of TPP curves of different compositions allows one to compare their efficiencies on the stage of intumescence as well as on the final stage when formations of the protective char cap is finished. Oblique of the last part on the TPP curve gives one the char ablation rate, i.e. its thermal stability (one possible criterion for a classification). More, some others standard tests, based on the detecting of the temperature response on the backwall of the sample under action of fixed external heat flow (e.g. British standards[40]) may be used. Unfortunately, TPP tests are very rarely used to investigate intumescent systems. Nevertheless, comparison of TPP for some intumescent systems[41] indicates that polymer matrix very slightly affect char ablation rate despite efficiency of intumescence is significantly different. Investigation on the part played by the polymeric matrix is illustrated by the study of mixture of melamine salt of derivative of phosphoric acid with PER (DFM) in PP, PE and PS matrixes (Figure 6).

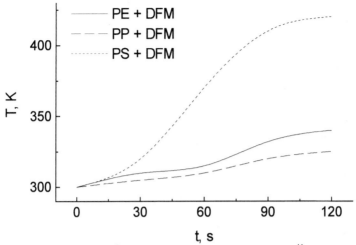

Figure 6. *TPP for some intumescent systems*[41].

It has also been shown that thermal stability of the intumescent chars depends on the chemical structure of the char, particularly the presence of the cyclic structures increase char thermal stability[16].

Studies of the weight loss during degradation at a constant temperature and during high temperature pyrolysis give more useful information, because they allow one not only to compare the thermal stability of various chars, but also to obtain kinetic parameters of char degradation. So using high temperature pyrolysis, when experiments proceed under the spherical geometry, degradation of the char can be described vs. time τ in the terms of ablation coefficient k and sample diameters: d (current) and d_0 (initial) by the Sreznevsky law[42]:

$$d^2 = d_0^2 - k\tau$$

Unfortunately, only qualitative discussions exist about pyrolysis of char forming polymers[43-44]. To improve this point, we present now some results on high temperature pyrolysis studies of some intumescent compositions. Figure 7 compares weight losses of the intumescent compositions: epoxy resin (ER) with APP-PER (curve *a*) and CFR with APP-Sorbitol (curve *b*) during high temperature pyrolysis. This study shows that char produced by ER-based formulation is more stable than this produced by CFR-based samples. Computing ablation coefficients, the thermal stability of the chars may be compared numerically.

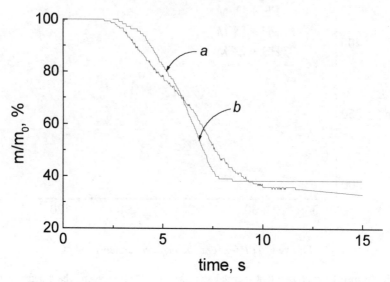

Figure 7. *High temperature pyrolysis of intumescent compositions ER with APP-PER (curve a) and CFR with APP-Sorbitol (curve b).*

Factors affecting char thermal stability have now to be discussed. The chemical structure of the char has to be, of course, first reported: as already referred, presence of cyclic structures improve thermal stability, strength of the chemical bonds, number of cross-linking bonds are also important factors. Physical structure is important too, as an example, presence of graphite structures increases thermal stability of chars (Aseeva et al.[45]). However, it should be noted that some authors disagree with this assumption and believe that arrangement does not affect oxidation processes[46]. Next significant factors are the conditions of pyrolysis, such as oxygen contents, temperature. In fact, it is known that in the oxygen-enriched atmosphere, char oxidation and degradation proceed in great extent via oxygen absorbed by the surface (Stuetz et al.[47]). Now, the remaining question is: "does oxidation proceed only on the surface or also in the bulk?". Bulk oxidation may be provided by the presence of "holes" in the char cap structure.

5. MECHANICAL STABILITY

One of the most important applied aspect of the commercial intumescent compositions is the mechanical strength of the intumescent chars. This factor is significant because protective layer should not be destroyed under action of wind, of mechanical action of fire or of convective air flows. Aspects of mechanical stability of intumescent chars sometimes appeared in the technical literature in discussions about the nature of bound(s) which provides char stability[11, 17-18, 48].

Previous studies of the mechanical properties of the foamed chars[49-50] have shown that char behaviour during mechanical destruction is generally unlike to classical rigid foams behaviours and, therefore, may not be characterised by the destruction force. More, they propose a criterion for the stability, i.e. work for the destruction of unit area under action of plane plate in the direction of the intumescence divided by char cap height. Studies of the mechanical properties of the CFR matrix with various intumescent additive contents and under different pyrolysis conditions[49-50] showed that mechanical stability depends on the contents of high-molecular fraction in the residue, the higher this fraction, the higher the stability.

Other results dealing with mechanical property of the char are reported in another article in this Book[51].

We may assume that chemical structure is one of the main factors affecting mechanical stability. moreover, physical structure data (as thickness of the walls inside the char cap, micro-density of the char, density of the char material) should also be taken into consideration. Other factors as fullness of the pyrolysis affect only the char chemical structure and so do not need a separate discussion. Finally, mechanical stability probably depends on the direction of external action and its type (direct, rotation, vibration), but these last factors are not investigated yet.

6. THERMO-PHYSICAL PROPERTIES

Thermo-physical properties of intumescent chars are needed generally in the purpose of modelling intumescent behaviour in fires. Evaluation of such parameters, such as thermal conductivity and heat capacity, is very complex and frequently is not reliable enough because of the complexity of char physical structure[52] and the ability of char to sorb gases on its internal surface[53] which leads to the time scale dependence of these parameters.

As an illustration, we have tested heat capacity vs. time during pyrolysis of a polycarbonate (PC) matrix with aromatic sulphone sulphonate additive which provides intumescence (Figure 9; heat capacity was measured using DSM-3 differential scanning calorimeter). Heat capacity curve shows a minimum at approximately 20 min on the time scale. Such behaviour is explained by two simultaneous processes. The first one consists in the desorbtion of volatile degradation products, and the second to absorption of air and water from the atmosphere which are sorbed instead of desorbed products.

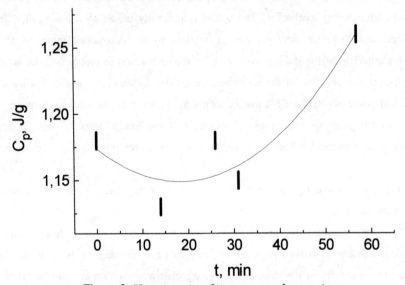

Figure 9. *Heat capacity of intumescent char vs. time*

Contrarily to char heat capacity, thermal conductivity of intumescent chars has been previously widely discussed[54-55] because it is the main factor used for the thermo-physical modelling of intumescent behaviour[6]. Practically in all works, thermal conductivity of the foamed char was computed as in study of a composite material with high air contents[56]. However, it has been shown that, due to radiant heat transfer inside the char cap, approach from the positions of the "effective coefficient of thermal conductivity" may be incorrect[57]. Theoretical investigation of the TPP of intumescent chars[23] indicated that it significantly depends on the char physical structure and that this dependence can be written:

$$T_b = T_0^* - \frac{Q}{\lambda_{eff}^*} h + h_r d$$

where T_b is the backwall temperature, d the average pores diameter, h the char cap height and Q the external heat flux.

Another problem in experimental measurements of the thermal conductivity of the foamed chars consists in impossibility to provide good thermal contact with an external plate because of the very low strength of the char material. So, steady methods are practically unsuitable. A possible solution may consist in the use of non-steady, pulse methods[57]. The only disadvantage of these methods is that they give thermal diffusivity data which can be easily converted to needed values.

To conclude, the intumescent char investigations, not yet solved, wait for the investigators attention.

References

1. L. Mashlyakovsky, A. Lykov and V. Repkin, in *"Organic Coatings with Lowered Flammability"*, Khimia, Leningrad, 1989.
2. H. Vandersall, *J. Fire and Flammability.*, **2** (1971) 97.
3. G. Camino, L. Costa and G. Martinasso, *Polymer Deg. & Stab.*, **23** (1989) 359.
4. M. Kay, A. Price and I. Lavery, *J. Fire Retardant Chem.*, **6(1)** (1979) 69.
5. R. Delobel, M. Le Bras, N. Ouassou and F. Alistiqsa, *J. Fire Sci.*, **8(2)** (1990) 85.
6. I. Reshetnikov, A. Antonov and N. Khalturinskij, *Combustion, Explosion & Shock Waves [Fizika Goreniya I Vzryva]*, **33(6)** (1997) 18.
7. S. Brauman, *J. Fire Retardant Chem.*, **7(2)** (1980) 61.
8. Ye. Gnedin, N. Kozlova, R. Gitina, O. Fedoseeva, M. Sevost'yanov and S. Novikov, *Vysokomol. Soed.*, **33** (1991) 1568.
9. R. Delobel, M. Le Bras, N. Ouassou and F. Alistiqsa, *J. of Fire Sci.*, **8** (1990) 85.
10. S. Bourbigot, M. Le Bras, L. Gengembre and R. Delobel, *Appl. Surf. Sci.*, **81** (1994) 299.
11. S. Bourbigot, M. Le Bras, R. Delobel, P. Bréant and J.-M. Trémillon, *Carbon*, **33** (1995) 283.
12. S. Bourbigot, M. Le Bras, P. Bréant, J.-M. Trémillon and R. Delobel, *Fire & Materials*, **20** (1996) 145.
13. G. Camino, L. Costa and M. P. Luda, *Makromol. Chem., Macromol. Symp.* **74** (1993) 71.
14. L. Costa, L. Rossi di Montelera, G. Camino, E. Weil and E. Pearce, *Polymer Deg. & Stab.*, **56** (1997).
15. L. Mashlyakovskii, M. Aleskerov, N. Kuzina, E. Khomko and V. Krasheninnikov, *Russ. J. of Appl. Chem.* 67 (1994) 248.
16. Gy. Marosi, Gy. Bertalan, P. Anna, A. Tohl, R. Lagner, I. Balogh, and P. La Manita, *J. of Thermal Anal.*, **47** (1996) 1155.

17. Gy. Marosi, Gy. Bertalan, I. Balogh, A. Tohl, P. Anna and K. Szentirmay, *Flame Retardants'96, Proceedings. of Flame Retardants'96 Conf.*, London (1996).

18. Gy. Marosi, P. Anna, I. Balogh, Gy. Bertalan, A. Tohl and M. Maatoug, *J. of Thermal Anal.*, **48** (1997) 717.

19. K. Gibov, B. Zhubanov and L. Shapovalova., *Polymer Sci. [Vysokomol. Soed.]*, **26(2B)** (1984) 108.

20. K. Gibov, L. Shapovalova and B. Zhubanov, *Fire & Materials*, **10** (1986) 133

21. C. E. Anderson, D. E. Ketchum and W. P. Mountain, *J. Fire Sci.*, **6** (1988) 390.

22. I. Reshetnikov and N. Khalturinskij, *Chem. Phys. Reports [Khimicheskaya Fizika]*, **16(3)** (1997) 499.

23. I. Reshetnikov and N. Khalturinskij, *Chem. Phys. Reports [Khimicheskaya Fizika]*, **16(10)** (1997) 102.

24. I. Reshetnikov, M. Yablokova and N. Khalturinskij, *Appl. Surf. Sci.*, **115** (1997) 199.

25. E. Gnedin, S. Novikov and N. Khalturinskij, *Makromol. Chem., Macromol. Symp.*, **74** (1993) 329.

26. Ye. Gnedin, R. Gitina, S. Shulyndin, G. Kartashev and S. Novikov, *Polymer Sci. [Vysokomol. Soed.]*, **33** (1991) 621.

27. S. Brauman, *J. Polym. Sci., Polymer Chem. Ed.*, **15(6)** (1977) 1507.

28. S. Brauman, *J. Fire Retardant Chem.*, **6(4)** (1979) 266.

29. S. Brauman, *J. Fire Retardant Chem.*, **7(3)** (1980) 119.

30. I. Reshetnikov, T. Rudakova and N. Khalturinskij, *Plasticheskie Massy*, (1996) 24.

31. K. Butler, in: *"Polymer Foams: Science and Technology"*. *ASC Symp. Ser. 669*, May 1997.

32. Ye. Gnedin, N. Kozlova, R. Gitina, O. Fedoseeva, M. Sevast'yanov and S. Novikov, *Polymer Sci. [Vysokomol. Soed.]*, **33(7A)** (1991) 1568.

33. M. Le Bras, S. Bourbigot, Y. Le Tallec and J. Laureyns, *Polymer Deg. & Stab.*, **56** (1997) 11.

34. J. Kipling and P. Shooter, *Carbon* **4(1)** (1966) 1.

35. T. Davlichin, B. Zhubanov and K. Gibov, in *"Chemistry and Physical Chemistry of Polymers"*, Alma-Ata, Nauka, **49** (1979).

36. M. Yablokova, E. Potapova, I. Reshetnikov and N. Khalturinskij, *2nd Int. Seminar "Fire and Explosion Hazard of Substances"*, Moscow, (1997) 129.

37. M. Abdul, V. Bratnagar and J. Vergnaud, *.J. Thermal Anal.*, **29(5)** (1984) 1107.

38. T. Hirata, S. Kawamoto and A. Okuro, *J. Appl. Polym. Sci.*, **42(12)** (1991) 3147.

39. N. Khalturinskij and Al. Berlin, in: *Degradation and Stabilization of Polymers*, H. Jellinek ed., Elsevier, New York, 1983.

40. British Standards BS 476, Part 20 (1987).

41. I. Reshetnikov, A. Antonov, T. Rudakova, G. Aleksjuk and N. Khalturinskij, *Polymer Deg. & Stab.*, **54** (1996) 137.

42. F. Williams, *"The Theory of the Combustion"*, Nauka, Moscow, 1971.

43. A. Gal'chenko, Doctoral Dissertation, Moscow, 1983.

44. A. Antonov, V. Dubenko, I. Reshetnikov and N. Khalturinskij, *2nd Int. Seminar "Fire and Explosion Hazard of Substances"*, Moscow (1997) 124.

45. R. Aseeva and G. Zaikov, *"Combustion of Polymeric Materials"*, Nauka, Moscow,1980.

46. S. Pokrovsky, *Dep. 209, R. Zh. Khim., 24T.*

47. D. Stuetz, A. Diedwardo, F. Zitomer and B. Barnes, *J. Polym. Sci., Polym. Chem. Ed.*, **13(3)** (1975) 585.

48. Gy. Marosi, Gy. Bertalan, P. Anna and I. Rusznak, *J. Polym. Eng.*, **12(1-2)** (1993) 33.

49. I. Reshetnikov, M. Yablokova, E. Potapova, N. Khalturinskij, V. Chernyh and L. Mashlyakovsky, *Appl. Polym. Sci.*, under press.

50. Al. Berlin, I. Reshetnikov, M. Yablokova and N. Khalturinskij, in *"Proceedings. of 6th European Meeting on Fire Retardancy of Polymers"*, Lille (1997) 35.

51. I. Reshetnikov and N. Khalturinskij, in *Proceedings. of Int. Meeting "Advances in Computational Heat Transfer"*, Cesme, Izmir (1997).

52. A. Pivkina, P. van der Put, Yu. Frolov and J. Schoonman, *J. Europ. Ceram. Soc.*, **16** (1996) 35.

53. J. Buckmaster, C. Anderson and A. Nachman, *Int. J. Eng. Sci.*, **24(3)** (1986) 263.

54. E. Gnedin, S. Novikov, in *"Proceedings. of 1st Int. Conf. on Polymeric Materials with Lowered Flammability"*, Volgograd (1990) 198.

55. K. Kanary, *Denki sikinse hose daigaku*, **176** (1973).

56. I. Reshetnikov and N. Khalturinskij, in *"Proceedings. of 6th European Meeting on Fire Retardancy of Polymers"*, Lille (1997) 47.

57. F. Camia, *"Traité de Thermo-cinétique Impulsionnelle"*, Dunod, Paris, 1967.

Acknowledgements

Authors would like to thank Prof. V. Chernyh (Moscow State Food Academy) for the "Structurometer ST-1" device, providing for experiments and for funding of the mechanical measurements. Also authors are very thankful to Dr. V. Svistunov (Institute of Synthetic Polymeric Materials) for his kind permission to present calorimetric measurements data.

SOME ASPECTS OF MECHANICAL STABILITY OF INTUMESCENT CHARS

Al. Al. Berlin, N. A. Khalturinskij

Semenov Institute of Chemical Physics,
4 Kosygina street, Moscow, 113977, Russia

I. S. Reshetnikov and M. Yu. Yablokova

Polymer Burning Laboratory
Institute of Synthetic Polymeric Materials,
70 Profsoyuznaya street, Moscow, 117393, Russia

1. INTRODUCTION

It is known, that the main aspect of high efficiency of intumescent fire retardant materials is their high thermo-protection properties, provided by the high thermo-protection properties of the foamed char cap[1-2]. As a consequence efficiency of fire protection depends on the char formation. This point has been discussed in detail[3-4]. Some aspects of char properties have been already discussed[5 and Refs. therein]. In particular it has been shown[6] that flammability of intumescent materials depends on the structure of carbonaceous cap. Bourbigot et al.[7] investigated surface and bulk chemical structure of the char. Chemical structure of foamed chars produced under different pyrolysis conditions has been discussed by Mashlyakovsky et al.[8]

The aim of present work is to discuss the mechanical properties of foamed chars, in particular their mechanical strength, parameter which allow to point out change in stability. In fact, efficiency of fire protection depends on these parameters. Indeed in the conditions of fire, char destruction can proceed not only by means of ablation and heterogeneous surface burning but also by means of an external influence such as wind. So the question is: how to create intumescent fire retardant system with high mechanical stability. Unfortunately, this parameter has been very poorly discussed in previous works. It has been proposed that bond(s) in chemical char structure will provide mechanical strength[9], but without further explanation. Earlier a simple criteria[10] was proposed for the mechanical stability of the intumescent chars: mechanical properties of foams depend both on their structure, porosity and mechanical properties of the foam material. So considering foamed chars produced during burning, a similar dependence may be expected.

2. EXPERIMENTAL

2.1.Materials

Intumescent fire retardant formulation on the base of carbamide-formaldehyde resin (CFR; Russian grade CFRLTCS) was used. Different blends of ammonium polyphosphate (Russian grade TU 6-18-22-101-87; APP), d-sorbitol ("NEOSORB" grade, supplied by Roquette Corp., France), pentaerythritol (Russian grade TU 6-09-3329-78; PET) and citric acid (CA) were used as intumescent additives (IA). Tetraetoxysilane (TES), γ-aminopropylvynilsilane (AGM), and a mixture of silane-containing oligomers (OM) were used as modification agents. Compositions were prepared by mixing of components with subsequent addition of a catalytic agent (phosphoric acid). After mixing, composition were put on a wood substrate. Cure was carried out at 20 °C, in air during 24 hours. The thickness of the final coatings was 1 mm.

2.2.Pyrolysis

Experiments were carried out under a constant heat flow from a radiation panel under air. Heat flow near the panel was 5 W/cm^2, value corresponding to the heat flow from the flame in fires[11]. In experiments carried out under one heat flow, pyrolysis of all samples was made simultaneously.

2.3. Mechanical strength measurements

All measurements were carried out at the constant speed 100 mm/min using device "Structurometr ST-1" developed in Moscow State Food Academy[10]. A_{destr} will represent the work necessary to obtain a full destruction per unit of area of the tested plate.

3. RESULTS AND DISCUSSION

Char structure and coke yield significantly depend on the ratio of the IA components, as previously reported[10], and on the IA content in the matrix. Our study deals now with dependence of char structure and strength on the IA contents in the polymer matrix (Figure 1). CFR is a "self-intumescent" system[13]. Foaming ratio for the CFR without additives is about of 15. For low IA concentrations, intumescence is provided by the matrix and foaming ratio is quite the same that this of pure CFR. Same foaming ratios are obtained with high IA contents: intumescence is in this case generally provided by the additive. The foaming ratio of the additive (7-10) is comparatively low and increase of IA contents leads to the corresponding foaming ratio value. Behaviour on the intermediate part of the IA contents scale is very interesting. Intumescence is inhibited when IA contents are about 20%. In this case, gaseous reaction products make the carbonising material structure not strong enough to create foamed structure, amount of cross-linked structures being too low. These cross-linked structures, produced during pyrolysis of the matrix, are more insufficiently strong owing to defects resulting from the additive. A synergetic effect and, as a consequence, a maximal foaming ratio are observed when the IA contents are about 30%.

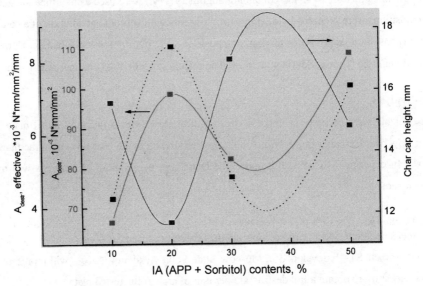

Figure 1. *Char mechanical strength and foaming ratio vs. intumescent additive (APP and sorbitol, 7:3) contents.*

Foaming ratio clearly correlates with char cap strength. On the graph the total work of destruction is presented in order to compare with destruction work recomputed on the unit of char cap height - criteria value of mechanical stability (dotted line). The figure shows that foamed chars with high ratio coefficient present low mechanical strength. In fact, it is obvious that thin wall is less strong than thick one and stress required to broke two thin walls may be assumed weaker than stress requires to broke one wall with double thickness.

Increase of char strength when IA contents increase may be readily explained by the formation of phosphorus-contained glass-likeness structures formed during IA degradation. So use of 10% and 50% IA contents leads to quite the same foaming ratio but mechanical properties of the last composition is much more better than this of the first one. Similar behaviours are observed using other IA, based on PET and CA as shown by Figures 2 and 3.

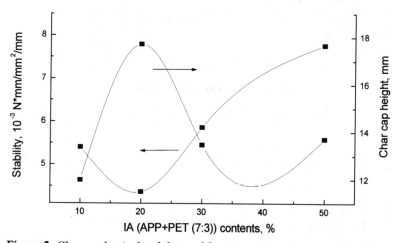

Figure 2. *Char mechanical stability and foaming ratio vs. intumescent additive (APP and PET, 7:3) contents.*

Previous works[10] show that silicone-based additives increase char strength. To verify this action, different silicone-containing additives in CFR matrix with 30% of APP-sorbitol additive are investigated. Introduction of TES in concentrations characteristic of surface active agents (~0.1-0.5%) leads to significant increase of the foaming ratio and of mechanical strength (Figure 4). When its concentration increases TES in fact plays the role of an inert filler and its active action compete with its antagonistic effect which may be related to isolation of APP active centres. Moreover an increase of TES contents to concentrations typical for common additives leads to an increase of the char cap height. This may be explained by formation of water from reactions of the OH groups of the TES and the corresponding increase of the total amount of volatile products.

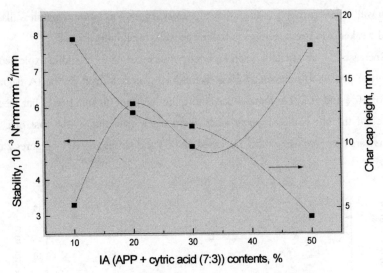

Figure 3. *Char mechanical stability and foaming ratio vs. intumescent additive (APP and CA, 7:3) contents.*

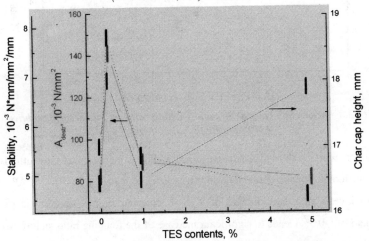

Figure 4. *Char mechanical strength and foaming ratio vs. modification agent (TES) contents.*

Such a behavior is not observed using other silicone-containing additives. For example, action of AGM and OM in the tested systems are illustrated in Figures 5 and 6. Introduction of AGM and OM practically does not affect char stability, even decreasing it in some extent.

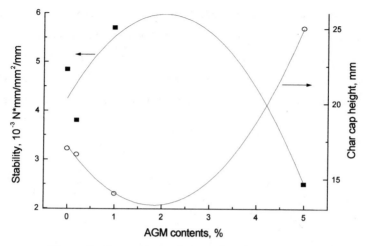

Figure 5. *Char mechanical strength and foaming ratio
vs. modification agent (AGM) contents.*

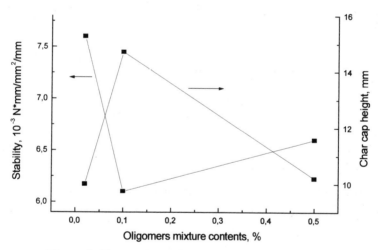

Figure 6. *Char mechanical strength and foaming ratio
vs. modification agent (OM) contents.*

It is known[14] that structure of foamed char produced by APP-based materials depends on the APP particle size. Mechanical properties of compositions with different APP particles sizes are shown in Figure 7. The mechanical stability presents a maximum when APP particle size is about 0.09 mm. An further increase of this size leads to an increase of average pores diameter and, as a sequence, a decrease of the mechanical stability.

It has been previously shown[10] that integral foamed char cap cannot be characterised by existing criteria for rigid polymer foams[12] (force of destruction). The question is why and how this force is distributed through the char cap - from the surface layer and up to "frozen" pyrolysis zone. This distribution depends on char cap physical and, in less extent, chemical structures. Typical dependence of destruction force vs. distance is shown in the Figure 8.

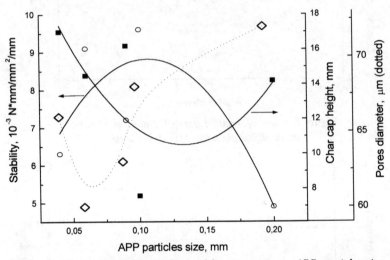

Figure 7. *Char mechanical stability and foaming ratio vs. APP particles size.*

In the upper cap layer, char strength is very low and despite destruction force slightly growth, its value remains very small. This trend is different in the pyrolysis zone (9-11 mm on distance scale) where a significant growth of the destruction force is observed. This may be explained by an increase of the amount of residual non-degraded polymer. In "frozen" pyrolysis zone (12-13 mm) the char behavior is similar to this of rigid polymeric foam and may be characterised by the destruction force value.

The structure, showed on the Figure 9, gives a model approach. Pyrolysis zone in this approach is characterised by a regular structure with small pores. Further, in "pre-pyrolysis" zone, pores diameters increase due to burning out of high molecular polymer residue. It is clear, that in this last zone mechanical strength has to decrease.

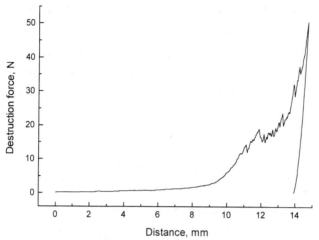

Figure 8. *Typical distribution of the destruction force vs. distance from the surface.*

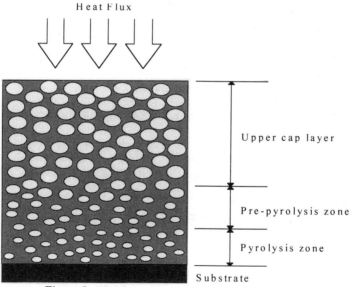

Figure 9. *Model structure of the intumescent char.*

4. CONCLUSIONS

Mechanical properties of different intumescent chars, produced during burning have been investigated and discussed. Our study shows that mechanical strength depends both on the chemical and the physical structure of the foamed char cap. In particular, char stability depends on the amount of cross-linked glass-like structures (e.g. phosphorus containing materials). Introduction of silicone-containing additives sometimes leads to and improvement of char properties, but all the silicone-containing additives do not give such an effect.

Finally, this study, considering physical structure of char, shows that one of the main basic aspects of mechanical stability is the porosity of the char: the smaller pores the better char strength.

References

1. G. Camino, L. Costa, and G. Martinasso, *Polym. Deg. Stab.*, 1989, **23**, 359.
2. I. Reshetnikov, A. Antonov, T. Rudakova, G. Aleksjuk and N. Khalturinskij, *Polym. Deg. Stab.*, 1996, **54**(2-3), 137.
3. S. Bourbigot, M. Le Bras, R. Delobel, R. Decressain and J.-P. Amoureux, *J. Chem. Soc., Faraday Trans.*, 1996, **92**.(1), 149.
4. I. S. Reshetnikov, M. Yu Yablokova., N. A. Khalturinskij, *Applied Surface Sci.*, 1997, **115**, 199.
5. I. S. Reshetnikov, M. Yu. Yablokova and N. A. Khalturinskij, *this Book.*
6. E. V. Gnedin, S. N. Novikov and N. A. Khalturinskij, *Makromol. Chem., Macromol. Symp.*, 1993, **74**, 329.
7. S. Bourbigot, M. Le Bras, L. Gengembre and R. Delobel, *Appl. Surface Sci.*, 1994, **81**, 299.
8. L. N. Mashlyakovskii, M. M. Alekserov, E. V. Khomko and V. A. Krasheninnikov, *Russ. J. Appl. Chem.*, 1994, **67**(2), 248.
9. S. Bourbigot, M. Le Bras, R. Delobel, P. Bréant and J.-M. Trémillon, *Carbon*, 1995, **33**, 283.
10. I. S. Reshetnikov, M. Yu. Yablokova and N. A. Khalturinskij, *Appl. Polym. Sci., 1998*, **67(10)**, 1427.
11. N. A. Kahlturinsky, Al. Al. Berlin, '*Polymer Combustion.*' in: '*Degradation and Stabilization of Polymers'* (Volume 2), H. H. G. Jellinek ed., Elsevier, Oxford,. 1989.
12. A. A. Berlin, F. A. Shutov, in '*Polymer Foams on the Base of Reactable Oligomers'*, Khimia, Moscow, 1978.
13. H. Vandersall, *J. Fire and Flammab.*, 1971, **2**, 97.
14. N. A. Khalturinskij, I. S. Reshetnikov and M. Yu. Yablokova, *this Book.*

Acknowledgements

Authors team would like to thank Aleuron Ltd. Company for the device "Structurometer ST-1", granted for the experiments. Also authors are very thankful to the Dr. Muzafarov (ISPM) for his funding of work with silicone-containing additives.

MODELLING FIRE RETARDANT INTUMESCENT POLYMERIC MATERIALS

V. Sh. Mamleev and K. M. Gibov

A. B. Bekturov Institute of Chemical Sciences, Kazakh Academy of Sciences,
Valikhanov street 106, Almaty 480100, Kazakhstan

1. INTRODUCTION

Despite the obvious practical interest in the creation of intumescent fire retardant paints, coatings and polymeric formulations, the development of quantitative theory of the intumescence process progresses very slowly.

As far as we know, there exist only several works[1-4] dedicated to thermo-physical models of intumescence (foaming). The most detailed one of them was suggested by Clark et al.[1]. Clark has assumed that decomposition of the coating material takes place as a result of two reactions. Only the first one of these reactions causes an intumescence. Thus, the zone where this reaction takes place corresponds to zone of intumescence within the coating. If all gas released in the first reaction were retained in the foam, Clark's model would be reasonable enough. However, in reality, as a rule, only a small part of gas is kept in the foam. Therefore, an extent of foaming is determined rather by viscoelastic properties of the material than by amount of gas isolated.

The work of Wauters and Anderson[2] is practically a reproduction of the model of Clark. When describing decomposition of the material, the authors ignored the laws of chemical kinetic. Instead of using the Arrhenius equations they approximated a curve of decomposition by Fourier series. In contrast to the model of Clark, according to which foaming takes place only within some region of the coating, the model of Wauters and Anderson[2] implies that foaming occurs in each coating layer undergoing decomposition up to the surface of the coating. At the same time, it is quite obvious that foaming must cease when the melted (liquid) polymer turns into a solid coke. We suppose[5,6] that probably the end of the foam formation occurs even earlier, when the material hardens as result of some curing reaction in the polymer.

Buckmaster et al.[3] apparently took into account the inaccuracies of Wauters and made assumptions being, in a sense, opposite to those of Wauters; namely, they assumed that the foaming occurs in an extremely narrow temperature range. This model allows one to explain some peculiarities of dynamics of heating the coatings[3,4], but being very formal it reflects by no means the chemical processes of foaming.

It is reasonable to assume[5,6] that in systems with intensive foaming the reactions of curing and gas formation should be synchronised. Earlier[5,6] we have offered a numerical procedure calculating factor of expansion of a coating via an amount of gas held in the foam within certain range of variable viscosity of the material. This procedure is illustrated for an artificial example, when the heating of a material takes place by the linear law[6].

The goal of the present paper is to consider a model, in which local heating of a substance in a coating obeys the heat conduction equation. Such a model allows one to investigate how kinetic parameters of above reactions influence a final behavior of a coating under conditions of actual heating by a flame or other source of heat.

2. EQUATIONS OF MASS CONSERVATION

Foam formation result in the widening of local portions of the substance, i.e. in a decrease of density of the material. Current x and initial ξ coordinates of the moving material point (Figure 1) are connected by the equation:

$$x(\xi,t) = \xi + \int_0^t u(\xi,\theta)d\theta$$

where $u(\xi,t)$ is the linear velocity of the movement of the material point. ξ and x are respectively the Euler and Lagrange coordinates.

The expansion coefficient connected with the difference of the velocities of two points, being at the initial moment in the positions ξ and $\xi + d\,\xi$, is expressed:

$$E(\xi,t) = \partial x/\partial \xi = 1 + \int_0^t [\partial u(\xi,\theta)/\partial \xi]d\theta$$

where

$$\partial u(\xi,t)/\partial \xi = \partial E(\xi,t)/\partial t$$

Let us consider the velocity of the substance flow at a distance x from the back-wall (the substrate):

$$dx/dt = v(x,t) \qquad (1)$$

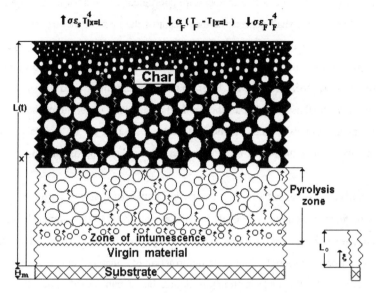

Figure 1. *Model of an intumescent coating.*

It is noteworthy that velocities $u(\xi,t)$ and $v(x,t)$ are not equal, since in equation (1) it is not determined what portion of the substance at the moment t reaches the x position. However, let us choose a certain portion, fixing its initial coordinate:

$$x_{|t=0} = \xi, \quad \partial x/\partial \xi_{|t=0} = 1 \quad (2)$$

Now the velocity value is determined at characteristic $x(\xi,t)$[7], i.e.

$$v(x,t) = v(x(\xi,t),t) = u(\xi,t)$$

A differentiating equation (1) by ξ leads to the equation:

$$d/dt(\partial x/\partial \xi) = \partial v/\partial x(\partial x/\partial \xi) \quad (3)$$

Solving with respect to $\partial x/\partial \xi$ an ordinary differential equation (3) under the initial condition (2), we obtain:

$$E(\xi,t) = \partial x/\partial \xi = \exp\{\int_0^t [\partial v(x(\xi,\theta),\theta)/\partial x] \, d\theta\} (4)$$

By means of the simple transformations, equation (4) turns into an identity:

$$\exp[\int_0^t \partial v/\partial \xi(\partial \xi/\partial x)d\theta] = \exp[\int_0^t (\partial \ln E/\partial \theta) \, d\theta] \equiv E$$

To calculate the change of a local density of the coating we have to do a number of assumptions. First, we shall neglect a temperature coefficient of expansion of polymer mass of the coating. Second, we shall neglect the movement of the mass of the material connected with gravity. Third, we shall consider that only gases isolated in each elementary layer of the coating take part in the foam formation inside this layer. As to the gases isolated in the lower layers, these penetrate freely through the layer under consideration toward the surface of the coating. Fourth, we shall assume that all the gas in the coating is under atmospheric pressure P. The latter assumption is required to not include a correction for an excess pressure in bubbles when using the equation of Clapeyron-Mendeleev. This approximation needs to be explained.

An Laplacian excess pressure is essential in bubbles of a size ~ 1 μm. At the same time, gas in the foam is accumulated mainly in bubbles of a size $\sim 0.01 - 1$ mm. Therefore, the Laplacian component of the excess pressure may be neglected at once. A reactive component of the excess pressure connected with plastic deformation of a material in course of growth of bubbles is small whilst the material remains liquid, but it can sharply rise after gelation (solidification) of the polymer. It has been shown[6] that the foam formation begins to occur at a viscosity ~ 300 Pa×s. At a viscosity ~ 3000 Pa×s the material is still able to viscous flow. However, moment of time of gelation practically coincides with the moment of reaching the viscosity ~ 3000 Pa×s, i.e., an amount of gas that is retained in the foam when changing a viscosity from 300 Pa×s to 3000 Pa×s is much greater than that being held in the foam when changing a viscosity from 3000 Pa×s to ∞. In other words, the gas bubbles, which cross over a very narrow zone of gelation, harden, having no time to grow. After gelation the excess pressure can lead to a rupture of bubbles walls, but, since the material is a solid one, the hardened bubbles conserve their volume and form (merely closed hollows turn into open ones). We note that the latter fact is confirmed by investigation of slits of the coatings. Thus, when calculating a change of the expansion coefficient one may ignore a correction for the excess pressure in the equation of Clapeyron-Mendeleev up to the moment of gelation and consider that the expansion coefficient reaches a limit (stationary) value after the moment of gelation.

With the assumption made, in the chosen layer of the coating (Figure 2) a density decreases due to the foam formation and due to decrease in the layer mass resulting from chemical decomposition of the material.

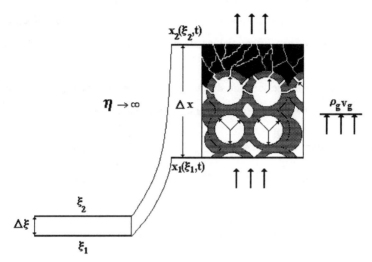

Figure 2 *Evolution of a small fragment of a coating.*

Depending on the position of a layer under consideration, gases isolated upon decomposition can leave this layer either due to disintegration of bubbles (the region of a liquid foam) or due to filtration through the system of micropores in the solid material (the region of a solid foam).

To apply the laws of formal kinetics, one should consider the gas isolation as being a result of decomposition of several individual components. Let us consider first, to simplify, one-component media. If the component undergoes chemical decomposition, the continuity equation in Euler and Lagrange coordinates has the following form:

$$\partial \rho / \partial t + \partial \rho v / \partial x = - d\Gamma/dt \qquad (5)$$

$$d\rho/dt + \rho \partial v / \partial x = - d\Gamma/ dt \qquad (6)$$

where ρ is a density of the component, $- d\Gamma/dt$ is a source of a loss of mass due to the decomposition reaction. If the component does not decompose, then $d\Gamma/dt = 0$. The function Γ depends both upon x (or ξ) and upon t , however, partial derivative of Γ with respect to x (or ξ) will be nowhere used below, and so, in an expression for Γ, the x (or ξ) variable can be considered as a parameter. That is why in equations (5) and (6) we use full derivative of Γ with respect to time.

In equation (6) we used known designation[7]:

$$d\rho/dt = \partial\rho/\partial t_{|\xi=const} = \partial\rho/\partial t_{|x=const} + v\partial\rho/\partial x$$

Indeed, if the functions $\partial v/\partial x$ and $d\Gamma/dt$ are formally considered as known ones, equation (6) becomes an ordinary differential equation with respect to ρ, in which ξ is some parameter and not an independent variable.

To determine a form of the function $d\Gamma/dt$, let us consider the change of mass of the component Δm within a segment between two characteristics (Figure 2):

$$d\Delta m/dt = A d/dt \int_{x_1(\xi_1,t)}^{x_2(\xi_2,t)} \rho dx = A\left(\rho v_{|x=x_2} - \rho v_{|x=x_1} + \int_{x_1(\xi_1,t)}^{x_2(\xi_2,t)} (\partial\rho/\partial t)dx\right)$$

where A is the surface area under consideration that is perpendicular to the flow.

Using equation (5) it is easy to show that:

$$d\Delta m/dt = -A \int_{x_1}^{x_2} (d\Gamma/dt)dx \qquad (7)$$

The function $\Delta m/\Delta m_0$, where Δm_0 is the initial mass of the component, conforms to the laws of formal kinetic. It is noteworthy that this mass ratio does not depend on absolute values of the masses, i.e. $\Delta m/\Delta m_0 = m/m_0$. The function $m(t)/m_0$ (the curve of the mass loss) is usually measured by means of thermogravimetric analysis. It can be expressed generally by the following differential equation:

$$d(m/m_0)/dt = -k_c(m/m_0)^n \exp(-E_c/RT)$$

where E_c is an activation energy, T is temperature, k_c is a rate constant, n is an order of reaction, R is the gas constant.

Integration gives:

$$m/m_0 = f(\tau) = [k_c\tau(n-1) + 1]^{-1/(n-1)}, \quad (n > 1)$$

$$m/m_0 = f(\tau) = \exp(-k_c\tau), \quad (n = 1)$$

where

$$\tau[T(\theta)]_{|\theta=t} = \int_0^t \exp[-E_c/RT(\theta)]d\theta \qquad (8)$$

is the reduced reaction time.

We put the double brackets in the left-hand side of equation (8), assuming that τ is a functional and not a function of temperature. Further we shall use the expression $\tau = \tau(t)$, bearing in mind, however, that $\tau(t)$ is unambiguous function of time only with a fixed temperature-time dependence $T(t)$. Note that $\Delta m_0 = A\rho_0\Delta\xi$, where ρ_0 is the initial density of the component. Thus, from equation (7) we obtain:

$$\rho_0\,\Delta\xi\,df/dt \approx -\,\Delta x\,d\Gamma/dt$$

Going to the limit at $\Delta\xi \to 0$ $(\Delta x \to 0)$ we have:

$$d\Gamma/dt = -\,[\rho_0/E(\xi,t)]df/dt \quad (9)$$

Substituting expression (9) into equation (6) and integrating with respect to $\rho(\xi,t)$ the ordinary differential equation (6) under the initial condition $\rho(\xi,0) = \rho_0(\xi) = \rho_0$, we obtain:

$$\rho(\xi,t) = \rho_0\exp(-I)\left\{1 + \int_0^t\,[\exp(I)/E(\xi,\theta)](df/d\theta)d\theta\right\}$$

where

$$I = \int_0^t\,[\partial\,v(x(\xi,\theta),\theta)/\partial\,x]d\theta$$

Taking into account equation (4), one can readily see that:

$$\rho(\xi,t) = [\rho_0/E(\xi,t)]\,f(\tau)$$

The latter equation may be deduced from equations (6) and (7) by one more manner:

$$E[d\rho/dt + \rho\,\partial v/\partial\xi(\partial\xi/\partial x)] = Ed\rho/dt + \rho\partial E/\partial t = d(E\rho)/dt = d(\rho_0\,f)/dt$$

from here:

$$E\rho = \rho_0\,f$$

For going to the case of a media consisting of N components it is sufficient to use the rule of additivity to give:

$$\rho(\xi,t) = \sum_{i=1}^N\,\rho_i(\xi,t) = [\rho_0/E(\xi,t)]\,\sum_{i=1}^N\,w_i\,f_i\,(\tau_i)\;(10)$$

$$d\Gamma/dt = -\,[\rho_0/E(\xi,t)]\,\sum_{i=1}^N\,w_i\,df_i\,(\tau_i)/dt$$

where $w_i = \rho_{0,i}/\rho_0\,(1 \le i \le N)$ are mass fractions of the components.

Being calculated per gram of the coating, the mass of gaseous products q isolated to the moment t at the point with the initial coordinate ξ constitutes:

$$q(\xi,t) = \sum_{i=1}^{N} w_i \, [1 - f_i \, (\tau_i)] = 1 - \sum_{i=1}^{N} w_i \, f_i \, (\tau_i)$$

The volume of this mass is equal to $RT(\xi,t)q/PM$, where M is the molecular weight of the gas.

Thermogravimetric dependence (the curve of the mass loss) for multi-component system is expressed as:

$$m[T(\theta)]_{|\theta=t} \, /m_0 = \sum_{i=1}^{N} w_i \, f_i \, (\tau_i \, [T(\theta)]_{|\theta=t})$$

The algorithm of the $E(\xi,t)$ calculation, proposed earlier[5,6], is based on the hypothesis that the gas released is retained in the foam only in a certain range of rheological state of the material, when its dynamic viscosity varies within the limits $[\eta_m,\infty)$. At the point of coating with the initial coordinate ξ the time interval $[t_{min}(\xi), t_{max}(\xi)]$ corresponds to the viscosity interval $[\eta_m ,\infty)$. Thus, a gas volume kept in the foam is equal to:

$$(RT/PM)\tilde{q}(\xi,t) = \begin{cases} 0, \, t < t(\eta_m(\xi)) = t_{min}(\xi) \\ (RT/PM)[q(\xi,t) - q(\xi,t_{min})], \, t_{min} \, < t < t_{max} \\ (RT(\xi,t_{max})/PM)q_{max}, \, t > t(\eta(\xi) \to \infty) = t_{max}(\xi) \end{cases} \tag{11}$$

Initially gram of the coating occupies a volume of $1/\rho_0$. Due to the decomposition of the material this volume decreases. If we assume that the decrease in the volume is proportional to the lost mass, then the volume of polymer mass under consideration will constitute $(1 - q)/\rho_0$. The overall volume of the foam is a sum of the volume of polymer mass and the volume of isolated gas. Hence, the ratio of the volume of the foam to the initial volume of the coating is equal to:

$$E(\xi,t) = [(1 - q)/\rho_0 + \tilde{q} \, (\xi,t)RT(\xi,t)/PM]/(1/ \, \rho_0) =$$

$$= 1 - q + \rho_0 \, \tilde{q} \, (\xi,t)RT(\xi,t)/PM , \, t_{min}(\xi) < t < t_{max}(\xi) \tag{12}$$

If presuming that the sole result of the material gasification is the increase in porosity, it is easy to express the volume fractions of the retained gas ω_g and the polymer mass ω_p as follows:

$$\omega_g = (E - 1 + q)/E, \quad \omega_p = (1 - q)/E \quad (13)$$

The actual density of the material is equal to:

$$\rho = \omega_p\rho_0 + \omega_g\rho_g \quad (14)$$

where ρ_g is the gas density. The formulas (10) and (14) differ, because equation (10) describes the change of a partial density. However, since $\rho_0 \gg \rho_g$, the equations (10) and (14) give close results.

Let us derive the equation of a material balance for the gaseous phase, considering again the layer between two characteristics (Figure 2). Taking account of equation (7), one can write:

$$\rho_g v_g|_{x=x_2} - \rho_g v_g|_{x=x_1} = \int_{x_1}^{x_2} (d\Gamma/dt)dx - d/dt \int_{x_1}^{x_2} \omega_g \rho_g \, dx \quad (15)$$

where v_g is the velocity of gas movement. When substituting the expression for ω_g from equation (13) into equation (15) and using integration along Lagrange coordinate, we obtain:

$$\rho_g v_g|_{x=x_2} - \rho_g v_g|_{x=x_1} =$$

$$= \rho_0 \int_{\xi_1}^{\xi_2} (dq/dt) \, d\xi - d/dt \int_{\xi_1}^{\xi_2} E \rho_g \, d\xi + d/dt \int_{\xi_1}^{\xi_2} (1 - q)\rho_g \, d\xi \quad (16)$$

While the foaming takes place the equality $dq/dt = d\tilde{q}/dt$ should be fulfilled. Using equations (11) and (12) for the stage of foam formation from equation (16) we find:

$$\rho_g v_g|_{x=x_2} - \rho_g v_g|_{x=x_1} = 0$$

In other words, at this stage all the gas isolated within the layer under consideration is retained in the foam. Note that the latter result is a natural consequence of our initial assumption.

At the limit $\Delta\xi \to 0$ equation (16) gives:

$$\partial \rho_g v_g /\partial\xi + d (E - 1 + q) \rho_g /dt = \rho_0 dq/dt \quad (17)$$

We used in equation (17) the full derivatives with respect to time, just like in equation (6), keeping in mind that all the functions in this equation are given in Lagrange coordinate system. We note that equation (17) describes balance of gas within layers which move with velocity of polymer substance. However, a velocity of gas differ from a velocity of polymer $(v_g \neq v)$, therefore, equation (17) is not the continuity equation for gas.

By expressing the gas density as $\rho_g = PM/RT$, one may rewrite equation (17) as follows:

$$\partial \rho_g v_g / \partial \xi = \rho_0 dq / dt - \rho_g d(E+q)/dt + (PM/RT^2)(E - 1 + q) \ dT/dt \quad (18)$$

The second and third terms in the right-hand part of equation (18) correspond respectively to the retention of gas in the material due to foaming and to the gas flow connected with thermal expansion of gas in pores. Since the retention of gas in the foam usually is much lesser than the amount released on the whole, these terms can be omitted. Taking into account that the gas flow at the back-wall is equal to zero, one can write with good accuracy the gas flow as:

$$\rho_g v_g |_{x=x(\xi,t)} = \rho_0 \int_0^\xi (dq / dt) \ d\xi \quad (19)$$

Clark et al.[1] used equation (19) in their calculations, although more accurate equation has the form:

$$\rho_g v_g |_{x=x(\xi,t)} = \int_0^\xi G(\xi,t) \ d\xi$$

where

$$G(\xi,t) = \begin{cases} 0, t_{min} < t < t_{max} \\ \rho_0 \ dq/dt - d(E - 1 + q) \ \rho_g / dt, t < t_{min}, t > t_{max} \end{cases}$$

3. EQUATION OF ENERGY CONSERVATION

Equation of heat conduction in the Euler coordinate system is as follows:

$$(c_g \rho_g \omega_g + c_p \rho_p \omega_p)\partial T / \partial t + [c_g \rho_g v_g + v(c_g \rho_g \omega_g + c_p \rho_p \omega_p)]\partial T / \partial x +$$
$$+ (\rho_0 /E) \sum_{i=1}^N \Delta H_i \ w_i \ df_i (\tau_i)/dt = \partial / \partial x(\lambda \partial T / \partial x)$$

where c_g, c_p are the specific heat capacities of the gas and polymer mass; ΔH_i $(1 \leq i \leq N)$ are heat effects of decomposition reactions of the components; λ is the heat conductivity.

By taking account of the equality:

$$\partial T/\partial t_{|\xi=const} = \partial T/\partial\ t_{|x=const} + v\partial T/\partial x$$

let us go into the Lagrange coordinates

$$E(c_g\rho_g\omega_g + c_p\rho_p\omega_p)\partial T/\partial t + c_g\rho_g v_g\ \partial T/\partial\xi + \rho_0 \sum_{i=1}^{N} \Delta H_i\ w_i\ df_i\ (\tau_i\ )/dt$$
$$= \partial/\partial\xi[(\lambda/E)\partial T/\partial\xi] \quad (20)$$

The first term in the left-hand part of equation (20) takes into account the accumulation of heat in the coating, the second term takes into account the convective transfer by the decomposition gases, the third term takes into account the heat effect of the decomposition reactions, and, finally, the right-hand part corresponds to the heat exchange through the material of the coating.

If we neglect the accumulation of heat by gases being in the pores $c_g\rho_g\omega_g \ll c_p\rho_p\omega_p$ and take into account the equality $\omega_pE = 1 - q$, the left-hand part of equation (20) becomes independent of E. Thus, the effect of the retardancy of heating caused by intumescence can be explained[1] exclusively in terms of an effective heat conductivity: $\lambda_{effective} = \lambda/E$.

λ, according to the studies of Anderson et al.[8], is approximated satisfactory as:

$$\lambda = \lambda_g\lambda_p /(\ \omega_g\lambda_p + \omega_p\lambda_g) = \lambda_g\lambda_p\ E/\ [(E - 1 + q)\ \lambda_p + (1 - q)\lambda_g\ ] \quad (21)$$

where λ_g, λ_p are heat conductivities of gas and polymer mass respectively. Since $\lambda_p \gg \lambda_g$, even at low values of E (E ~ 1.5 - 2) the local conductivity of the coating is ractically stipulated by heat conductivity of gaseous products in the pores. When E $\gg$ 1, the veritable heat conductivity λ, in accordance with the formula (21), is limited by the heat conductivity of gas ($\lambda \approx \lambda_g$), and, consequently, the effect of deceleration of heat transfer in the coating is stipulated only by a local expansion: $\lambda_{effective} = \lambda_g/E$. If, furthermore, the coefficient E has approximately a constant value along the depth of the coating, then the heat-shielding effect is determined solely by the factor of increasing a thickness of the coating. At the same time, equation (21) seems to us as imperfect one. In the systems with large pores the heat conductivity, in particular, should certainly depend on the size of pores. To solve equation (20) it is necessary to set boundary conditions:

$$(\lambda/E)\ \partial T/\partial\xi_{|\xi=L_0} = \alpha_F[T_F - T_{|\xi=L_0}] + \varepsilon_F\ \sigma T_F^4 - \varepsilon_s\ \sigma T^4_{|\xi=L_0} \quad (22)$$

$$(\lambda/E)\ \partial T/\partial\xi_{|\xi=0} = c_m\rho_mH_m\ dT/dt_{|\xi=0} + \alpha_B[T_{|\xi=0} - T_0\ ] \quad (23)$$

and initial condition

$$T = T_0 \, , \rho = \rho_0 \, , t = 0 \ (24)$$

where T_0 is an initial temperature; T_F is the flame temperature; α_F is the coefficient of heat exchange between flame and coating surface; ε_F , ε_s are the coefficients of emission of the flame and the coating surface; σ is the Stephan- Boltzmann constant; L_0 is the initial thickness of the coating; c_m, ρ_m , H_m are heat capacity, density and thickness of the substrate; α_B is the coefficient of heat exchange between the substrate and outer media.

The equations (20) - (24) form the complete system to study numerically a dynamics of intumescent coating.

4. EXPERIMENTAL RESULTS AND DISCUSSION

4.1. Experimental

The intumescent coating was prepared[5] at 120°C by mixing 80 wt. % of phenol-formaldehyde resin (PhFR) and 20 wt. % of boron oxide (B_2O_3).

A composition layer of 1 cm thickness was put on an aluminum disk with the diameter of 15 cm and thickness of 0.3 cm. A propane-air flame, which was created by a gas injector situated at a distance of 10 cm from the initial position of the coating surface (Figure 3), served as a source of heat.

The measurements of a temperature were carried out using chromel-alumel thermocouples with the junction diameter of 200 μm. For avoidance of destruction of the thermocouples' material the surface of the thermocouples was covered by a film of silicon oxide. The temperature equilibrium between thermocouple and the flame is reached for ~ 10 - 20 s. This time is much less than that of the experiment.

The thermocouples were fastened at different heights from the back-wall (the substrate). A current position of the coating surface relatively to the back-wall was measured by a cathetometer.

Initially the thermocouples are located within the flame and, consequently, show the flame temperature ($\approx$ 1150°C), but, as the coating foams, the thermocouples penetrate in the depth of the material and, in so doing, display the temperature change in the interior layers of the coating in the Euler coordinate system.

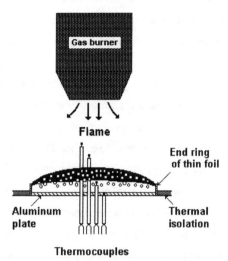

Figure 3. *Scheme of testing the coating.*

Temperature on the coating surface was measured by optical pyrometer. For time ≈ 5 min it reaches the stationary value, which is approximately equal for different regions of the surface and constitutes $\approx 850 - 870°C$.

4.2. Temperature Field in the Coating

Calculations carried out according to above model as well as a determination of the necessary parameters will be considered in detail elsewhere[9].

We adduce here (Figure 4 (a,b)) only results of calculations of indications for the thermocouples fastened at different heights (h_i - experiment, h_i' - theory) from the substrate and dependence upon time of the change of the coating height

$$L(t) = \int_0^{L_0} E(\xi,t)d\xi$$

The theoretical dependencies are depicted as points to show how a time step was being varied in the computations.

At the initial moment the thermocouples, which are above the coating show the flame temperature ($1150°C$). An algorithm computes temperature only inside the coating, i.e. the theoretical temperature-time dependence for above thermocouples contains a discontinuity between the flame temperature and a temperature on a surface of the coating. We showed these discontinuities by using rectangular fragments on the theoretical dependencies (dotted lines in Figure 4(a)). The experimental indications of these thermocouples are continuous functions on account of thermocouples' relaxation and a temperature gradient within the flame.

As foaming proceeds, the thermocouples penetrate into the depth of the material and, naturally, first the temperature decrease is registered. When the velocity of movement of the material of the coating with respect to the thermocouple position significantly decreases, the thermocouple registers the temperature increase. So, if the Euler coordinate is higher than the initial height of the coating, the corresponding temperature-time dependence passes through a minimum. The temperature change in the points located at the initial moment inside the coating, depends on the position of the point and character of foam formation. Usually a preliminary heating of material precedes intensive foaming. That is why one may expect that the temperature in the points situated near the coating surface at small time should increase to a maximum. Then, when an intensive foam formation begins to occur in the coating, the temperature in these points is decreasing. The form of the calculated temperature-time dependence (curve with $h_i' = 0.8$ cm in Figure 4(a)) for a thermocouple situated near the surface is probably typical.

Thermocouples located in the lower layers of the coating in the majority of cases show monotonous growth of the temperature (experimental curve with $h_i = 0$ cm in Figure 4(a)). However, in the calculations the temperature of the substrate passes through the maximum. This maximum is connected with the effect of heat dissipation on the back-wall into the external media (a room). In case of adiabatic back-wall ($\alpha_B = 0$) the temperature change should be of course monotonous.

The model describes the temperature field within the coating with an appreciable inaccuracy. It is more probable that the observed discrepancies of the model with the experiment are due to the fact that the coating is not actually one-dimensional.

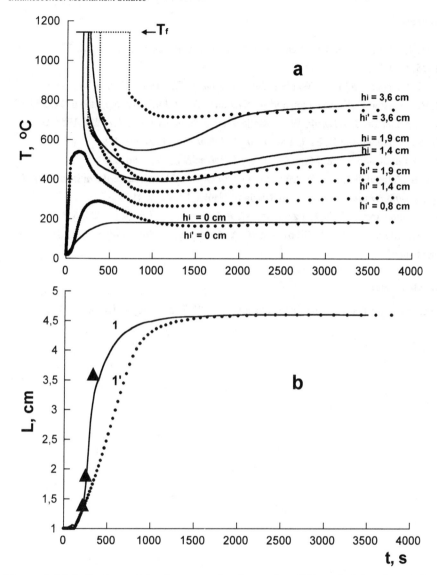

Figure 4 *(a) Indications of the thermocouples fastened at different heights (in cm) from the back-wall. The experimental (* h_i *) and numerically calculated (* h_i' *) results are shown by the solid lines and by points respectively. The dotted lines display a discontinuity between the flame temperature and a temperature on a surface of the coating. (b) The experimental (1) and calculated (1') dependencies of a height of the coating upon time. The points resulted from the indications of thermocouples at moments of their crossing with surface of the coating.*

References

1. D. E. Cagliostro, S. R. Riccitiello, K. J. Clark and A. B. Shimizu, *J. Fire & Flammability*, 1975, **6**, 205.
2. C. E. Anderson and D. K. Wauters, *Int. J. Engineer. Sci.*, 1984, **22**, 881.
3. J. Buckmaster, C. E. Anderson and A. Nachman, *Int. J. Engineer. Sci.*, 1986, **24**, 263.
4. C. E. Anderson, J. J. Dziuk, W. A. Mallow and J. Buckmaster, *J. Fire Sci.*, 1985, **3**, 161.
5. V. Sh. Mamleev and K. M. Gibov, *J. Appl. Polym. Sci.*, 1997, **66**, 319.
6. K. M. Gibov and V. Sh. Mamleev, *J. Appl. Polym. Sci.*, 1997, **66**, 329.
7. Ya. B. Zeldovich and A. D. Myshkis, in *"Elements of Mathematical Physics: A Medium of Non-interacting Particles"* (in Russian), Nauka, Moskow, 1973.
8. C. E. Anderson, J. D. E. Ketchum and W. P. Moumtain, *J. Fire Sci.*, 1988, **6**, 390.
9. V. Sh. Mamleev, K. M. Gibov and E. A. Bekturov, *J. Appl .Polym. Sci.*, 1998, to be published.

Acknowledgements

The work was fulfilled with the financial support of "INTAS" (Project INTAS - 93 – 1846).

MODELLING OF THERMAL DIFFUSIVITY DURING COMBUSTION – APPLICATION TO INTUMESCENT MATERIALS

S. Bourbigot and J.-M. Leroy

Laboratoire de Chimie Analytique et Physicochimie des Solides,
Ecole Nationale Supérieure de Chimie de Lille, U. S. T. L.,
BP 108, F-59652 Villeneuve d'Ascq Cedex, France
serge.bourbigot@ensc-lille.fr

1. INTRODUCTION

In recent years there has been considerable research into various aspects of fires and associated hazards in buildings, automotive or aeronautics. Because polymers are commonly used, it is crucial to study the flame retardancy of these materials and their fire behaviour. Intumescent technology is a method of providing flame retardancy to polymeric materials[1-4]. Fire retardant intumescent materials form on heating foamed cellular charred layers on the surface, which protects the underlying material from the action of the heat flux or of a flame. The proposed mechanism is based on the charred layer acting as a physical barrier which slows down heat and mass transfer between gas and condensed phase.

Table 1. *LOI values of PP-based intumescent systems compared to virgin PP.*

Formulation	LOI (%)	U.L. 94 Rating
PP	18	No rating
PP-APP/PER	32	V-0
PP-AP750	38	V-0

To predict the thermal performance of an intumescent material, we must first know the thermal environment to which the material is to be exposed. In previous work [5], we have studied the heat transfer of intumescent polypropylene (PP)-based materials during combustion. The intumescent formulations were a mixture ammonium polyphosphate (APP)/pentaerythritol (PER) and an intumescent commercial additive the Hostaflam AP750 of Hoechst[5]. The evaluation of the fire proofing performances showed that the formulation

containing AP750 has better performances than this one with APP/PER (table 1) but that in
the two cases a UL-94 test V-0 rating was achieved.

These results was confirmed by a cone calorimeter [7] study (Figure 1). The presence of
intumescent systems in the PP decreases strongly the rhr (rate of heat release) values
compared to the virgin polymer. Moreover the rhr values of PP-AP750 reach only 80 kW/m²
whereas these ones of PP-APP/PER reach 400 kW/m².

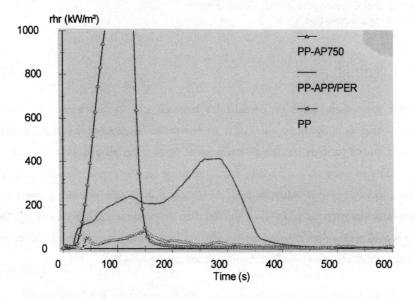

Figure 1. *rhr (rate of heat release) curves of PP-based intumescent systems
in comparison with the virgin PP (loading = 30 wt.-% ; thickness of the sheet = 3 mm).*

The comparison between the two systems showed therefore that the formulation PP-
AP750 is comparatively very efficient and that the intumescent coating developed from PP-
AP750 implied another mode of protection than that for PP-APP/PER. This was explained by
the protection of the intumescent coating developed from PP-AP750 which is still efficient at
high temperature (450-550°C) and which is developed faster. The measurement of
temperature profiles in the conditions of the cone calorimeter showed that the temperatures
reached with PP-AP750 were lower than the ones obtained with PP-APP/PER (Figure 2).
Moreover it was shown that the temperature of PP-AP750 remained always lower than the
temperature of the beginning of the fast degradation rate of PP which implied that there was

never a fast degradation of the polymeric matrix. Finally, we computed the position of the degradation front[8] versus the conversion degree of the degradation of the material and we proposed that an intumescent system enhances its performance when the degradation front occurs at low temperature and when it moves into the material at a rate high enough to restore the intumescent shield and so, stop the degradation of the polymeric matrix.

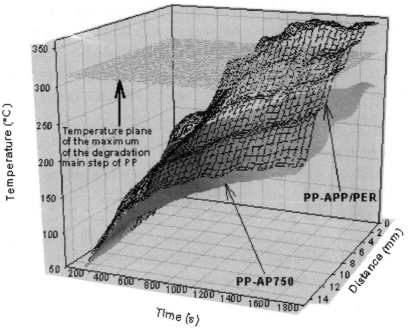

Figure 2. *temperature profiles of PP-APP/PER and PP-AP750 (temperatures measured in the conditions of the cone calorimeter under a heat flux equalling 50 kW/m² and with a sheet of 2cm thickness).*

The aim of the work is to complete the above study by considering first the heat transfer in the material during forced combustion (condition of the limiting oxygen index). Then we apply the "surface" model or "ignition" model (static model) developed by Albert et al. [9] to determine thermo-physical parameters of the materials formed during the combustion of the flame retarded polymers. Finally, a "bulk" model (dynamical model) developed in our laboratory is proposed to study the development of the intumescence and will be compared with the "ignition" model.

2. EXPERIMENTAL

2.1. Material

Raw materials were PP (Polypropylene supplied by Solvay), PER (Aldrich R.P. grade), APP $((NH_4PO_3)_n$, n=700, Hoechst Hostaflam AP422, soluble fraction in H_2O: < 1wt%) and AP750 (ammonium polyphosphate with an aromatic ester of tris(2-hydroxyethyl)-isocyanurate and bound by an epoxy resin [5]). The study has been carried out using the APP/PER mixture for the ratio APP/PER = 3 (wt/wt). In the particular case of the polyethylenic materials, the fire-retardant properties are maximum for this ratio [4]. The additive(s) was(were) incorporated at 30 % (wt/wt) in the polymer. Initial mixture(s) was(were) first prepared by ballmilling after mechanically grinding and sifting (200 μm) the raw materials. Sheets (100 x 100 x 3 mm^3) were then obtained using a Darragon press at T = 190°C and at a pressure of 3 MPa.

2.2. Temperature profile

Thermocouples (K type; diameter = 0.25 mm to avoid to modify the combustion conditions of the sample) are put into the sample each 3 mm from the top and the temperature versus time response is recorded (one point per second) at LOI+4 (LOI+4 chosen in order to get a continuous forced combustion, does not correspond to a traditional standard). Limiting Oxygen Index (LOI: Standard Test Method for Measuring the Minimum Oxygen Concentration to Support Candle-like Combustion of Plastics) was previously measured using a Stanton Redcroft instrument on sheets (120 x 6 x 200) mm^3.

3. RESULTS AND DISCUSSION

3.1. Temperature profile

Temperature profiles versus time of the combustion of PP-APP/PER and PP-AP750 in the condition of LOI+4 are shown in figures 3 and 4 respectively. Two different behaviours are observed. In the case of PP-APP/PER, the temperature versus time increases up to temperatures between 575 and 650°C. In the case of PP-AP750, the temperatures go over maxima (from 175 to 325°C) because the material extinguishes at about 250s.

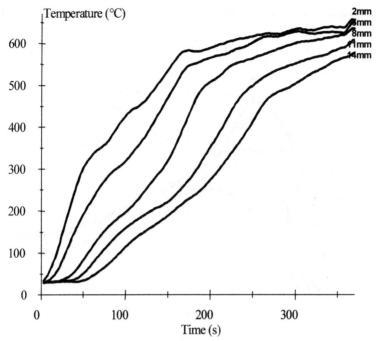

Figure 3. *temperature profiles measured in the conditions of LOI+4 of PP-APP/PER..*

Visually, the formation of the intumescent chars are very different. In the case of PP-APP/PER, the coating seems to be brittle and the amount of char produced is comparatively low. In the case of PP-AP750, the yield of char is very high and the degradation rate is faster than in PP-APP/PER. This confirms, therefore, our assumption about the relation FR performance-degradation rate. Indeed we have proposed in the case of the intumescent systems that a fast degradation rate at the beginning of the development of the intumescence is needed to give good performance [5,10-11]. Moreover, it is to be noted that when the sample is burning, the flame front is located in the polymer-char interface and moves progressively. This proves also that the intumescent shield has good thermal insulative properties. Indeed between 150 and 250s, the temperatures of the highest thermocouples decrease (Figure 4) even though the material is burning.

From this comparison, it may be concluded that the formulation PP-APP/PER is comparatively permeable to oxygen diffusion. This may lead to efficient oxidation reactions and then, to a fast degradation of the protective shield. Conversely, the system PP-AP750 shows great efficiency because of the fast degradation of the material leading to the formation of an intumescent shield impermeable to oxygen and having good thermal insulating properties.

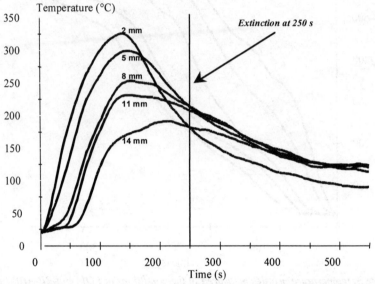

Figure 4. *temperature profiles measured in the conditions of LOI+4 of PP-AP750.*

3.2. Ignition model

The ignition model allows to determine thermo-physical parameters in the conditions of a fire using the cone calorimeter [9]. The model is one-dimensional which assumes that the vaporisation at the surface of a material occurs at a given temperature. The formula for the case of purely radiative losses from the surface provides the ignition time (t_{ig}) versus the external heat flux (q''_{ext}):

$$t_{ig}^{-1/2} = \left[\frac{2}{\sqrt{\pi k \rho c_p}\left(T_{ig} - T_\infty\right)} \right] \cdot q''_{ext} - \left[\frac{0.64 \times \dot{q}''_{cr}}{\sqrt{\pi k \rho c_p}\left(T_{ig} - T_\infty\right)} \right]$$

where $\dot{q}''_{cr} = \varepsilon.\sigma.\left(T_{ig}^4 - T_\infty^4\right)$ is the critical irradiance, ε is the emissivity (we will assume in the following that $\varepsilon=1$), σ is the Stephan-Boltzman constant, T_{ig} is the ignition temperature, T_∞ is the surrounding temperature, c_p is the heat capacity, ρ is the density and k is the thermal conductivity.

If $\dot{q}''_{ext}$ is plotted versus $t_{ig}^{-1/2}$, the equation is that of a straight line y =ax + b and a and b can be computed. Figures 5 and 6 show the curves obtained by plotting the external heat flux versus $t_{ig}^{-1/2}$. The curves are straight lines as demonstrated in the model. From these straight lines, the values in table 2 can be deduced. In order to compute the thermal conductivity k and the thermal diffusivity $\alpha=k/\rho c_p$ of the materials, density (ρ) and heat capacity (c_p) were measured by extinguishing the external flux after different characteristic times of the rhr curves, the on the different residues. From this study, it is shown that ρ and c_p remain constant from ignition to the end of the combustion (values in table 2). This means therefore that when the intumescent shield is formed, the chemical transformations occurring in the coating do not modify the thermal parameters.

Table 2 shows that the ignition temperatures of the two systems are very similar. This result suggests that the ignition is governed by the polymeric matrix. On the other hand it can be noticed that the density of PP-AP750 is twice as high as this one of PP-APP/PER. Thus that the thermal conductivities and the storage energies (ρc_p) are very different. These properties may explain the FR performances. If k is low, the heat transfer in the intumescent shield is slowed down, the temperature rise in the substrate remains low and the degradation rate of the material after the formation of the intumescent coating is reduced. Moreover, if the coating stores energy, then it is not given back to the surroundings. It may be therefore concluded that the differences of the thermal properties of the two intumescent shields explain their different fire behaviour.

Table 2. *thermo-physical parameters measured and computed of the formulations PP-APP/PER and PP-AP750*

Formulation	T_{ig} (°C)	$\dot{q}''_{cr}$ (kW/m²)	$k\rho c_p$ (kJ²/m⁴.s.K²)	ρ (kg/m³)	c_p (kJ/kg.K)	ρc_p (kJ/m³.K)	k (W/m.K)	α (m²/s)
PP-APP/PER	298	5.6	0.79	200	2.7	540	1.5	$2.7.10^{-6}$
PP-AP750	308	6.0	0.77	400	3.0	1200	0.6	$0.5.10^{-6}$

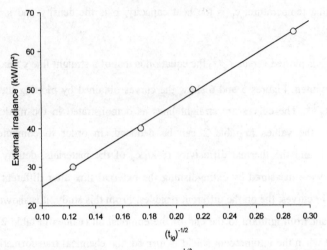

Figure 5. *External heat flux versus $t_{ig}^{-1/2}$ of the system PP-APP/PER.*

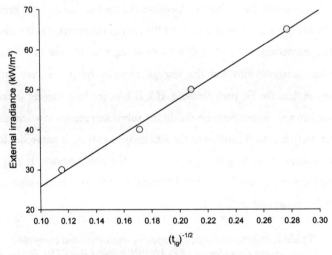

Figure 6. *External heat flux versus $t_{ig}^{-1/2}$ of the system PP-AP750.*

3.3. Dynamical method to determine the thermal diffusivity

The model described above provides only a "surface" view of the intumescent shield and it is interesting to develop a method which allows the measurement of thermal parameters in the bulk during the combustion. In the method, we use the temperature profiles recorded during the combustion of the samples under the conditions of the cone calorimeter[5].

The cone calorimeter allows to compute the total heat evolved (THE) versus time and by combining this value with the recorded temperatures, an apparent thermal diffusivity α can be defined versus time and versus the depth:

$$\alpha = \left(\frac{1}{T - T_0} \times \frac{\partial T}{\partial t} \right)^2 \times \frac{THE}{q''_{ext}}$$

with : α = thermal diffusivity (m²/s), T = temperature in the material (K), T_0 = room temperature (K), $\frac{\partial T}{\partial t}$ = temperature rate in the material (K/s), THE= total heat evolved (kJ)

and q''_{ext} = external heat flux (kW/m²).

Figures 7 and 8 present the curves obtained with the method. In the two cases, the overall behaviour is identical : α decreases when the time increases. When the material is under an external heat flux, the polymer melts, the intumescence develops and thus to the decrease of α. In the case of PP-APP/PER, the shape of the curves is complex. This phenomenon may be assigned as discussed in a previous work [5], to an equilibrium between the partial destruction of the intumescent coating and its reformation. This chaotic evolution becomes very pronounced in the upper thermocouple positions and is less variable in the lower positions because of the thermal gradient in the coating.

In the case of PP-AP750, only the two upper positions show irregular evolution. This result suggests that the intumescent shield undergoes significant stresses due to the flame only on the surface. Indeed if we assume that the intumescent coating is able to create a "skin effect" on the surface of the material, which slows down the heat transfer, then the internal layers of the char form beneath in a more stable fashion. Moreover it can be noted that at about t>400s, α becomes very low and explains the very low values of rhr (Figure 1).

It may be noted that the α values of PP-AP750 are always lower than those of PP-APP/PER which confirms the conclusions of the previous paragraph. Now if we compare the values computed with the ignition model with these ones computed with our model, it is observed that the values of the last model converge to the values of the first one after the initial chaotic zone. The ignition model assume that α is constant versus time, and so the values of the two models can only converge to the same values when the intumescent shield is formed and the material reaches a pseudo-steady state of combustion.

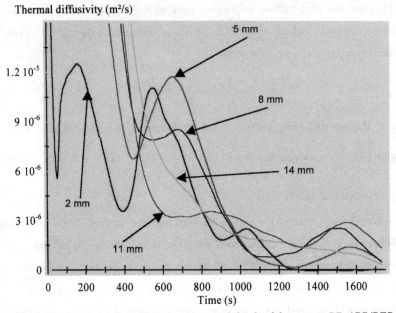

Figure 7. *thermal diffusivity versus time and depth of the system PP-APP/PER.*

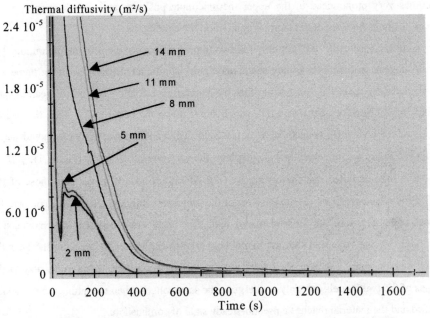

Figure 8. *thermal diffusivity versus time and depth of the system PP-AP750.*

Our model agrees, therefore, with the ignition model and confirms the conclusions of the previous paragraph about the relation between FR performance and thermo-physical properties. It shows further that it provides information about the thermal parameters of the intumescent shields during their formation (unsteady state) and their behaviour during the combustion of the materials.

4. CONCLUSION

In this work we have studied the fire behaviour and the heat transfer in the intumescent formulations PP-APP/PER and PP-AP750 and we have proposed a method to compute the thermal diffusivity of materials during combustion. We have shown that the fire resisting properties of PP were strongly improved using the commercial additive AP750 in comparison with the model system APP/PER. This is explained by the different thermal properties of the intumescent coatings. The shield developed from the PP-AP750 system has comparatively low thermal diffusivity and high storage energy which allows it to protect the substrate longer and at higher temperature.

References

1. H. L. Vandersall, *J. Fire & Flamm.*, 1971, 2, 97.
2. G. Montaudo, E. Scamporino and D. Vitalini, *J. Polym. Sci. Polym. Chem.*, 1983, 21, 3361.
3. G. Camino, in « *Actes du 1er Colloque Francophone sur l'Ignifugation des Polymères* », Edited by J. Martel, Saint-Denis (France), 1985, pp. 36.
4. R. Delobel, M. Le Bras, N. Ouassou and F. Alistiqsa, *J. Fire Sci.*, 1990, 8(2), 85.
5. L. Morice, S. Bourbigot and J.M. Leroy, *J. Fire Sci.*, 1997, 15(5),358.
6. E. Jenewein and W. D. Pirig, *European patent EP 0 735 119 A1, assigned to Hoechst A.G.*, 1996.
7. V. Babrauskas, *Fire and Materials*, 1984, 8(2), 81.
8. S. Bourbigot, M. Le Bras and R. Delobel, *J. Fire Sci.*, 1995, 13(1), 3.
9. R.L. Alpert and J. de Ris, *NISTR 89-4188.*, 1989.
10. S. Bourbigot, R. Delobel, M. Le Bras and Y. Schmidt, *J. Chim. Phys.*, 1992, 89, 1835.
11. C. Siat, S. Bourbigot and M. Le Bras, *Polym. Deg. & Stab.*, 1997, 58, 303.
12. J.E. Staggs et R. H. Whiteley, in « *Interflam '96* », London (1996), pp. 103-112.

SPECIAL FEATURES OF BUBBLE FORMATION DURING INTUMESCENT SYSTEMS BURNING

I. S. Reshetnikov, M. Yu. Yablokova

Polymer Burning Laboratory, Institute of Synthetic Polymeric Materials,
70 Profsoyuznaya street, Moscow, 117393
Russia

N. A. Khalturinskij

Semenov Institute of Chemical Physics,
4 Kosygina street, Moscow, 113977
Russia

1. INTRODUCTION

In the scope of investigation and development of intumescent fire retardant materials[1-3] the practical approach plays the prevalent role during long time. It means that the search for the most effective compositions was made by means of look over of the different components and their concentration and frequently without discussion of the physical processes which take place during intumescence. Works on the modelling of the intumescent polymers burning also represent an attempt to describe the intumescence process.

    Now mathematical models are needed which should allow one to predict intumescent behaviour in fires and estimate factors affecting efficiency of intumescence. There exist a great number of works devoted to the numerical description of the intumescent systems burning processes.[4 and Refs. therein] and, in particular, works dealing where physical structure of the foamed char layer and the processes of its formation are simulated.[5-7]

2. THEORY OF BUBBLE GROWTH IN POLYMER FOAMS

Obviously, the structure of the foamed char layer will depend, above all, on the bubble growth process. Under spherically symmetric conditions, the equilibrium radius of a vapour bubble that has nucleated in a single-component Newtonian fluid which contains no dissolved gases, is:

$$R = 2\sigma / [p_v(T) - p]$$

where $p = p_s - \rho g x$ (p_s: surface pressure, ρ: liquid density, g: gravity, σ: surface tension of the liquid, p_v: pressure inside the bubble).

After the bubble has nucleated, the primary obstacle to initial radial growth is the opposite action of the surface tension. The next stage controlled by liquid inertia, is followed by an intermediate growth stage in which both inertia and thermal effects dominate. Finally thermal effects alone control the growth process.

So, bubble growth is a complex process which depends on the rheology of polymer melt, the surface tension of the melt and/or the rate of outgasing reactions. Because each of these processes have some peculiarities in the conditions of intumescence, bubble formation processes also has some special features. But bubble growth is the constituent part of a number of theories - theory of polymer foams and/or bubble formation during boiling. Let us refer to some results of these mentioned theories.

Previous experimental investigation[8] of bubble formation during boiling of different liquids (such as water or benzene) on a metallic substrate has shown that radius r is:

$$r = \sqrt{2\beta\frac{\lambda\Delta T}{r\rho}\tau} \qquad (1)$$

where λ is the thermal conductivity of the liquid, r the heat of vaporisation, ρ the density of the vapor, τ the growth time, β a numerical coefficient (in the article value 6 was used), and $\Delta T \equiv T - T_v$ (T_v and T are vapour and surrounding temperature respectively). This expression corresponds with the theoretical one obtained for the final stage of bubble growth, when thermal effects control the growth process[9]. Similar expression was used for theoretical model which describes the in-depth effect of bubbles on the steady-state transport of volatile gases from the surface of thermoplastic material, subjected to an incident heat flux.[10].

For an over-saturated liquid-gas solution, the expression for the bubble radius [11] (using assumption that dissolved gas concentration satisfies a spherically symmetric diffusion equation) is:

$$\dot{r} = DS\left(\frac{P_0}{P_c} - 1\right)\left[\frac{1}{r} + \frac{1}{\sqrt{\pi D\tau}}\right] \qquad (2)$$

where D is the diffusion coefficient, S the gas solubility and P_0 and P_c respectively the initial supersaturating pressure and the minimal critical pressure for bubble inflation. In this expression, convection from bubble expansion is neglected, the initial concentration is assumed constant and the concentration at the bubble interface equal for a saturated solution. Later it has been proved that this expression is valid for the growth of bubble in elastomers[12]. Moreover, this expression has been previously used for modelling of intumescence processes[5].

Bubble growth is one of the main point of the polymer foams theory[13]. The rate of bubble growth in this theory strongly depends on the conditions of the foam formation. In the case of a diffusion-controlled bubble formation we can use de Vries expression[14] for the bubble radius r:

$$\dot{r} = 2\sigma \frac{DSRT}{rP_0\theta} \qquad (3)$$

where σ is the surface tension of the fluid and θ the distance from the nearest bubbles. For non-elastic fluids this expression transforms to equation (2). If assumed that bubbles shells do not break during foaming and that the bubble growth proceeds due to rise of internal pressure and bubbles confluence, solution of above differential equation gives the relationship between kinetic characteristics of bubble growth with physico-chemical parameters of polymeric matrix and external conditions[13]:

$$r = r_0 + N\sqrt{\tau}\left(1 - \frac{N}{L}\right)\frac{L^2\tau}{2r_{cr} + L\sqrt{\tau}}, \quad L = 0.5\left(N + \sqrt{N^2 + 4M}\right)$$

where N and M are the expression of external parameters :

$$M/2 = DS'(P_0^| / P_{cr} - 1) - 2\sigma(2SRT / \Theta P_0)$$
$$N/2 = DS'(P_0^| / P_{cr} - 1)\sqrt{\pi D}$$

where super-index dash corresponds to the conditions of oversaturating. This expression was used, for example, for modelling of destruction of polyethylene and foaming by nitrogen[15, 16]. It should be noted, however, that this theory may be only used for foams based on rigid thermoplastic polymers which under action of external forces display only plastic deformations.

Polymer foams with filler is a particular class of foamed materials. Bubble growth in such systems also has some peculiarities. The first one consists in the position of the filler particles: there exist two different possible positions: particles can be either a component of the cellular structure or, if particle is large enough, located in the cavities[17].

The criterion of micro-heterogeneity can be written as:

$$\delta_c \approx (2F_c/\rho g)^{1/3},$$

where δ_c is the particle size, ρ the particles density and F_c the force of the single contact between particles. This expression implies that characteristic particle size is not a constant and depends on the properties of the particles material. However, despite it was mentioned that filler particles play the role of nucleation sites, detailed mathematical description of the processes of bubble growth on the filler particles is not developed yet.

3. BUBBLE GROWTH IN INTUMESCENT SYSTEMS

Hitherto it was common practice to consider bubble formation processes during intumescence in the same manner as polymer foam formation. In particular it was arbitrarily stated that gas is initially solved in polymer melt. Of course it is verified for some intumescent systems but the validity of this assumption has to be verified with all intumescent compositions.

Two main classes of intumescent systems may be considered. In the first one, the polymers is able to produce foamed char by itself under an heat flow. Typical examples are polycarbonate-based materials[18]. The second class includes the blends of polymer matrix (such as polypropylene) with special intumescent additives. These additives consist generally in mixtures of a polyol (e.g. pentaerythritol) and an acidic agent (ammonium polyphosphate, for instance)[2]. The following discussion will concern both the two mentioned types.

Case 1. Polymer takes part in the intumescence processes.
Describing physical aspects of intumescence in such systems, bubbles may be assumed formed from the gases initially dissolved in the matrix, because isolation of volatile products occurs due to rearrangements in the main polymer chain. As a matter of fact, above referred expressions for bubble growth in an over-saturated liquid-gas solutions may be proposed valid, at least in a qualitative aspect. It means that bubble size depends on the polymer melt properties all in all. Estimation of diffusion coefficient and gas solubility may be difficult because there were no attempts to measure these parameters from polymer melt during degradation. Unfortunately, the mentioned parameters have not been previously specified, even in works dealing with intumescence modelling where expression for bubble growth in liquid with dissolved gases was used[5].

Case 2. Intumescence is provided by the special multi-component additive

Two subclasses may be in this case emphasised. The first one consists in materials where additive is uniformly located in the polymer matrix. This case usually takes place when all components of intumescent additive are liquid and thermodynamically compatible with each others and the polymer matrix. The modelling is apparently the same that the modelling of the "self-intumescent polymers".

The second subclass is more general and more complex. Let us illustrate it by an example. Standard intumescence additive contains a dry powder (e;g ammonium polyphosphate), which is poorly solved in water. This component may be stable up to temperatures (215 - 360°C) where intumescence occurs. It is obvious that theory of bubble growth from the dissolved gases is absolutely inadequate for modelling. Some rough expressions for bubble growth on the particle is discussed in this text.

Subclass 1. In this case, a number of assumptions allows a simple discussion. First, reactions of gasification take place only on the interface particle-melt. Second, these reactions start only when the temperature rises up to a critical value T_g (here we shall not use any particular value for T_g and discuss its nature). The third assumption is that the particle and bubble become of a spherical shape. And the last one is that the particle only gasified without formation of high molecular or cross-linked residue on the surface. This assumption allows us to avoid description of the diffusion process through the particle surface char layer. On the basis of these assumptions, a relationship between bubble radius R and time τ may be studied. At the initial time ($\tau = 0$) particle radius is r_0; then the temperature equals T_g and the reactions starts. Expression for hetero-phase reaction of gasification may be written[19]:

$$\frac{dC}{d\tau} = kSD$$

where k is the rate constant, S the area of interaction, D the coefficient (which takes account of the diffusion of the melt into the particle, particle and melt densities, and so forth). For further simplification, it may be supposed that temperature variations are small as compared with T_g. In the following discussion, only value T_g is used. Contact area is a function of time, because of changes in bubble and particle radii, i.e. $S = S(\tau)$. Moreover, the values k and D may be considered in first approximation as constants.

So, the total amount of the volatile products is:

$$C = kD \int_0^\tau S(\xi)d\xi$$

Char   Bubble

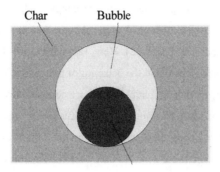

Solid particle

Figure 1. *Configurations of the particle in growing bubble when only gas isolation takes place on the interface.*

At the initial time, particle is surrounded by the melt and contact area is equal to $S = 4\,\pi\,r_0^2$. Instantaneously (during time interval $\delta\tau$, characteristic diffusion time) this value decreases sharply because bubble radius increases, when particle radius otherwise decreases. Figure 1 shows the corresponding particle location. Bubble radius R may be expressed versus the amount of isolated gases C:

$$R = \sqrt[3]{0.75\pi(\alpha C + v)},$$

where $v = v(\tau)$ is the current particle volume, α a coefficient which connects mass and volume of the gas. Note that in general α is not a constant but a function of R because the pressure inside the bubble is equal to $4\sigma/R$, σ being the surface tension of the surrounding liquid. Considering that the contact area is the area where the distance between spheres surfaces is lower than the characteristic diffusion distance δr, the relationship between contact area and current bubble and particle radii may be expressed:

$$R\left(1 - \cos \arctan \frac{a}{R}\right) - r\left(1 - \cos \arctan \frac{a}{r}\right) = \delta r.$$

a^2 corresponds in order of value to the contact area. Solving together above equations from initial time to $\tau = t$, corresponding to the zero particle radius, numerical relationship between bubble radius and time is obtained. The choice of the constants in the expressions depends on the nature of the particles, the density of the gas and so on, and must be made using real values for each particular case.

Subclass 2. In this case , gaseous products are not the only products of the reactions on the particle melt interface; i.e. during contact with the bubble wall high-molecular residue can appear. At the end of the reaction, this residue can play the role of a "support" which connects the bubble wall and the particle (Figure 2). Here, the question of the nature of the support-bubble wall connection is neglected and the behavior of the particle is only considered.

In the initial stage, particle and bubble sizes are comparable and a negligible influence of these "supports" may be assumed. Motion of the particle so is stochastic and very similar to the "Brown's motion". Neglecting the forces acting on the particle during interaction with the walls, in thirst approximation, these contacts may be considered free collisions. Frequency of collisions f_c for bubble radius R and particle radius r may be written as:

$$f_c = \frac{v_m}{R - r}$$

where v_m is the characteristic particle velocity. Computing the amount of gases isolated during single collision $\delta C(R, r)$ as referred in subclass 1, the total amount of isolated gases can be computed (note that R and r are not constants but functions of the time) using:

$$C = \int_0^\tau f_c^{-1} \delta C \, d\xi.$$

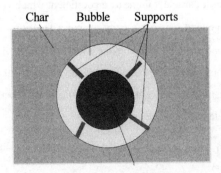

solid particle

Figure 2. *Configurations of the particle in growing bubble, when high-molecular reaction products appear during interaction on the interface.*

For the second limit case (r < R) influence of "supports" may not be neglected. However because the particle size is small, the particle may be assumed a point suspended from some springs inside the sphere. In this case Lagrange function[20] for the particle can be written as:

$$L = \frac{m(\tau)}{2}v^2 - \frac{1}{2}\sum_i k_i(|x - s_i|-l_i)^2,$$

where x represents the position of the particle, s_i the attachment point for support number i, l_i is the free length of the support, k_i is the strength coefficient of the support number i and $|y|$ the length of y vector.

Gravitation and non harmony of the supports in the present model are neglected. While particle is far from the wall, its movement represents a free oscillation (more exactly it is only finite movement.).

If amplitude of these "oscillations" is small compared with R, the bubble stops its growth. Otherwise in some time collision occurs. During collision (duration is $2\delta r/v$), a constant force F proportional to the amount of isolated gases δC, acts on the particle in the direction of the bubble center. The resolution of the problem is the same as for motion in the conditions of external uniform potential field. A new term U_e (increasing of the system potential energy due to action of external force) has to be introduced in the Lagrange function considering that the force acts only in small neighbourhood of the bubble walls:

$$L = \frac{m(\tau)}{2}v^2 - \frac{1}{2}\sum_i k_i(|x - s_i|-l_i)^2 + U_e\Theta(R - \delta r+|x|),$$

where Θ is the Heavy-Side function.

Solving motion equations based on this function with assumption that s_i, R, v, m and (may be) k_i are functions of time we can expect to get the dependence of bubble radius at final stage of the bubble growth.

For the intermediate stage, it seems impossible to adequately describe particle motion in the terms of classical mechanics. The main problem is that particle must be considered as a rigid body and so rotation of the particle can not be neglected. Moreover, if in the initial stage zero number of "supports" may be assumed and in the final stage number of supports stay a constant, in intermediate stage processes of "support" formation and possible destruction should be described.

For these reasons we suggest to get numerical solution only for large ($\sim R$) and small ($<<R$) particle sizes, when middle part can be deduced from an extrapolation in assumption that practically all (~ 90 %) particles are gasified.

4. LIMITING FACTORS

Considering above bubble growth, we use background assumption that there is no external factor which affect bubble growth. However it is not so. Non-regularity of the particle can lead to the increase of diffusion into the particle volume and, as a consequence, to the increase of growth rate. Moreover, low surface tension of polymer melt can induce the destruction of the bubble and/or two or more bubbles can begin to simultaneously rise on one particle. In the initial stage of bubble growth (particle and bubble radii are comparable and gas isolation rate is large enough) viscosity of the melt may be very significant.

Temperature variations which are neglected in present discussion are also very important: firstly, because they affect the rate of gasification. and secondly, gas volume considerably depends on the temperature. But, in turn, temperature depends on the radii of bubbles of the upper layer, on the migration of the bubble due to viscosity and temperature gradient, and on other factors. It seems very complex to take these effects into consideration with enough level of accuracy.

5. CONCLUSIONS

Bubble growth process during intumescence systems burning has been discussed considering various types of intumescent systems; Bubble growth proceeds in significantly different ways according to the system type.

First this review reports literature results about bubble formation from the dissolved gases. Then a simple mathematical model dealing with bubble growth on the particle has been developed and problems which appear during mathematical modelling was discussed. The reported theoretical considerations need experimental validations. In this purpose, we present now some comparison of theoretical results obtained from suggested model with earlier experimental results. So for the final bubble radius R in the case of small initial particles size we can predict as follows:

$$R \sim r \sqrt[3]{\frac{\rho_s}{\rho_g}} \sim 2 \div 3 \ r.$$

To check this point, average pores diameters as a function of APP particles size in the carbamide-formaldehyde resin (CFR) with 30% of intumescent additive (APP - sorbitol, 7:3 wt./wt.) have been investigated (Figure 3).

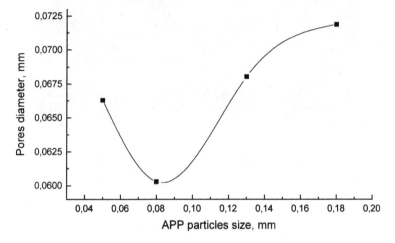

Figure 3. *Average pores diameter vs. APP particles size*

"Apparent" pores diameters is 2-3 times smaller than initial particle size. To explain it, one more experimental point may be readily tested, i.e. the assumption that particle inside the bubble is attached to the walls when gaseous products do not only appear during reactions on the melt-particle interface. Electron scanning microscopy of formed char cap (Figure 4) shows a particle (outlined) of ammonium polyphosphate attached to the walls with high-molecular "supports". These "supports" play, in fact, the role of the walls (at least, they look like walls on the picture). So, "apparent" pore diameter may be smaller than predicted and coefficient of proportionality correlate with coordinating value which, in turn, is about of 6. Total pore diameter under such assumption will be 2-3 times smaller than initial APP particles size, as predicted.

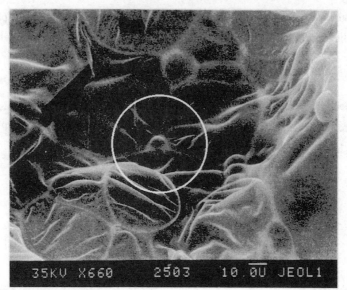

Figure 4. *Structure of char pore. Residue of APP particle is emphasised.*

To conclude, proposed approach of modelling of the bubble growth during intumescence is correct enough, at least in qualitative aspects. In combination with different others model, such as model of heat transfer in porous media at high temperatures, model of bubble migration in the melt and/or model of the polymer melt viscosity (versus temperature) and others, presented model of bubble growth may be used as a consistent part of a full mathematical model of intumescence.

References

1. H. Vandersall, *J. Fire and Flamm.*, 1971, **2**(4), 97.
2. G. Camino, L. Costa and G. Martinasso, *Polym. Degr. and Stab.*, 1989, **23**, 359.
3. S. Bourbigot, M. Le Bras and R. Delobel, *J. of Fire Sci.*, 1995, **13**, No. 1, 3.
4. I. Reshetnikov, A. Antonov and N. Khalturinskij, *Comb., Explosion and Shock Waves.*, 1997, **33**(6), to appear.
5. K. Butler, *"Physical modelling of intumescent fire retardant polymers (Chapter 15)"* in *"Polymer foams: science technology"*, *ASC symp. ser.* 669, ASC Book. 1997.
6. R. Pehrson and J. Barnett, *J. Fire Prot. Eng.*, 1996. **8**, 13.
7. I. Reshetnikov and N. Khalturinskij, *Rus. J. Chem. Phys.*, 1997, **12**(3), 102.

8 . V. Golovin, B. Kol'chugin and E. Zakharova, *High Temperature,* 1966, **4**(1), 147.

9. G. Birkhoff, R. Margulies and W. Horning, *Phys. Fluids,* 1958, **25**, 493.

10. I. Wichman, *Comb. and Flame,* 1986, **63**, 217.

11. P. Epstein and M. Plesset, *J. Chem. Phys.,* 1950, **18**, 1505.

12. A. Gent and D. Tompkins, *J. Appl. Phys.,* 1969, **40**, 2520.

13. A. Berlin and F. Shutov, *"Chemistry and Technology of Foamed Polymers"*, Nauka, Moscow,1980.

14. A. De Vries, *Rubber. Trav. Chim.,* 1958, **77**, No 2, 81.

15. P. Durill, R. Grisley, *Amer. Ind. Chem. Eng. Journ.,* 1966, **12**, 1147.

16. A. Gal'chenko, N. Khalturinskij and Al. Berlin, *Vysokomol. Soed.,* 1980, **22A**(1), 16.

17. O. Tarakanov, I. Shamov and V. Al'pern, *'Filled Cellular Plastics'*, Khimia, Moscow, 1989.

18. A. Antonov and S. Novikov, *Vysokomol. Soed.,*1993, **35A**(9), 1442.

19. N. Emanuel and E. Denisov, *"Chain Reactions of Oxidation in Liquid Phase"*, Nauka, Moscow, 1965.

20. L. Landau and E. Lifshits, *'Theoretical Physics. Mechanics'*, Nauka, Moscow, 1988.

THE ROLE OF RADIATION OVER INTUMESCENT SYSTEMS BURNING

I. S. Reshetnikov

Polymer Burning Laboratory
Institute of Synthetic Polymeric Materials,
70 Profsoyuznaya street, Moscow, 117393 Russia

N. A. Khalturinskij

Semenov Institute of Chemical Physics,
4 Kosygina street, Moscow, 113977 Russia

1. INTRODUCTION

Since Tramm et al. got first patent on intumescent fire retardant composition[1] (1938), this process began to develop with high extent. Although the consumption of these materials in present remains a small fraction of fire retardant materials used in industry, the impact of the needs and application is large.

It would be wrong to say that intumescent polymers are poorly investigated. Chemical and physical properties of such systems have been described in detail in technical literature[2-5]. However, these experimental results commonly describe how intumescence occurs instead of why it occurs and, as a consequence, there exist only isolated works devoted to the physical structure of foamed char layer[6-7]. In the same time, practically all authors agree that thermo-protection of foamed char layer is the main aspect of high efficiency of intumescent fire retardant systems and investigation of heat transfer mechanisms in the char cap is one of the main tasks in the investigation of intumescence.

Up to present intumescent char layer was considered as an uniform medium with "effective" coefficient of thermal conductivity[8-9]. This coefficient, as a rule, is calculated as for composite or air-filled material with known porosity. Anderson et al.[10] presented the only attempt to treat foamed char layer as a complex non-uniform system. In this work, Authors suggest to treat cellular char cap as a number of layers: char layer - air layer. In this assumption thermal resistance may be assumed the sum of thermal resistances of each layer.

Finally in accordance with theory of heat transfer[11], total thermal conductivity k_{eq} versus thermal conductivity of solid (k_s) and vaporous (k_v) fractions may be written:

$$\frac{1}{k_{eg}} = \frac{V_s}{k_s} + \frac{V_v}{k_v},$$

where V_s and V_v are volume fractions of solid and vaporous materials respectively.

When porosity is not very high, heat transfer may be calculated as supposed in the model of intumescence developed by Butler, Kashiwagi et al.[7]. In their work, heat transfer through the medium with pores (bubbles) has been calculated using the model where bubble plays the role of disturbance in the solution of the heat transfer equation.

There exist four main ways of heat transfer - conduction through the char material; conduction, convection and radiation through the pores gases. In most real tasks, convection may be neglected, because pores diameter is, as a rule, less than critical value. But when radiation is neglected, it can be expected that predicted value of thermal conductivity will be ten or more times less than experimental. The aim of this paper is the discussion of the role of radiation in the intumescent systems burning processes, considering in particular processes of heat transfer in protective foamed char layer.

2. EXPERIMENTAL

Thermo-protection properties (TPP) were investigated using the laser beam device. The sample, located on the asbestos table (Figure 1), with the thermocouple in its back side, was subjected to a continuous heating flow. The latter was generated using a CO_2-laser (10.6 μm). Power of laser beam was 7 W/cm^2. Pt-Pt+10%Rh thermocouple, 50 μm wire diameter was used. The developed method allowed one to record the power, time of treatment and the temperature response at the back-wall thermocouple.

As a model system, carbamide-formaldehyde resin (Russian grade CFRLTCS) with intumescent additive system: ammonium polyphosphate (Russian grade TU 6-18-22-101-87) and sorbitol (grade "NEOSORB-70", produced by "Roquette Corp. ", France) was used. Ammonium polyphosphate - sorbitol ratio was 3:1, contents of intumescent additive in polymeric matrix was 30 wt. %.

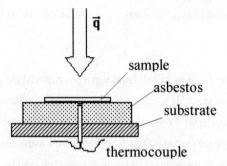

Figure 1. *Objective table for investigation of thermo-protection properties.*

3. THORETICAL PART

Predicting the exact geometrical structure of the char is an exceedingly difficult task. However, it is not unfair to state that the char consists of a considerable volume fraction of gas trapped within a carbonaceous-like material. All real chars have one common feature - contents of the trapped gas is significantly high and as a rule is more than 80% of the volume (in some cases it can be up to 99%)[6].

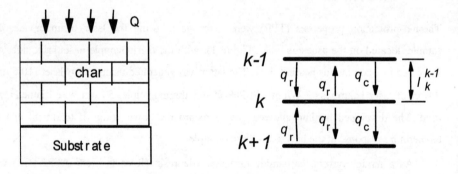

Figure 2. *Model structure of foamed char cap.*

To study heat transfer phenomena in the systems with high air contents we suggested to use a simple model, based on the physical structure of foamed char[12-13]. In this model, char cap is constituted of a number of planes, separated by the air streaks (Figure 2). Surface of the model cap is affected by the continuous heat flow from the external source. Heat flux to each layer is combined from the radiant heat fluxes (denoted as q_r on the figure) from the upper and lower planes and conductive heat flux (q_c) from the upper, more heated plane. Heat losses of the layer are radiation from the plane surfaces and conduction to the lower less heated plane. All calculations was made in real time scale, time step was 0.01 s. Heat losses through substrate was taken into consideration via expression $\gamma(T_b-T_0)$, where γ is the coefficient of heat losses, T_b the temperature of lowest char layer and T_0 the environmental temperature.

4. RESULTS AND DISCUSSION

As previously shown[12], experimental and theoretical results for the response of thermocouple (located on the back side of the sample) are in good agreement. We will now consider the most significant mechanism taking the part in the total heat transfer. Figure3 present under various assumed conditions the values of the temperature at the lowest layer. Dotted curve corresponds to experimental data, solid curve 1 to when all heat transfer mechanisms are allowed, curve 2 radiation heat exchange is prohibited, curve 3 to theoretical curve when radiation heat exchange is prohibited only in volume, heat losses via radiation from the surface are taken into consideration. Comparison of the curves shows that when radiant part is ignored, the temperatures values are comparatively low.

Generally in fires, char cap temperature on the back side is approximately 400°C when the surface temperature can be up to 1000°C, i.e. temperature gradient is approximately 100°C/mm. Char cap height is supposed equal to 6 mm, pore diameter $l = 0.1$ mm. It means that on the single pore temperature difference between upper and lower sides ΔT is approximately 10°C. Radiant heat flow in this case, may be estimated (using surface temperature: 800°C):

$$q_r = \varepsilon\sigma T^3 \Delta T \sim 1\cdot5.67\cdot10^{-8}\cdot800^3\cdot10 \sim 300 \text{ W/m}^2$$

when the heat flux due to air thermal conductivity is:

$$q_c = \frac{\lambda\Delta T}{l} \sim 50\cdot10^{-3}\cdot10/10^{-4} \sim 500 \text{ W/m}^2.$$

with ε is the char blackness, σ the Stephan-Boltzman constant, T the average pore temperature and λ the thermal conductivity of the air.

q_r and q_c values are of the same order and, thus, it may be expected that neglecting of radiant part of heat transfer would lead to great errors.

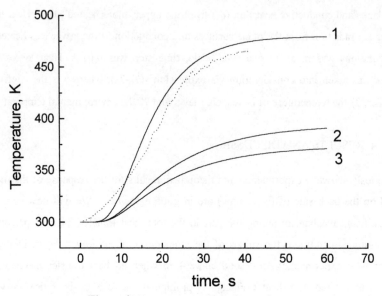

Figure 3. *Temperature of the lowest layer.*
1 - all heat transfer mechanisms are allowed, 2 - radiation is prohibited,
3 - radiation prohibited only in volume, radiative heat losses from the surface allowed.

Let us compare further contributions of radiant and conductive heat fluxes to the total heat flow on the layer in the model char cap in the conditions of maximal heating rate (i.e. at the point of maximum rate of the back-side temperature growth). Figure 4 presents these contributions computed using the model mentioned above. Total radiant heat flow to the layer was calculated as a sum of radiant heat flows from the upper and lower layers minus radiant heat losses of the considered layer. In the whole temperature range, the values of the radiant heat flux to the layer are always higher than those of the conductive heat flux.

As discussed earlier in the theory of thermal conductivity of rigid polymer foams and thermo-insulating materials[14] radiation plays a significant role in heat transfer. A number of other theories about heat transfer (Loeb's theory[15], as an example) try to take radiation into account through an empirical coefficient. But physical nature of this "empirical coefficient" is not clear and as a consequence, practical use of these theories remain limited.

More, previous experimental studies on heat transfer in glass fibers vs. density of the material and its temperature[16] has shown that radiant part becomes very significant when materials density is low (less than 0.01 g/cm^3) and the temperatures are high. This result allows to expect that radiant part of heat flux is very great in conditions of intumescence, when temperature in bulk is higher than 300°C (surface temperature: at least 700°C).

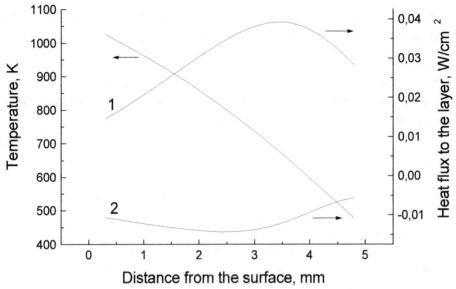

Figure 4. *Temperatures and heat fluxes inside of foamed char cap:*
1 - radiation, 2 - conduction through the air.

We proposed previously[13] a simple empirical relationship between back-side temperature T_b and char structural parameters (cap height h and pores diameter d):

$$T_b = T_0^* - \frac{Q}{\lambda_{eff}^*} h + h_r d ,$$

where Q is the external heat flow; T_0^*, λ_{eff}^* and h_r are some constants. In this Work, we assumed that the coefficient h_r corresponds to the radiant heat flow.

Let us now check this point. The only model parameter related with radiation, is the blackness of the char (ε). Our analysis indicates that h_r is reverse proportional to the char blackness (in the neighbourhood of 1). Of course such a relationship has no physical meaning because, discussing model varying material blackness, we change only the amount of radiation from the surface but we assume that all dropped energy is absorbed. Nevertheless, obtained results allow to draw one conclusion: char blackness is an important parameter and cannot be ignored.

4. CONCLUSIONS

Radiant heat transfer in porous char cap has been discussed. Particularly, we show that radiant heat transfer inside the pore is of the same order of value as conduction through the gas. Considering the total heat flux to the pores wall, it was predicted that conductive heating flow from the upper wall and conductive heat flow to the lower layer are practically equal and so, resulting conductive heat flow to the horizontal pore wall is very small. Concurrently, difference in radiant heat fluxes is comparatively high because of a difference of neighbour horizontal walls temperatures. The main conclusion of our work is that the heating of the char takes place mainly mainly because of the radiant part of total heat flow.

References

1. H. L. Tramm, *US patent* No 2 106 938 (assigned to Ruhrchemie Aktiengesellchaft), 1938.
2. H. Vandersall, *J. Fire Flammab.*, 1971, **2**(4), 97.
3. G. Camino, L. Costa and G. Martinasso, *Polym. Deg. Stab.*, 1989, **23**, 359.
4. S. Bourbigot, M. Le Bras and R. Delobel, *J. Fire Sci.*, 1995, **13**(1), 3.
5. I. Reshetnikov, A. Antonov, T. Rudakova, G. Aleksjuk and N. Khalturinskij, *Polym. Deg. Stab.* 1996, **54**(2-3), 137.
6. Ye. Gnedin, N. Kozlova, R. Gitina and O. Fedoseeva, *Vysokomolek. Soed.*, 1991, **33A**, 1568.
7. K. Butler, *"Physical modelling of intumescent fire retardant polymers* (Chapter 15)" in *"Polymer foams: science technology"*, ASC Symp. Ser. 669, ASC book. May 1997.
8. D. Cagliostro, S. Riccitiello, K. Clark and A. Shimizu, *J. Fire Flammab.*, 1975, **6**, 205
9. V. Zverev, G. Isakov, V. Nesmelov and V. Nazarenko, *J. Polym. Mat.*, 1993, **20**(1-2), 91.
10. C. Anderson, D. Ketchum and W. Mountain, *J. Fire Sci*, 1988, **6**, 390.
11. K. Kanary, In *"Thermal conductivity of high polymers"*. *Denki sikinse hose daigaku*, 1973, **176**.
12. I. Reshetnikov and N. Khalturinskij, *Chem. Phys. Reports*, 1997, **12**(3), 499.
13. I. Reshetnikov and N. Khalturinskij, *Chem. Phys. Reports*, 1997, **12**(10), 104.
14. V. Cherepanov and I. Shamov, *Plasticheskie massy*, 1974, **10**, 53.
15. A. Loeb, *J. Amer. Ceram. Soc.*, 1954, **37**, 96.
16. O. McIntire and R. Kennedy, *Chem. Eng. Prog.*, 1948, **44**, 27.
17. J. Vaschoor, P. Greebier and N. Mainville, *Trans. ASTM*, 1952, **74**, 961.

STUDIES ON THERMAL DEGRADATION AND CHARRING OF POLYMERS BY XPS/TGA *TO RE-EXAMINE THE TRANSITION TEMPERATURE T_{GRL} OF THE GRAPHITE-LIKE STRUCTURE FOR POLYMERS ON HEATING*

J. Wang

National Laboratory of Flame Retarded Materials
School of Chemical Engineering & Materials Science
Beijing Institute of Technology, 100081 Beijing, China

1. INTRODUCTION

In the study of flame retardancy and smoke emission of polymers, the thermal *degradation* and *charring* are two processes of central importance. A great effort in the characterization of polymers, which are subject to heat, has attracted ubiquitous interests. It is well known that the study of charring process is too complicated to be illustrated simply and many parameters are required to describe the char in its integrity within the entire range of temperature. This intractable difficulty lies normally in the black colored char residues with highly crosslinked network.

More than twenty years ago van Krevelen[1] found that there is a very significant relationship between the limiting oxygen index (LOI) and the pyrolysis residue (%) of the polymer. A further considerable attention to this relation leads to many applications in the area of flame retardancy of polymers.

The pyrolysed carbon matrix is, itself, reported to have a wide range of properties depending sensitively on its structures. It is natural to think of the problem : what about the chemical structures of the char residues of polymers. It is wondered if some structural parameters exist which can be used for the appraisal of flame retardancy of polymers. The relation between the structural parameters and the flame retardancy has not been completely established yet so far. In the laboratory we have investigated this problem with the aid of XPS (X-ray Photoelectron Spectroscopy) for years[2-5]. Very recently, an additional parameter, the transition temperature T_{GRL} of the graphite-like structure for polymers on heating has been detected and fortunately proved to be one of the parameters capable of being a clear-cut criterion for the description of the charring process. In the present work a set of experimental data on T_{GRL} has been presented. A correlation between T_{GRL} and flame retardancy and/or smoke emission has been observed.

2. T_{GRL} AND ITS SIGNIFICANCE IN CHARRING

Most organic materials that do fuse during heat treatment will form mesophase. Increasing ordering of organic matter towards graphite during metamorphorphism is widely studied whereas the transition behavior of charring of organic polymers on heating which is of vital importance in flammability within the range of ambient up to say, 600°C is poorly known. There is in general no distinct boundary between thermal degradation and charring. As mentioned above the T_{GRL} is defined as the transition temperature of graphite-like structure from mesophase to graphitic phase and has been characterized by XPS technique[3]. Unlike the conventional techniques, such as UV, IR, NMR, and so on, the characteristic parameter, T_{GRL} measured by XPS gives information about the pseudo-phase transformation *in its integrity* rather than the *local* structural features. Therefore T_{GRL} could be employed as a clear-cut criterion for the description of the pseudo-phase transformation of charring residues. It was found[3-5] that the T_{GRL} of organic polymers is very much lower (generally, $T_{GRL} < 500°C$)than the temperature of graphitization (normally, in the range of 2500 –3000°C)[5].

2.1. (Ethylene – vinyl acetate) copolymers (EVA)

The parameter T_{GRL} was firstly tested using EVA copolymers as models on the premise of simultaneous fulfillment of the following stipulations :

1) *no charging* observed in C1s spectra, that is, the binding energy of C1s approaching to the graphite-like structure;

2) 2) *tailing* appearing in C1s spectra;

3) 3) a *typical valence band* of graphite-like structure taking place.

The value of T_{GRL} for EVA(45%) had been determined to be something in-between 450 –500 °C (470°C) through interpolation as seen in Table 1[2].

The VA content, atmosphere, and flame retardant additives are all the factors affecting T_{GRL} as shown in Table 2. A very small amount of oxygen (0.5%) present in nitrogen can decrease T_{GRL} of EVA copolymers by 40°C (see table 2). It is well known that a very important role is played by oxygen in the emergence of charring of polymers as evidently confirmed by the dramatic reduction of 70°C in T_{GRL} in air (see Table 2). That is to say, generally, the lower the T_{GRL}, the earlier the charring takes place.

Table 1. *Characteristic Changes in C1s Spectra and Valence Bands of EVA (45%VA) as a Function of Temperature in Argon Atmosphere.*

Spectrum Temperature °C	C1s			Assignment in Valence Band
	Binding energy eV	Charging eV	Shape	
Ambient	287.1	2.6	Symmetric	EVA Copolymer
360	287.2	2.7	Symmetric	Polyethylenic Backbone
460	287.2	2.7	Symmetric	Polyethylenic Backbone
500	284.9	0.3	Unsymmetric Tailing	Graphite-like High-crosslinked

Table 2. *Factors Affecting the Average T_{GRL} of EVA Copolymers.*

Copolymer	Average Transition Temperature, T_{GRL}, °C		
	Ar	N₂*	Air
EVA (15%VA)	470	430	400
EVA (45%VA)			
EVA (60%VA)			
EVA (15%VA) + P_x	480	470-480	460
EVA (45%VA) + P_x			
EVA (60%VA) + P_x			
EVA (15%VA) + APP	460	460	400
EVA (45%VA) + APP			
EVA (60%VA) + APP			

* Industrial grade (containing 0.5 % oxygen)

However, exceptions have often been detected, e.g. in the case of red phosphorus, which is frequently recognized as a highly effective flame retardant for polyolefins. It increases the T_{GRL} of polyolefins by +10°C, indicating a different mechanism of thermal decomposition and/or imparting flame retardancy rather than the earlier charring. It was really the case as reported by Peters[6] for polyethylenes that the P_x increases the thermal oxidative stability of the base resin by scavenging oxygen at the surface of the polymer via forming a protective coating which physically separates the surface underneath from the attack of oxygen and heat transfer from surroundings[3]. In the case of ammonium polyphosphate (APP) an obvious reduction in T_{GRL} is self-explanatory, as expected, meaning an early charring even in argon, usually from the effective dehydration.

2.2. Polyethylenes

The T_{GRL} values for high density polyethylene (HDPE, HDPE/P_x and HDPE/APP) and low density polyethylene formulations (LDPE, LDPE/P_x, and LDPE/APP) were nearly identical at 470°C in nitrogen, except for LDPE/APP T_{GRL} which was 10 °C lower than the others. A considerable influence of oxygen on T_{GRL} has been found, e.g. a decrease of 140°C in T_{GRL} induced by oxygen in air for HDPE and LDPE. The addition of P_x and APP to polyethylenes caused an additional increase of 20°C and a decrease of $10 - 20$°C in T_{GRL}, respectively in the similar way as that in EVA copolymers[3].

3. RE-EXAMINATION OF THE ONSET OF T_{GRL} - THE LIMITING T_{GRL} (LT$_{GRL}$)

The measurement of T_{GRL} was based on the simultaneous fulfillment of three stipulations. In some cases difficulties may occur in precise determination. The first question encountered was the ambiguous definition of T_{GRL}. The so-called graphite-like structure is a general and nonspecific nomenclature denoting structural changes occurring in a wide range of temperature. The aim of this study is to define the onset temperature of the pseudo-phase transformation, i.e. the lowest temperature at which the structural transformation just emerges. By analogy with Limiting Oxygen Index (LOI), the limiting transformation temperature (LT$_{GRL}$) term may be used.

Secondly, the binding energy of C1s spectrum of the limiting graphite-like structure may still suffer from ambiguity in the presence of heteroatoms, e.g. halogen, oxygen, sulfur, etc. which may cause poor conductivity of the samples. Thirdly, the typical valence band of graphite-like structure may subject to the overlapping in the presence of some elements; it may also sustain a serious contamination because of the higher inherent surface sensitivity of valence band than C1s core level.

In order to overcome drawbacks built-in the previous definition and improve the precision of measurements, some additional requisites must be invoked. Polyvinyl alcohol (PVA) and polyvinyl chloride (PVC) and their related composites were selected as models for this aim.

3.1. Poly (vinyl alcohol) and PVA/ KMnO$_4$[7]

The reason for PVA to be chosen as a model polymer is twofold :

(i) PVA has been proposed to be acting as a halogen-free char-former used for improving the charring propensity of other polymers;

(ii) the study of the synergy between the char-former PVA and Nylon-6 (in terms of LT$_{GRL}$) is of particular interest. New types of ecologically safe flame retardant composition based on PVA and PVA oxidized by KMnO$_4$ have been recently proposed for nylon 6,6 by Zaikov[8].

The precise determination of LT$_{GRL}$ by XPS technique seems to be much complicated than expected. For example, upon oxidation the valence band of the oxidized top layers, particularly oxidized by the strong oxidative agent tends normally to be poor conductor as expected and becomes so involved that it is not suitable for precise measurement. Poly(vinyl alcohol) (PVA) contains -OH group as substituent within the polymer. It is, therefore, important to have detailed knowledge of C1s binding energy of the graphite-like structure in the presence of oxygen. It has been argued that C1s binding energy of graphite - dominated materials is noticeably lower than that of polymeric hydrocarbons. Barr et al.[9] have reported that the C1s binding energies of O$_3$-treated graphite was 284.9 eV relative to 284.7 eV for the non-oxidized specimen. It is noted indeed that the presence of oxygen in the surface layers does increase the C1s binding energy.

Following up the above analysis the C1s binding energy of 285.0 eV seems to be a reasonable value recommended for the measurement of LT$_{GRL}$ for PVA in the present paper.

Determination of LT$_{GRL}$ from C1s bending energy can be achieved as follows from Figure 1: an horizontal line parallel to the X-axis is drawn through a fixed point at the coordinate depending upon the specific polymer involved, 285.0 eV for PVA. The horizontal line intersects the curve of C1s binding energy vs. temperature at a point through which the LT$_{GRL}$ can be readily determined from abscissa.

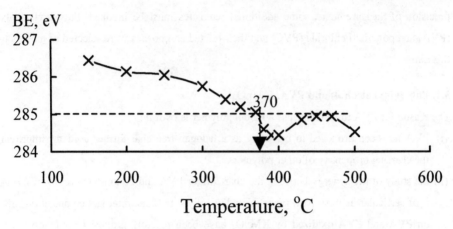

Figure 1. *The Determination of LT$_{GRL}$ for PVA.*

The LT$_{GRL}$ values for PVA and PVA oxidized by KMnO$_4$ were measured to be 370 and 305°C, respectively. Similarly, data for some PVA-containing systems are given in Table 3. Unlike HDPE, LDPE, EVA copolymer the poly(vinyl alcohol) shows a value of 370°C, that is 100°C lower than the value for polyolefins (LT$_{GRL}$ = 470°C). It indicates clearly that the char formation from PVA does begin in the condensed phase at a temperature lower than PE, EVA by 100°C. This explains the use of PVA as a char-former. As a consequence, our result explains again why hydroxy groups linked to a polyethylenic backbone, or somewhere else, are often used as a carbonific source to impart flame retardancy to polymers which on their own form no char at all while burnt. Compared with PVA alone a further reduction of 65°C in LT$_{GRL}$ for PVA oxidized by KMnO$_4$ shows that the charring propensity of PVA can be surely accelerated by KMnO$_4$[7].

To verify the oxidation part played by chelated metal (Mn, Ni), the system PVA/H$_2$O$_2$ was tested. Reduction in LT$_{GRL}$ by a certain amount (-10°C) was found for system of PVA/H$_2$O$_2$. Evidently, the influence of H$_2$O$_2$ on the charring of PVA proved to be insignificant when compared to PVA/KMnO$_4$. That is to say, attention must be focused on the chelated metal, i.e. manganese in KMnO$_4$ and nickel in NiCl$_2$ (see Reference 7). In fact, a distinctive decrease of 40°C in LT$_{GRL}$ for PVA/NiCl$_2$ (from 370 to 330°C) (Table 3) verifies that nickel chloride (as does KMnO$_4$) is useful for PVA to be charred.

Table 3. *The Transition Temperature LT$_{GRL}$ for Some Systems.*

System	PVA	PVA/H$_2$O$_2$	PVA/NiCl$_2$	PVA/H$_2$O$_2$/NiCl$_2$	PVA/KMnO$_4$
LT$_{GRL}$ (°C)	370	360	330	325	305

El-Shahawy[10] had recently studied the structural state of Ni (II) in heat-treated PVA-NiCl$_2$ composites by optical absorption technique and reported that the addition of NiCl$_2$·6H$_2$O to PVA seems to ***retard*** the process of the conjugated polyene formation along the polymer chain by thermal degradation. It may be therefore proposed that there is not an inevitable connection between the polyene formation and the charring process.

3.2. Nylon-6 and Nylon-6/PVA

Table 4 presents more LT$_{GRL}$ values for PA-6 and PA-6-containing systems. The addition of PVA (20% by weight) to PA-6 does induce a further reduction of 20°C (from 380 to 360°C) in LT$_{GRL}$. Obviously, a synergistic charring between PA-6 and PVA is verified by the lowest LT$_{GRL}$ of nylon-6/PVA system[7].

Table 4. *The Transition Temperature LT$_{GRL}$ for Some Systems.*

System	Nylon 6	Nylon 66	Nylon 6/PVA (8:2 by weight)	Nylon 6/PVA/KMnO$_4$ (Nylon-6/PVA=8:2 by weight)
LT$_{GRL}$ (°C)	380	365	360	350

3.3. Polyvinyl Chloride (PVC) and Energy Loss (Plasmon) in C1s Spectra[11]

PVC is one of the most important bulk plastics. Its inherent flame retardancy stems from the presence of chlorine atoms attached to the backbone. To see the role of chlorine on charring in terms of LT$_{GRL}$ is the point of our interest. The utilization of the plasmon loss peak ΔE_L in C1s spectra deserves to be mentioned in respect of improving the measurement. Both binding energy and plasmon energy loss ΔE_L (denoting the separation between the loss peak and the main peak) in C1s spectra are depicted in Figure 2. In an opposite direction to the binding energy the ΔE_L grows up along with the temperature increase. Based upon the same assumption as in PVA, the ΔE_L in C1s spectrum of PVC of 25.4 eV was determined corresponding to the binding energy of 285.0 eV. We have now an additional supplementary parameter ΔE_L besides the fulfillment of three stipulations mentioned above: no charging, tailing, and typical valence band.

Following the above procedure the LT_{GRL} of PVC was ascertained to be 370°C, i.e. 100°C lower than polyethylenes due to the presence of chlorine atom. In other words the chlorine atom enables the charring of PVC to be started earlier by 100°C relative to PE; namely, for charring the chlorine atom imparting flame retardancy to PVC plays the same part as the hydroxyl group in PVA.

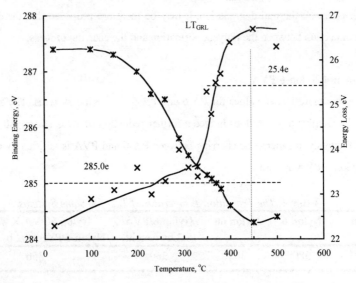

Figure 2. *Binding Energy and Energy Loss ΔE_L observed from C1s spectra of PVC vs. temperature (XPS measurement).*

3.4. Influence of Transition Metals on the Charring Processes of PVC

The application of the transition metals and their compounds to PVC in order to control its flammability and smoke suppression has been widely documented[11].

Some of the LT_{GRL} values for systems of PVC/transition metals were selected and shown in Table 5 accompanied with their plasmon loss symbol ΔE_L. It is noted that the cuprous oxide with a specific molar ratio of Cl/M (100/4) has the lowest LT_{GRL} (310°C) in agreement with the most effective capability of smoke suppression. All the systems of PVC/metallic oxide exhibit lower LT_{GRL} values than pure PVC. It implies again that all metallic compounds shown in Table 5 would be expected to be effective additives for smoke suppression due to earlier charring. This conclusion can also be drawn from the correlation between LT_{GRL} and char residue measured by TGA at 600°C (Figure 3).

The nature of the loss spectra ΔE_L for carbon-containing systems implies to some extent the potential π density in the carbon system. When the π density in the carbon system is maximized (graphite) it becomes an interlaced collective network of electrons that form a near perfect, free electron plasmon with $\Delta E_L \cong 31$ eV[9]. Based on our XPS data the loss peak ΔE_L for ascertaining the onset of graphite-like structures for PVC and systems of PVC/transition metal at a level of Cl/M=100/4 was determined to be 25.4 eV and 24.0 - 24.7 eV, respectively. The charring for systems of PVC/transition metal has occurred earlier as shown by the difference of 0.7-1.4 eV relative to PVC. This shows once again the catalytic influence of the transition metals and of their compounds on charring (details in reference 11).

The influence of LT_{GRL} on the Reducing Smoke Efficiency (RSE) is depicted in Figure 4, where RSE = [(TSP)$_{PVC}$ - TSP] 100 %/ (TSP)$_{PVC}$; TSP represents Total Smoke Production tested by Cone calorimeter[11].

In Figures 3 and 4 the cuprous oxide LT_{GRL} deviates far from the linear regression and presents a very special position when charring or smoke emission is concerned. This question should merit really further investigation.

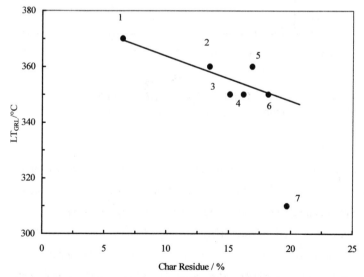

Figure 3. *Correlation between LT_{GRL} and char residue measured by TGA at 600 °C: 1-PVC, 2-FeOOH, 3-MoO$_3$, 4-Fe$_2$O$_3$, 5-CuC$_2$O$_4$, 6-Cu(HCOO), 7-Cu$_2$O.*

Table 5. *LT$_{GRLS}$ for Systems of PVC/Transition Metals (M) accompanied with ΔE_L in C1s Spectra.*

System	PVC	MoO$_3$ 100/4(Cl/M)	Cu$_2$O 100/4(Cl/M)	Fe$_2$O$_3$ 100/4(Cl/M)	FeOOH 100/4(Cl/M)
LT$_{GRL}$/°C	370	350	310	350	360
LT$_{GRL}$/°C	---	20	60	20	10
LT$_{GRL}$/°C	25.4	24.1	24.0	24.4	24.7

* Separation of loss peak from main peak in C1s spectra.

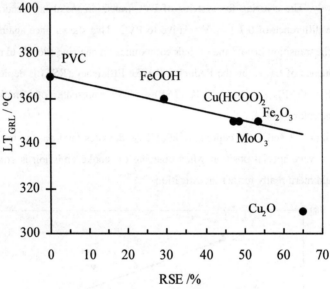

Figure 4. *Effect of LT$_{GRL}$ on RSE (Reducing Smoke Efficiency, from cone calorimeter study).*

3.5. Relation between LT$_{GRL}$ and Thermal Degradation of PVC

In XPS technique the absolute intensity of C1s spectrum can be used to trace the thermal degradation of polymers and the change in crosslinking[3]. The results for systems of PVC and PVC/transition metal using pseudo-in-situ technique are shown in figures 5 to 8)[11]. The data in the coordinates are all expressed as relative intensity , (i.e. the ratio of [(CPS)$_T$ - (CPS)$_{20}$]/(CPS)$_{20}$, where CPS represents the absolute intensity abstracted from C1s spectra).

As tested by Cone calorimeter the system of PVC/Cu$_2$O/MoO$_3$ manifests itself to be of more effectiveness in smoke reduction and char formation.[11] However, no synergistic effect has been observed on the basis of the LT_{GRL} (see Figure 8). Figures 5 to 8 show that the charring happening in the mixed system may suffer from a complicated interaction between Cu_2O and MoO_3 in the condensed phase. The flame retardancy and/or smoke reduction of the system arise likely from some different routes besides char formation.

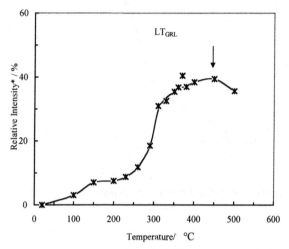

Figure 5. *Relative Intensity vs. Temperature of PVC.*

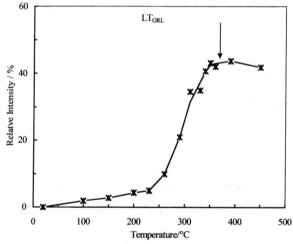

Figure 6. *Relative Intensity vs. temperature of PVC/MoO₃ (from XPS spectra).*

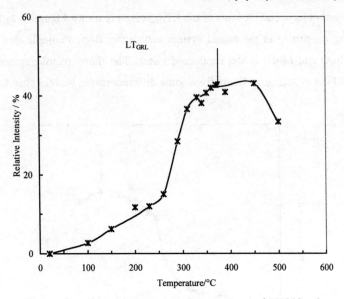

Figure 7. *Relative Intensity vs. Temperature of PVC/Cu₂O*
(from XPS spectra)

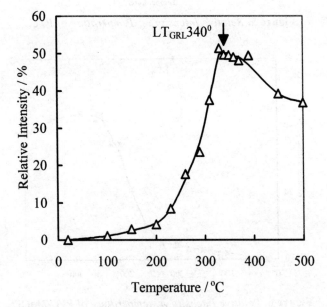

Figure 8. *Relative Intensity vs. Temperature of System of PVC/Cu₂O/MoO₃*
(from XPS spectra).

4. CONCLUSIONS

An improved parameter, LT_{GRL} (the limiting transition temperature of graphite-like structure) through the re-examination of T_{GRL} determined by XPS has been introduced as a clear-cut parameter for the precise description of thermal degradation and charring. A systematic assessment has been performed in terms of LT_{GRL}. Some conclusions can be drawn as below:

(1) the magnitude of the supplementary parameter, LT_{GRL} depends primarily on the structure of the repeat unit of polymer and some other factors, such as atmosphere, heating rate, loading of the additives, and so on.

(2) the binding energy of 285.0 eV in C1s spectra was recommended as a reasonable value in the determination of the LT_{GRL} values for PVA, PVC and their compositions. The plasmon energy loss ($\Delta E_L \geq 24.0$ eV) turns out to be quite useful for the precise measurement of LT_{GRL} as exemplified by PVC and PVC based composites. The energy loss ΔE_L could also be extended to other polymers.

(3) the loss peak ΔE_L for the onset of graphite-like structures for PVC and related systems of PVC/transition metal at a level of Cl/M=100/4 was determined to be 25.4 eV and 24.0 - 24.7 eV, respectively. As seen, the charring for systems of PVC/transition metal starts earlier as shown by the difference of 0.7-1.4 eV relative to PVC. This shows once again the catalytic influence of the transition metals and their relevant compounds on charring.

(4) similar decrease in LT_{GRL} (ΔLT_{GRL}=-100°C) for both PVA and PVC relative to polyethylenes indicates that the char-forming propensity of OH group and Cl atom is comparable.

(5) good correlations have been established : (i) between LT_{GRL} and char residue (Figure 3); and (ii) between LT_{GRL} and RSE (Figure 4). It is noted that cuprous oxide gives LT_{GRL} far deviated from the linear regression.

(6) the synergistic effect between nylon 6 and PVA and system of nylon6/PVA/KMnO$_4$ has been confirmed by their LT_{GRL} values. However, the improvement of charring may be ascribed to the enhancement from the metal complex through the pre-oxidation.

(7) inspecting through Figures 5 to 8 the charring occurring in the system of PVC/Cu$_2$O/MoO$_3$ may suffer from a complicated interaction between Cu$_2$O and MoO$_3$ in the condensed phase. The flame retardancy and/or smoke reduction of the system is likely conducted by some different mechanisms besides char formation.

References

1. D. W. Van Krevelen, *Chimia*, 1974, **28**, 504.
2. J. Q. Wang and H. B. Tu, *"Proceedings of the 2nd Beijing International Symposium - Exhibition on Flame Retardants"*, Beijing, China, 1993.
3. J. Wang, H. Tu and Q. Jiang, *J. Fire Sci.*, 1995, **13**, 261.
4. J. Q. Wang, '*Fire & Polymers II, Materials and Tests for Hazard Prevention*', American Chemical Society, Washington, D.C., **518**, 1995.
5. J. Wang, *"Proceedings of the 7th Annual BCC Conference on Flame Retardancy"*, Stamford, 1996.
6. E. N. Peter, *J. Appl. Polym. Sci.*, 1979, **24**, 1457.
7. J. Wang, X.B. Huang and B. Li, to be submitted.
8. G. E. Zaikov and S. M. Lomakin, *"Fire & Polymers II, Materials and Tests for Hazard Prevention"*, American Chemical Society, Washington", **518**, 1995.
9. T. L. Barr and M. P. Yin, *J. Vacuum. Sci. Tech.* A, 1992, **10**, 2788.
10. M. A. El-Shahawy, *Polym. Degrad. Stab.*, 1994, **43**, 75.
11. J. Wang and B. Li, unpublished results.

New Intumescent Polymeric Materials

POLYMER COMBUSTION AND NEW FLAME RETARDANTS

T. Kashiwagi, J. W. Gilman, M. R. Nyden,

Building and Fire Research Laboratory, National Institute of Standards and Technology
Gaithersburg, MD 20899 USA

S. M. Lomakin

Guest researcher from Institute of Biochemical Physics,
Russian Academy of Sciences, Moscow, Russia

1. INTRODUCTION

A majority of polymer-containing end products (for example, cables, carpets, furniture) must pass some type of regulatory fire test to help assure public safety. Thus, it is important to understand how polymers burn and how to best modify materials to make them less flammable in order to pass such tests without compromising their uniquely valuable physical properties and also significantly increasing the cost of end products. This paper briefly describes chemical and physical processes occurring in the gas and condensed phases during the combustion of polymers and methods to reduce their flammability.

Combustion of polymer materials is characterized by a complex coupling between condensed phase and gas phase phenomena. Characteristics of the critical role in each phase are briefly described below.

1.1. Condensed Phase

In order to burn a polymeric material, thermal energy must be added to it to raise its temperature sufficiently to initiate degradation. This energy could be from an external source, in the case of an ignition event such as a match, or from an adjacent flame as heat feedback in the case of flame spread and burning. Thermal radiation is the primary mode of energy transfer from the flame to the polymer surface as discussed later except for small samples (roughly less than 15 cm diameter).

When temperatures near the surface become high, thermal degradation reactions occur and these evolve small gaseous degradation products. The majority of the evolved products from polymers is combustible. Depending on the nature of the polymer, thermal degradation reactions may proceed by various paths. Since there are several excellent books and articles describing thermal degradation chemistry in detail[1-3], only an extremely brief discussion is presented here. It has been accepted that the majority of vinyl polymers degrade thermally by a free radical chain reaction path. Free radical chain reactions consist of random or chain-end initiated scission, depropagation, intermolecular or intramolecular transfer, and termination reactions. Polyethylene, PE, is a typical example of a polymer that undergoes scission at random locations on the main chain to yield many smaller molecular fragments. Polystyrene, PS, polypropylene, PP, and polymethylacrylate, PMA, belong to this group. Polymethylmethacrylate, PMMA, undergoes a reversal of the polymerization reaction after the initial breakage and yields mainly monomer molecules. Polyoxymethylene, poly-α-methystyrene, and polytetrafluoroethylene belong to this group. These two groups of polymers undergo almost complete degradation while leaving hardly any char (carbonised polymer residue). Polymers with reactive side groups attached to the backbone of a polymer chain may degrade initially as a result of interactions or instabilities of these groups; such reactions may then lead to scission of the backbone. Polyvinylchloride, PVC, and polyvinyl alcohol, PVA, are examples of such polymers. This group tends to undergo cyclization, condensation, recombination or other reactions which ultimately yield some char. Diene polymers, polyacrylonitrile, and many aromatic and heterocyclic backbone polymers also belong to this char-forming group. Common to the pyrolysis of all these polymers is the formation of conjugated multiple bonds, transition from a linear to a cross-linked structure, and an increase of the aromaticity of the polymer residue[4]. For polymers containing aromatic carbon- and/or heterocyclic links in the main chain of the polymer structure, general features of their pyrolysis and char yield have been derived[5,6].

As described above, the type of polymer structure, thermal properties, and the amount of heat transferred to the polymer determine the depth over which the polymer is heated sufficiently to degrade. Since the boiling temperatures of some of the degradation products are much less than the polymer degradation temperatures, these products are superheated as they form. They nucleate and form bubbles. Then, these bubbles grow with the supply of more small degradation products by diffusion from the surrounding molten polymer[7]. Since the polymer temperature is higher near the surface than deeper, the polymer sample is more degraded there and its molecular weight, M, is lower. Since the viscosity of the molten polymer, η, depends strongly on molecular weight and temperature (for example[8], $\eta = cM^{3.4}$

or $\eta = \exp\{-M/(E(T-T_g))\}$, the viscosity near the surface is much less than that in the interior. The net result is a highly complex generation and transport of bubbles containing small molecules from the interior of the polymer melt outward through a strong viscosity gradient that heavily influences bubble behavior. A qualitative description of this complex transport process and its effect on gasification rate has been previously given[9].

The transport of the degradation products through the molten polymer layer near the surface is not well understood. Understanding of transport processes is important if intumescent char layer or barrier layer formation is used as a flame retardant approach. Very little study has been conducted to understand these transport processes except a capillary transport study through a well-controlled char layer[10].

1.2. Gas Phase

The heat release rate is one of the key quantities characterising the hazard of a material. However, the heat from oxidation reactions in a flame is released in two components; one is convective and the other is radiative. The fraction of each component, the convective fraction X_c and the radiative fraction X_r (normalised by the idealised heat of combustion of the material), depends strongly on the chemical structure of the material. Typical results for a pool flame configuration as a function of fuel mass flux are shown in Figure 1 and Figure 2 for methane (and natural gas) and acetylene, respectively[11]. The term X_l is the fraction of the idealised heat release, which is fed back to the fuel surface. In these flames, the flame becomes taller and larger with an increase in mass flux.

For large size methane (and natural gas) flames, roughly 80% of the heat release is convected away and roughly 20% of the heat release is radiated. A small fraction, about 2-3% of the total heat release, is fed back to the fuel surface. However, for small flames, the radiative fraction of the heat release becomes quite small due to the smaller flame size and the feedback fraction, X_l, increases. For acetylene flames, the radiative fraction increases up to slightly above 30% and the convective fraction decreases to as low as 45%. Combustion efficiency, X_a (the measured chemical heat release ($X_c + X_r$) divided by the idealised heat release), decreases to about 65% with an increase in the fuel mass flux. Unsaturated materials and aromatic materials tend to have similar characteristics as acetylene and their radiative fraction tends to be between 30 and 40% due to an increase in soot particle concentration in their flames. These results show clearly that heat release characteristics and heat feedback

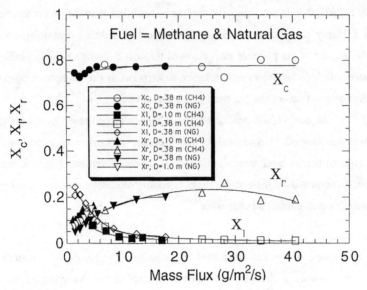

Figure 1. *Fractions of idealised heat release dissipated by convection (X_c), radiation (X_r) and feedback to the fuel surface (X_l) with respect to fuel mass flux.*

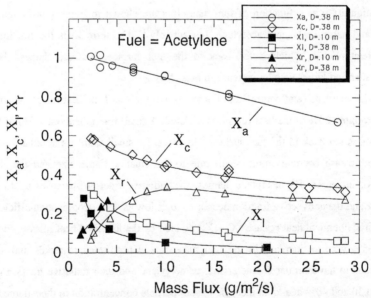

Figure 2. *Fractions of idealised heat release dissipated by convection (X_c), radiation (X_r) and feedback to the fuel surface (X_l) with respect to fuel mass flux, X is combustion efficiency.*

rates depend not only on the chemical structure of the materials but also the diameter of a pool flame and the fuel mass flux.

Radiation from the flame to the sample surface is a major heat feedback mode when the diameter of a pool flame becomes large (roughly more than 15 cm). The radiant flux from flame to the sample surface was measured using a miniature radiant flux gauge at the surface of 30 cm diameter methanol, heptane, and toluene pool flames [12]. Although the methanol flame is blue and does not generate soot particles, there is still a significant amount of radiative feedback by CO_2 and H_2O band emissions. The radiative heat feedback has a non-uniform spatial distribution. The fraction of radiation in the total heat feedback is about 80% at the center and gradually decreases to about 10% at the edge of the methanol pool flame as shown in Figure3. In this figure, the radiative component of heat feedback flux, Q_r , is Normalized by the local net heat feedback flux, Q. For the sooty toluene flame, however, this fraction is nearly constant (about 100%) across the pool surface [12]. It appears that the radiative feedback flux from a pool flame might not increase with pool diameter beyond a certain size due to absorption of radiation from the flame to the fuel surface by the vaporised fuel and particulates near the pool surface [13].

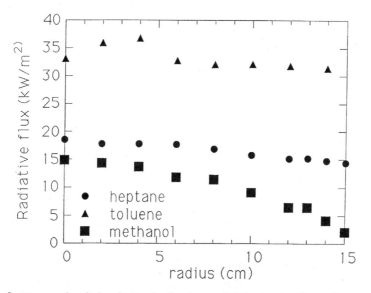

Figure 3. *Measured radial radiative feedback rate distribution for 30 cm diameter pool flames with three different liquid fuels.*

2. FLAME RETARDANTS

The fire safety of materials can be enhanced by increased ignition resistance, reduced flame spread rates, lesser heat release rates and reduced amounts of toxic and smoke products, preferably simultaneously within reasonable costs.

The most practical approach to enhance fire safety performance is the use of flame retardant additives to inexpensive and large volume commodity polymers such as PE, PP, PS, PVC, and so on. Unfortunately, the majority of these polymers have low to medium thermal stability and high heat of combustion. The additives must have a minimum impact on physical properties and product cost. Although halogenated flame retardants generally lower the heat of combustion and reduce the heat feedback rate from a flame to the polymer surface and are highly effective for reducing the heat release rate of commodity polymers, the future use of these retardants is unclear. Public perception of the environmental impact of combustion of certain halogenated flame retardants during incineration has become an issue in Europe[14,15].

Although there are many possible approaches to non-halogenated flame retardancy such as the use of aluminum trihydrate or magnesium hydroxide (both generate significant amount of water during degradation and act as a heat sink). Another approach is the formation of char. There are three mechanisms whereby the formation of char reduces flammability: (1) part of the carbon (and hydrogen) stays in the condensed phase, thus reducing the amount of gaseous combustible degradation products evolved; (2) the low thermal conductivity of the char layer over the exposed surface acts as thermal insulation to protect the virgin polymer beneath[16]; and (3) a dense char acts as a physical barrier to gaseous combustible degradation products[17].

The majority of commodity polymers do not form char during their combustion. This char forming approach is most successful if the polymer chars rapidly and early in the burning process. To be useful the charring process must be designed so that it occurs at a temperature greater than the processing temperature but before the polymer decomposition has proceeded very far. The physical structure of char has significant effects on polymer flammability. It is generally preferable to form an intumescent char (swollen char) having a cellular interior structure consisting of pockets of trapped gas[18]. The dominant protective role of an intumescent char is mainly via its thermal insulating capability[17,18] rather than an obstacle to the passage of volatile and low-viscosity products into the gas phase because low-viscosity polymeric melts can rise through an intumescent char layer due to capillary forces[10].

One possible such approach is the use of phosphorous based compounds whose effective flame retardant performance is well known[19,20]. However, it appears that the mechanism of flame retardancy depends on the polymer resin and the type if phosphorus compounds. It has been reported that the flame retardant operates in the condensed phase by forming char for rigid polyurethane[21] but, for polystyrene, phosphorous flame retardants act primarily in the gas phase[22]. The use of phosphorous compound was extended to determine flame retardant effectiveness of phosphine oxides, various hydrolytically stable aromatic phosphine oxides were chemically incorporated into nylon 6,6, PET, and epoxy polymers.

2.1. Phosphine Oxide

All phosphine oxide copolymers were synthesised by J. McGrath's group at Virginia Polytechnic Institute and State University and the synthetic methods used are described in refs. 23-25. Samples were prepared by compression molding and their size was 10 cm square with about 3 mm thickness. The flammability properties of these samples were measured by the Cone Calorimeter (ASTM E1354) at external flux of either 35 kW/m^2 or 40 kW/m^2 in air. The sample was wrapped in a thin aluminum foil except the irradiated sample surface and mounted horizontally on a calcium silicate board as an insulation material. A heavy metal container used in the standard test procedure was not used in this study to avoid heat loss to the container. The effect of incorporation of triphenylphosphine oxide, TPO, into nylon 6,6 as a copolymer on the heat release rate per unit surface area is shown in Figure4 for three different levels of phosphine oxide from 10 mol. % to 30 mol. %.A significant decrease in heat release rate is observed as the amount of the TPO co-monomers is increased. Piloted ignition delay time decreases slightly with increasing amount of the TPO. This is consistent with the slight decrease in the onset of thermal degradation temperature (from 410°C for nylon 6,6 to 402°C for the 30 mol. % of the phosphine oxide sample[24]) seen in the TGA data (thermal gravimetric analysis) in air. The mass burning flux is calculated from the transient sample weight loss rate divided by the initial sample surface area and the results are shown in Figure 5. The mass loss flux decreases with the amount of TPO, but this trend is much less than that of the heat release rate. The heat of combustion, ΔH_c, is calculated from the transient heat release rate divided by the transient mass loss rate at the same instance. The ΔH_c results shown in Figure 6 indicates that the heat of combustion decreases with increase in TPO. A reduction in the heat of

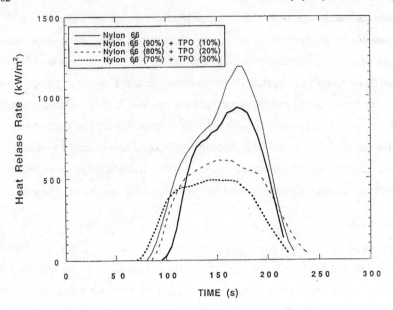

Figure 4. *Comparison of heat release rate per unit surface area of nylon6,6 and copolymer samples of nylon6,6/TPO at external flux of 40 kW/m².*

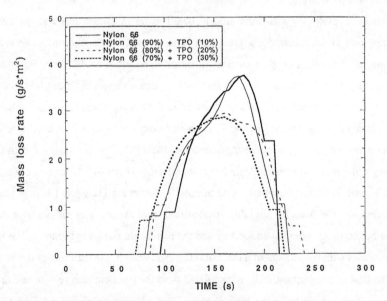

Figure 5. *Comparison of heat release rate per unit surface area of nylon6,6 and copolymer samples of nylon6,6/TPO at external flux of 40 kW/m².*

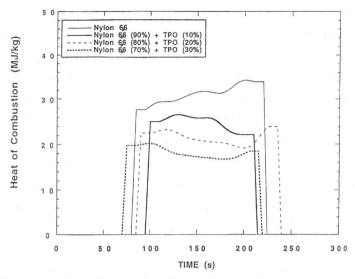

Figure 6. *Comparison of specific heat of combustion of nylon6,6 and copolymer samples of nylon6,6/TPO at external flux of 40 kW/m².*

combustion is about 40% from nylon 6,6 to the copolymer of TPO (30 mol. %) and nylon 6,6. However, the yield of char after the test is from 2.3 ±0.2 % for nylon 6,6 to 8.7 ±0.8 % for the copolymer sample. These trends indicate that there is some flame retardant activity in the condensed phase but it appears that the majority of the flame retardant activity is in the gas phase. This is confirmed by significant increases in specific extinction area, shown in Figure 7.

The specific extinction is calculated from the extinction measurement of a He-Ne laser beam passing through the exhaust duct of the Cone Calorimeter divided by the volume flow rate in the duct and the transient mass loss rate. This value indicates the concentration of soot particulates generated by the combustion of the sample. Since the effect of the TPO on the mass loss rate is relatively small, as shown in Figure 5, the overall rate of CO and soot particles formation increased with an increase in the phosphine oxide content in the copolymer sample.

The flammability properties of these samples are summarised in Table 1. A small increase in char yield (from 0 % to 8.5 %) with an increase in the TPO is also observed in the TGA study at 750°C[24]. Although the physical properties of the copolymer tends to be better than that of the blended sample, the cost of the copolymer sample might be higher than that of the blended sample. Another flammability study was carried out to compare the flame retardant effectiveness of the copolymer and a blend material. TPO (10 mol. %) was blended with nylon 6,6 and the flammability properties of this blend are compared with those of nylon 6,6 and of the nylon 6,6/TPO copolymer. The comparison of heat release rate among the three samples is shown in Figure 8. The heat release rate of the blended sample does not differ significantly from that of the copolymer but the ignition delay time of the blended sample tends to be shorter than that for the copolymer sample. There are no significant differences in burning rate, heat of flammability properties of nylon 6,6 with TPO as a blend or as a copolymer are not significantly different. Other copolymers based on polycarbonate, PET, and epoxy (Epon 828) with TPO were synthesised to examine the effects of polymer chemical structure on flame retardant effectiveness of the phosphine oxide. The results are similar to those for nylon 6,6. The heat release rates of these polymers are reduced by the incorporation of TPO as a copolymer but an increase in the amount of CO and soot particulates was also observed.

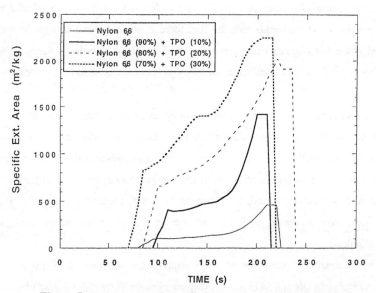

Figure 7. *Comparison of specific extinction area of nylon6,6 and copolymer samples of nylon6,6/TPO at external flux of 40 kW/m².*

The increase in the formation of soot particulates and CO, by the incorporation of TPO, could be caused by the combustion of pendant benzene groups from TPO. When benzene is a part of the polymer backbone, it tends to participate in formation of char [5]. However, pendant benzene groups do not always promote char formation. This is the case in polystyrene where the pendant benzene groups tend to generate soot particulates instead of char. In order to confirm this hypothesis, a new copolymer sample was synthesised with the pendant benzene replaced with methyl.

Sample	Peak heat release rate (kW/m²)	Total heat release (MJ/m²)	Heat of combustion (MJ/kg)	Char yield (%)	CO yield (%)	Specific extinction area (m²/kg)
nylon 6,6	1190 ±150	95 ± 10	31 ± 3	2.3 ± 0.2	1.4 ± 0.2	177 ± 30
nylon 6,6+-TPO(10 %)	930 ± 120	72 ± 7	25 ± 2	6.1 ± 0.6	10 ± 1.5	700 ± 100
nylon 6,6 + TPO(20 %)	610 ± 90	62 ± 6	21 ± 2	7.5 ± 0.7	15 ± 2	1120 ± 150
nylon 6,6 + TPO(30 %)	490 ± 70	50 ± 5	18 ± 2	8.7 ± 0.8	16 ± 2	1480 ± 200

Table 1. *Effects of TPO incorporation on the flammability properties of nylon 6,6 at an external flux of 35 kW/m².*

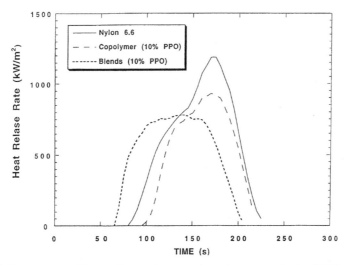

Figure 8. *Comparison of heat release rate per unit surface area of nylon6,6, the blended sample and the copolymer sample of nylon6,6/TPO at external flux of 40 kW/m².*

The heat release rate of copolymer samples of diphenylphosphine oxide, DPO, with nylon 6,6
is not significantly different from that of copolymer samples of TPO with nylon 6,6 as shown
in Figure 9.

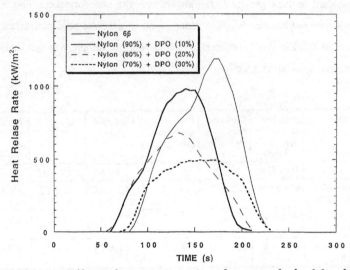

Figure 9. *Comparison of heat release rate per unit surface area of nylon6,6 and copolymer*
samples of nylon6,6/TPO at external flux of 40 kW/m².

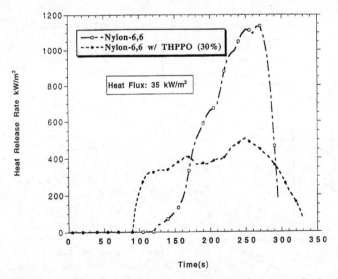

Figure 10. *Comparison of heat release rate per unit surface area of nylon6,6 and the*
nylon6,6/THPPO (30 wt. %) blended samples at external flux of 35 kW/m².

The mass loss rate, heat of combustion, CO yield, and specific extinction area of copolymer samples of DPO/nylon 6,6 are not significantly different from those of copolymer samples of TPO/nylon 6,6.

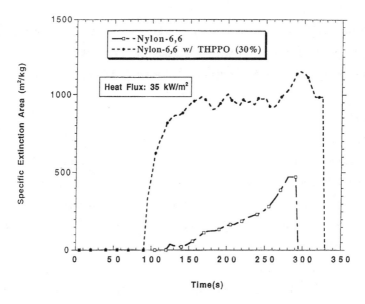

Figure 11. *Comparison of specific extinction area of nylon6,6 and the nylon6,6/THPPO (30 wt. %) blended samples at external flux of 35 kW/m².*

These results indicate that the pendant benzenes do not enhance the formation of CO and soot particulates. To determine if benzene in the copolymer backbone was contributing to the increase in CO and soot particulates, we examined an aliphatic phosphine oxide blended with nylon 6,6. Trihydroxypropylphosphine oxide, THPPO, was used as the aliphatic phosphine oxide. The blended sample has 30 wt. % of THPPO. The heat release rate of the blended sample with THPPO is compared with that of nylon 6,6 sample (This nylon 6,6 is a commercial sample whose thermal degradation characteristics might be different from that of the nylon 6,6 sample used for the copolymer study.). The results, shown in Figure 10 at an external flux of 35 kW/m², show a significant reduction in heat release rate similar to the copolymer samples with TPO. In addition, the specific extinction area of the blended sample is much higher than that of nylon 6,6, as shown in Figure 11. The char yield of the blended sample was 4.2 %. These results are similar to those of the copolymer samples with TPO. The results suggest that phosphorous is the major factor in controlling the reduction of the heat release rate and the increase in CO and soot particulates. There is evidence which suggests that, if phosphorous is released into the gas phase, then it acts as a radical scavenger of H-atoms[26,27].

On the other hand, the measurable char yield in the tested sample suggests that there is some activity in the condensed phase. If phosphorous stays in the condensed phase during combustion, phosphorous could be a significant char forming flame retardant.

2.2. Silica Gel

The intention of using silica gel with K_2CO_3 was to devise a method of in-situ formation of silicon based fire retardants during combustion. The reaction of silica gel and organic alcohols in the presence of metal hydroxides has been shown to give multi-coordinate organosiliconate compounds[28]. Instead of synthesising these materials and then combining them with various polymers to evaluate their effect on polymer flammability properties, we envisioned the reaction occurring in the condensed phase of the pyrolyzing polymer beneath the burning surface, by combining a polyhydroxylic polymer, e.g. PVA or cellulose, with silica gel and K_2CO_3. If the reaction between the polymer and the additives occurs, it should crosslink the polymer and might assist in forming a silicon-oxy-carbide, SiOC, type protective char during combustion. The flammability properties of these samples were measured in the Cone Calorimeter at an incident flux of 35 kW/m^2. The results are summarised in Table 2 for the polymers and polymers with the addition of silica gel and K_2CO_3[29]. Assuming all additives remained in the polymer residue after the test, the char yield was determined as (polymer residue weight - initial additives weight)/initial mass of polymer in the sample. The results show that the additives enhance the formation of carbonaceous char even if the original polymer does not generate any char such as PP, PS and PMMA. The increases in carbonaceous char yield for PVA and cellulose is nearly a factor of 10. It is not surprising that a significant increase in char yield was observed for PVA and cellulose but char was not expected to form for PP and PS which do not have any alcohol groups in their polymer structure. The reduction in peak heat release rate by the additives is quite significant, reaching about 50% for PP, PVA, cellulose, and nylon 6,6. A typical result for the reduction in heat release rate is shown in Figure 12 for PP. However, the heat of combustion is not significantly affected by the additives and also the concentrations of particulates and CO in the combustion products do not increase with the additives[29]. These trends are significantly different from those for halogenated flame retardant additives or even for the previously described copolymer samples of phosphine oxide. The results presented here clearly demonstrate that the flammability of a wide variety of polymers is dramatically reduced in the presence of relatively small concentrations of silica gel and K_2CO_3.

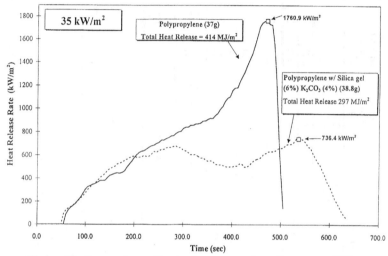

Figure 12. *Comparison of heat release per unit surface area of PP and the PP blended sample with silica gel/K_2CO_3 at external flux of 35 kW/m².*

All the data with the addition of silica gel and K_2CO_3 in the polymers described in Table 2 show trends of lower heat release rate, lower sample mass loss rate, and no significant effects on heat of combustion, yields of smoke and CO, and the formation/enhancement of carbonaceous char. These trends show that the site of flame retardancy of the additives is in the condensed phase.

This is more clearly demonstrated by the gasification study of PMMA with the additives in nitrogen[29]. The PMMA with silica gel and K_2CO_3 (mass ratio of 95:4:1) sample of about 75 cm diameter with about 0.6 cm thickness was exposed to external radiant flux of 41 kW/m² in our radiative gasification device.

Sample disk: 75mm x 8mm	Char Yield (%)	LOI (%)	Peak HRR (A) (kW/m²)	Mean HRR (kW/m²)	Mean Heat of Combustion (MJ/kg)	Total Heat Released (MJ/m²)	Mean Specific Ext. Area (m²/kg)	Mean CO yield (kg/kg)
PP	0	-	1,761	803	38	357	689	0.04
PP w/ 6%SG & 4%PC	10	-	736 (58%)	512	33	297	710	0.04
PS	0	18	1,737	1,010	25	277	1,422	0.07
PS w/ 6%SG & 4%PC	6	24	1,190 (31%)	725	25	246	1,503	0.07
PMMA	0	18	722	569	23	319	210	0.01
PMMA w/ 3%SG & 1%PC	15	25	420 (42%)	246	21	231	199	0.05
PVA	4	-	609	381	17	221	594	0.03
PVA w/ 10%PC	9	-	322 (47%)	222	17	145	571	0.03
PVA w/ 10%SG	29	-	252 (57%)	173	15	131	361	0.03
PVA w/ 3%SG & 1%PC	16	-	295 (52%)	232	16	166	447	0.03
PVA w/ 6%SG & 4%PC	43	-	194 (68%)	114	12	101	201	0.03
Cellulose	4	-	310	161	11	101	27	0.02
Cellulose w/ 6%SG & 4%PC	32	-	149(52%)	71	5.3	34	20	0.04
SAN	2	-	1,499	837	25	197	1,331	0.07
SAN w/ 6%SG & 4%PC	3	-	1,127 (25%)	772	23	169	1,301	0.06
Nylon 6, 6	1	30	1,131	640	23	108	234	0.02
Nylon 6, 6 w/ 4%PC	3	-	854 (25%)	570	25	103	342	0.03
Nylon 6, 6 w/ 6%SG	4	-	558 (51%)	365	24	111	164	0.02
Nylon 6, 6 w/ 3%SG & 2%PC	5	33	526 (53%)	390	22	105	171	0.02
Nylon 6, 6 w/ 6%SG & 4%PC	6	30	546 (52%)	370	24	102	185	0.02

Incident heat flux = 35 kW/m² ; SG = Silica Gel; PC = K_2CO_3

Table 2. Flammability properties of various polymers with silica gel and potassium carbonate in Cone calorimeter and Limiting Oxygen Index test. Uncertainties in peak heat release rate and in mean extinction area are ± 15 % and those of other quantities are ± 10 %.

The weight loss rate and gasification behavior of PMMA and PMMA with the additives were recorded by an electric balance and a video camera, respectively. Since the experiments were conducted in nitrogen, there were no flames and the results were solely based on the condensed phase process. The results show that the sample weight loss rate of PMMA with the additives started to become slower when slight char formation was observed. The colour of PMMA with the additives became darker with time and became black after 300 s exposure. Carbonaceous char yield of the PMMA with the additives sample was about 12 % (excluding the additives left in the residue) and the peak weight loss rate of the additive sample was about 30% less than that of PMMA without the additives. As expected, no carbonaceous char was observed at the end of the test for PMMA without the additives.

The envisioned flame retardant approach is the formation of silicon-oxy-carbide (SI-O-C) type protective char by crosslinking PVA with pentacoordinate organosilicate during combustion [28].

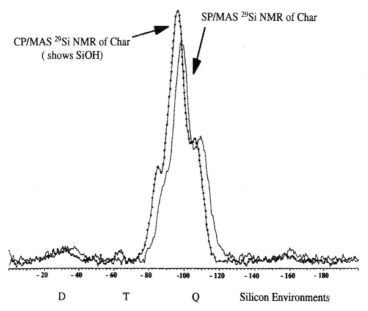

Figure 13. *CP/MAS and SP/MAS ^{29}Si NMR spectra of the char from combustion in the Cone Calorimeter of PVA with silica gel/K2CO3 additives (mass ratio, 90:6:4 respectively).*

In order to confirm this mechanism, the char from the combustion of PVA with silica gel/K_2CO_3 additives (mass ratio, 90:6:4 respectively) in the Cone Calorimeter was analysed using single pulse magic angle spinning (SP/MAS) ^{29}Si NMR and cross-polarisation (CP)/MAS ^{29}Si NMR which selectively enhances the signal intensities of Si nuclei near proton[30]. Their spectrum is shown in Figure 13. The SP/MAS ^{29}Si NMR spectrum shows a broad resonance from ~130 ppm to ~90 ppm, which indicates that the char may contain some silicate species. Comparison of this spectrum to the SP/MAS ^{29}Si NMR spectrum of the char from combustion of PVA with sodium silicate (mass ratio of 90:10) confirms this possibility, since both spectra show the majority of the silicons are of the Q^3 [(SiO)$_3$SiO(-)] (100 ppm to 110 ppm) and Q^4 [(SiO)$_4$Si] (110 ppm to 120 ppm) type. The CP/MAS ^{29}Si NMR spectrum of the PVA with silica gel/K_2CO_3 char reveals that there is still a significant fraction of Q^2 [-(SiO)$_2$-Si-(OH)$_2$] (85 ppm to 95 ppm) and Q^3 [-(SiO)$_3$-Si-OH] (95 ppm to 105 ppm) silanol functionality present after the combustion. Silanol is also present in the original silica gel structure. These spectra indicate that the majority of the silica gel original structure remains intact during the combustion and envisioned Si-O-C bonds are hardly observed in the spectra at T in the figure.

An alternative mechanism for these additives is through the formation of a potassium silicate glass as a surface barrier which insulated and slowed escape of volatile decomposition products. The latter might provide enough time for crosslinking and the formation of char among the degradation products and residues of polymer chains. At present, the flame retardant mechanism of silica gel/K_2CO_3 additives in not only hydroxylic polymers such as PVA and cellulose but also in non-hydroxylic polymers such as PP, PS, and PMMA is not understood. We are currently working to determine the effects of particle size, internal pore size and silanol content of the silica gel on flammability properties of PP and to understand the flame retardant mechanisms.

2.3. Another Flame Retardant Approach

Another approach is the in situ formation of a barrier layer near the polymer surface during burning. An inorganic additive such as silicon based particles is dispersed into a polymer sample and accumulation of the particles to form a layer to interfere in the transport rate of the thermal degradation products of the polymer to the gas phase or to act as a thermal insulation layer.

2.3.1. *Nanocomposites*

2.3.1a. Experimental Study

In the pursuit of improved approaches to fire retardant polymers, a wide variety of concerns must be addressed, in addition to flammability. Generally, the addition of inorganic or organic flame retardants into polymer tends to reduce mechanical properties of the polymer. However, nylon-6 clay-nanocomposites, first developed by researchers at Toyota Central Research and Development Laboratories, were developed to have unique mechanical properties when compared to conventional filled polymers[31]. The nylon-6 clay-nanocomposites (clay mass fraction from 2% to 70%) are synthesised by ring-opening polymerization of e-caprolactam in the presence of cation exchanged montmorillonite clay. This process creates a polymer layered silicate nanocomposite with either a delaminated hybrid structure (randomly dispersed silicate layers) or an intercalated hybrid structure (well ordered multilayer with spacing between the silicate layers of only a few nanometers). The mechanical properties of the nylon-6 clay-nanocomposites with 5% clay mass fraction show excellent improvement over pure nylon-6. The nanocomposite exhibits a 40% higher tensile, 68% greater tensile modulus, 60% higher flexural strength, 126 % increased flexural modulus, and comparable Izod and Charpy impact strengths. The heat distortion temperature is increased from 65 °C (nylon-6) to 152 °C (nylon-6 clay-nanocomposite)[32].

To evaluate the flame retardant effectiveness of the nanocomposite approach, we have measured the flammability properties of nylon-6 delaminated clay-nanocomposites with clay mass fractions of 2% and 5%, and compared them with pure nylon-6[33]. All nylon-6 clay-nanocomposites and nylon-6 were obtained from UBE industry and were used as received. The samples were prepared by compression molding into about 7.5 cm x 5 cm rectangular slab with about 1.5 cm thickness. The thermogravimetric analysis shows that there was no significant difference in weight loss rate history between nylon-6 and clay-nanocomposites with clay mass fraction of 5% [32]. The heat release rate curves from the Cone Calorimeter for nylon-6 and nylon-6 clay-nanocomposites (2% and 5%) when exposed to external radiant flux of 35 kW/m^2 are shown in Figure 14. The results show significant reduction in heat release rate and higher mass fraction of the clay reduces more heat release rate of nylon-6. Visual observation of the combustion experiments in the Cone Calorimeter reveals different behavior for the nylon-6 clay-nanocomposites compared to the pure nylon-6 from the very beginning of the thermal exposure. A thin char layer forms on the top of the all the samples in the first few minutes of exposure prior to ignition. In the case of pure nylon-6, this char layer fractures into small pieces early during the combustion. The char does not fracture with the nylon-6 clay-nanocomposites. This tougher char layer survives and grows throughout the combustion, yielding a rigid multicellular char-brick. The nanocomposite structure appears to enhance the toughness of the char through reinforcement of char layer. The TEM of a section of the residual char from the nylon-6 clay-nanocomposite (5% mass fraction) is shown in Figure 15. A multilayered silicate structure is clearly seen, with the darker, 1 nm thick, silicate sheets forming a large array of fairly even layers. This was the primary morphology seen in the TEM of the char, however, some voids were also present. The original nylon-6 clay-nanocomposite sample is mostly the delaminated structure[31,32], this implies that the layered structure seen in the Figure 15 formed during combustion. The delaminated hybrid structure, which subsequently collapses during combustion, may act as an insulator and a mass transport barrier, slowing the escape of the volatile products generated as the nylon-6 decomposes. The nanocomposite's low permeability for liquids and gases may slow the transport of volatile products and also molten polymers through the nanocomposite layers to the sample surface[34].

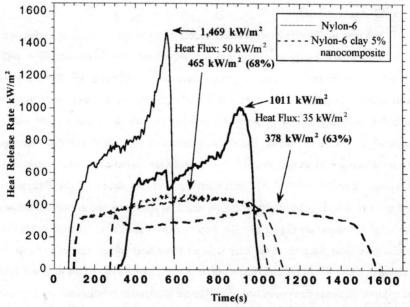

Figure 14. *Comparison of heat release rate for nylon-6 and nylon-6 clay nanocomposites.*

Figure 15. *TEM of combustion char sample of nylon-6 clay composite.*

Further X-ray and TEM analysis of the char and the original nylon-6 nanocomposite structure and gasification rate measurement of nylon-6 clay-nanocomposites in nitrogen and also in a 7% O2/93% N2 atmosphere without flaming in the radiative gasification apparatus are underway to better understand flammability behavior. Some new results are presented in our other paper[35] in this Book.

Comparison of nylon-6 clay-nanocomposites to other flame retarded nylon systems, such as the previously discussed nylon-6,6 triphenylphosphine oxide copolymer (nylon-6,6+TPO(30%)), where the flame retardant is also combined with the nylon at the molecular level, further illustrates the unique benefits the nanocomposite approach offers. Table 3 shows that the nylon-6,6+TPO(30%) copolymer gives similar reduction in heat release rate (58%) to that for the nanocomposites (63%) at a comparable level of incorporation of "flame retardants" (4% mass fraction of phosphorus). As described in the Section 2.1, the specific extinction area and CO yield (these values are normalised by mass loss rate) for the nylon-6,6 +TPO (30%) are much greater then that for the nylon-6,6. Even though the mass loss rate for the copolymer is about 50% lower than that for the nylon-6,6, the extinction by smoke is still four times greater, and the CO production rate is still 10 times greater, than that of nylon-6,6. Another additive flame retardant system for nylon, based on ammonium polyphosphate, requires 35% mass fraction of additive to significantly effect the flammability (measured by oxygen index) of nylon-6[36] and this results in as much as a 20% loss of mechanical properties. Finally, it should be noted that the nano-dispersed clay composite structure has a very different effect on the flammability of nylon than macro-or meso-dispersed clay-polymer mixtures. Le Bras and Bourbigot found, in their extensive study of clays in an intumescent polypropylene system, that montmorillonite clay, similar to the ion exchanged montmorillonite clay used to make the nylon composites, actually decreased the limiting oxygen index[37].

Other polymer silicate nanocomposites based on a wide variety of resins such as polystyrene, epoxy, poly(ethylene oxide), polysiloxane, polyesters, and polyphosphazenes, have recently been prepared via melt intercalation[38, 39]. These materials possess varying degrees of interaction between the polymer and the silicate layer and provide the opportunity to study the effect this variable has on flammability and to determine if the clay-nanocomposite approach is useful in reducing flammability of many polymers. We are continuing to investigate the mechanism of flame retardancy in clay and other nanocomposite materials and some of the results will be presented in our other paper in this Book.

Sample	Char Yield (%) ± 0.1	Peak HRR (Δ%) (kW/m²) ± 15%	Mean Heat of Combustion (MJ/kg) ± 10%	Total Heat Released (MJ/m²) ± 10%	Smoke Mean Extinction Area (m²/kg) ± 10%	Mean CO yield (kg/kg) ± 10%
Nylon-6	0.3	1011	27	413	197	0.01
Nylon-6 clay-nanocomposite 2%	3.4	686 (32%)	27	406	271	0.01
Nylon-6 clay-nanocomposite 4%	5.5	378 (63%)	27	397	296	0.02
Nylon-6,6	0	1190	30	95	200	0.01
Nylon-6,6 -PO 4% Phosphorus	8.7	490 (58%)	18	50	1400	0.16

Table 3. *Cone Calorimeter data of nylon-6,6 clay nanocomposites and nylon-6 triarly phosphine oxide copolymer*

2.3.1b. Computer Simulations

A computer program, hereafter referred to as MD_REACT, that was developed in this laboratory to study the thermal degradation of polymers[40-43], is also being used to determine the mechanism for the increase in the thermal stability and fire resistance of polymer nanocomposites. The basis of MD_REACT is molecular dynamics. This technique consists of solving Hamilton's equations of motion for each of the 3N molecular degrees of freedom. The form of the Hamiltonian used in the calculations reported in this investigation was derived from the Consistent Valence Forcefield (CVFF) developed by Molecular Simulations, Inc. (MSI)[44]. A more detailed description of the force field used in the calculations can be found elsewhere[43].

The unique capability of MD_REACT is that it allows for the formation of new bonds from free radical fragments generated when bonds in the polymer break and, thereby, account for the chemical reactions which play a major role in the thermal degradation process. Some of these are: bond scission, depolymerization, hydrogen transfer, chain stripping, cyclization, crosslinking and radical recombination reactions. The depolymerization and intramolecular hydrogen transfer reactions, in particular, are modelled by introducing two new atom types (cf and ccf) into the CVFF forcefield to account for beta scission. The ccf atom type corresponds to an aliphatic carbon (c) bonded to a free radical carbon (cf). Once a free radical is formed, the dissociation energies of the c-ccf and h-ccf bonds (beta to cf) are reduced by 317 kJ/mol (to 51 kJ/mol and 137 kJ/mol, respectively) which corresponds to the difference between a carbon-carbon single and double bond.

The thermal degradation experiments were performed on polypropylene/graphite, rather than nylon-6/clay nanocomposites. In making this decision, we were motivated by the considerable body of experience and high level of confidence that we have acquired in using the CVFF forcefield to model hydrocarbon polymers and surfaces and we did not want to introduce any additional ambiguities into the interpretation of the computer simulations. The model of the polymer/graphite nanocomposite consisted of 4 chains of isotactic polypropylene each containing 48 propylene monomers and a graphite sheet with about 600 carbon and 80 hydrogen (used to terminate the edges of the surface) atoms.

A series of nanocomposite structures with the polymer intercalated between graphite layers which were separated by a variable distance, b, was obtained by annealing the model polymer and graphite inside of a unit cell with the following dimensions: a = 100, c = 30 and b = 25, 28, 30, 32 and 50 Å. The same model polymer was used in all of the structures. Thus, only the distance between the graphite sheets and, consequently, the density of the composite was allowed to change from one simulation to the next. The simulated annealing was performed by heating the polymer/graphite assembly to 500 K for 100 time steps and then relaxing it by performing 100 iterations of the Polak Ribiere conjugate gradient minimization[44]. The entire process was repeated until the potential energy of the fully optimized structure was lower than any of the values attained during the course of the simulated annealing procedure.

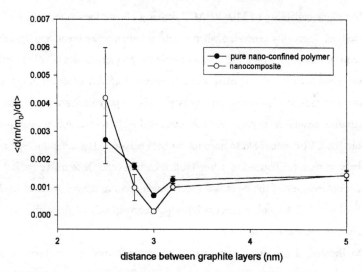

Figure 16. *The effects of the separation distance between the graphite layers on the average mass loss of PP with (open circles) and without (closed circles) the interaction between the layers and the polymer.*

The average rate of mass loss for both the pure nano-confined polymers and the nanocomposites are plotted as a function of the distance of separation between the graphite layers in Figure 16. The stabilization of the polymer is most pronounced in the b = 30 Å nanocomposite and approaches zero at b = 50 Å, when the graphite layers are too far apart for there to be a significant interaction between them. At these large distances of separation, the interactions are almost exclusively between the polymer and the graphite which should approximate what occurs in the delaminated nanocomposites where the graphite layers are individually dispersed in the polymer matrix. The observation that the thermal stability of the polymer increases when it is intercalated but is unaffected when the layers delaminate is consistent with recent experimental results that indicate that intercalated nanocomposites are more thermally stable than delaminated nanocomposites[45,46]. Indeed it was noted that the derivative thermogravimetric (DTG) curves corresponding to the delaminated nylon-6/clay nanocomposites were almost identical to the values obtained from pure nylon-6, whereas the DTG curves of intercalated polystyrene/clay nanocomposites were shifted to dramatically higher temperatures than what was observed for pure polystyrene[35].

A comparison of computer animations of the trajectories, corresponding to the nanocomposites and pure nano-confined polymers, corroborate the observations we made about the effects of the interactions between the polymer and the graphite from consideration of the mass loss data. In general, the polymers in the nanocomposites lost fewer fragments and retained their shape longer than the pure nano-confined polymers. We also observed that there was a tendency for the fragments that did form to collide with the graphite surface and bounce back into the central unit cell where they could undergo recombination reactions with other free radical polymer fragments, rather than escape from the melt as combustible fuel. The last frames from the animated trajectories of the pure polymer and the b = 30 Å nanocomposite are depicted in Figure 17. A comparison indicates that the fragments only escape from the sides of the nanocomposite (right), whereas they leave the pure nano-confined polymer from all directions (left).

Figure 17. *Animated trajectories of polypropylene with the separation distance of 30 Å between the graphite layers, without the interaction between layers and the polymer (left), and with the interaction (right).*

3. SUMMARY

The search for new effective flame retardants is becoming more efficient due to the availability of bench scale tools such as Cone Calorimeter for the measurement of various flammability properties relevant to fire performance, and the availability of various analytical tools to determine chemical structures of polymer residues collected at various stages of pyrolysis/combustion. The development of theoretical tools such as molecular modelling is acting a guiding tool for the development of new flame retardants. Thus, new promising flame retardants such as silica gel and nanocomposites are found and their flammability properties and characteristics are described in this paper, although their retardant mechanisms have not been fully understood. Albeit the characteristics of the new flame retardants need improvement to satisfy their environmental impact, processability, and cost, we are quite optimistic that more fire safe materials will be produced with new, more efficient flame retardants.

References

1. Mita, I., _"Effects of Structure on Degradation and Stability of Polymers"_ in _Aspects of Degradation and Stabilization of Polymers, Chapter. 6_, H..H.G. Jellinek, Ed., Elsevier Scientific, Amsterdam, 1978.
2. Kelen, T., _"Polymer Degradation"_, Van Norstrand Reinhold, New York, 1983.
3. Grassie, N., and Scott, G., _"Polymer Degradation Stabilization"_, Cambridge University Press, Cambridge, 1985.
4. Factor, A., _"Char Formation in Aromatic Engineering Polymers"_, in _"Fire and Polymers"_ (G.L. Nelson, Ed.), ACS Symposium Series 425, ACS, Washington, DC, pp.274-287, 1990.
5. Van Krevelen, D. W., _Polymer_ 16:615-620 (1975).
6. Aseeva, R. M., and Zaikov, G. E., _"Flammability of Polymeric Materials"_, _Adv. Polym. Sci._ 70:172-229 (1985).
7. Clift, R., Grace, J. R., and Weber, M. E., _"Bubbles, Drops, and Particles"_, Academic Press, New York, 1978.
8. Matsuoka, S., and Kwei, T.K., _"Physical Behavior of Macromolecules"_, in _Macromolecules_, F. A. Bovey and F.H. Winslow Eds., Academic Press, New York, p.346, 1979.
9. Kashiwagi, T., and Ohlemiller, T. J., _Nineteenth Symposium (International) on Combustion_, The Combustion Institute, Pittsburgh, pp.815-823 (1982).
10. Gibov, K. M., Shapovalova, L. N., and Zhubanov, B. A., _Fire Mater._ 10:133-135 (1986).

11. Hamins, A., Konishi, K., Borthwick, P., and Kashiwagi, T., *26th Symposium (Int.) on Combustion*, The Combustion Institute, Pittsburgh, pp.1429-1436 (1996).

12. Hamins, A. J., Fischer, S. J., Kashiwagi, T., Klassen, M. E., and Gore, J. P., *Combust. Sci. Tech.*, **97**:37-62 (1994).

13. Modak, A. T., *Fire Safety J.*, **3(2-4)**:177-184 (1981).

14. Nelson, G. L., *"Recycling of Plastics - A New FR Challenge"*, *"The Future of Fire Retarded Materials: Applications & Regulations"*, FRCA, Lancaster, PA, p.135, 1994.

15. Van Riel, H. C. H. A., *"Is There a Future in FR Material Recycling; The European Perspective"*, *The Future of Fire Retarded Materials: Applications & Regulations"*, FRCA, Lancaster, PA, p.167, 1994.

16. Anderson, C. E., Jr., Ketchum, D. E., and Mountain, W. P., *J. Fire Science*, **6**:390-410 (1988).

17. Camino, G., Costa, L., Casorati, E., Bertelli, G., and Locatello, R., *J. Appl. Polym. Sci.*, **35**:1863-1876 (1988).

18. Scharf, D., Nalepa, R., Heflin, R., and Wusu, T., *Fire Safety J.*, **19**:103-117 (1992).

19. Weil, E. D., *"Encyclopaedia of Polymer Science and Technology"*, Wiley-Interscience, New York, Vol.11, 1986.

20. Aaronson, A. M., *"Phosphorous Chemistry"*, *ACS Symposium Series* 486, Chapter 17, p.218, 1992.

21. Papa, A. J., *"Flame Retarding Polyurethanes"* in *"Flame Retardancy of Polymeric Materials"* (Volume 3), Kuryla, W. C. and Papa, A. J., Eds., Marcel Dekker, New York, pp.1-133, 1975.

22. Carnahan, J., Haaf, W., Nelson, G., Lee, G., Abolins, V., and Shank, P., *"Investigation into the Mechanism for Phosphorous Flame Retardancy in Engineering Plastics"*, *"Proceeding of 4th International Conference on Fire Safety"*, Product Safety Corp., San Francisco, CA, 1979.

23. Smith, C. D., Gungor, A., Wood, P.A., Liptak, S. C., Grubbs, H., Yoon, T. H., and McGrath, J. E., *Makromol. Chem., Macromol. Symp.*, **74**:185 (1993).

24. Wan, I. Y., McGrath, J. E., and Kashiwagi, T., ACS Symposium Series 599, *"Fire and Polymers II"*, edited by G. Nelson, Washington, D.C., p.29, 1995.

25. Wan, I.Y. and McGrath, J.E., *Polymer Preprints*, **36(1)**: 493 (1995).

26. Hastie, J.W. *"Molecular Basis of Flame Inhibition"*, *J. Research of NBS-A. Physics &Chemistry*, **77A**: 733-754 (1973).

27. Hastie, J. W. and McBee, C. L., *"Mechanistic Studies of Triphenylphosphine Oxide-Poly(Ethyleneterephthalate) and Related Flame Retardant Systems"*, NBSIR 75-741 (1975).

28. Laine, R. M., *Nature*, **353**:642 (1991).

29. Gilman, J. W., Ritchie, S. J., Kashiwagi, T., and Lomakin, S. M., *Fire and Materials*, **21**:23-32 (1997).

30   Gilman, J. W., Lomakin, S. M., Kashiwagi, T., VanderHart, D. L., and Nagy V., *Polymer Preprints*, **38(1)**:802-803 (1997).

31   Usuki, A., Kojima, Y., Kawasumi, M., Okada, A., Fukushima, Y., Kurauchi, T., and Kamigaito, O., *J. Mater. Res.*, **8**:1179 (1993).

32. Kojima, Y., Usuki, A., Kawasumi, M., Okada, A., Fukushima, Y., Kurauchi, T., and Kamigaito, O., *J. Mater. Res.*, **8**:1185 (1993).

33. Gilman, J.W. and Kashiwagi, T., *SAMPE J.*, **33**:40-46 (1997).

34. Giannels, E. and Messersmith, P., *J. Polym. Sci. A: Polym. Chem.*, **33**:1047, 1995.

35. Gilman, J. W., Kashiwagi, T., and Lichtenhan, J. D., *"Nanocomposites: A Revolutionary New Flame Retardant Approach"*, *"Proceeding of the 6th European Meeting on Fire Retardancy of Polymeric Materials"*, Lille (1997).

36. Levchik, S., Camino, L., Costa, L., and Levchik, G., *Fire and Materials*, **19**:1 (1995).

37. Le Bras, M., and Bourbigot, S., *Fire and Materials*, **20**:39 (1996).

38. Vaia, R., Jandt, K., Kramer, E., and Giannelis, E., *Macromolecules*, **28**:8080 (1995).

39. Giannelis, E., *Adv. Mater.*, **8**:29 (1996).

40. Nyden, M. R. and Noid, D. W., *Phys. Chem.*, **95**:940 (1991).

41. Nyden, M. R., Forney, G. P., and Brown, J. E., *Macromolecules* **25**:1658 (1992).

42. Nyden, M. R., Brown, J. E. and Lomakin, S. M., *Mat. Res. Soc. Symp. Proc.*, **278**:47 (1992).

43. Nyden M. R., Coley, T.R., and Mumby, S., *Polym. Eng. Sci.*, **37(9)** (1997).

44. *"Discover User Guide, Part 3"*, MSI, San Diego, (1995).

45. Lee, J., Takekoshi, T., and Giannelis, E., *Mat. Res. Soc. Symp. Proc.*, **457**:513 (1997).

46. Lee, J., and Giannelis, E., *Polymer Preprints*, **38**:688 (1997).

Acknowledgements

The authors are grateful to Professor J. McGrath and Dr. I. Y. Wan at Virginia Tech. for the sample preparation with phosphine oxides, Dr. Henry Yue at Dow Corning Corp. for ^{29}Si NMR analysis, Dr. C. Jackson for TEM analysis of polymer residues, Mr. Jack Lee and Michael Smith for the flammability measurements in Cone Calorimeter, and the FAA Tech. Center (Interagency Agreement DTFA003-92-Z-0018) for partial support of this Work.

NANOCOMPOSITES:

RADIATIVE GASIFICATION AND VINYL POLYMER FLAMMABILITY.

J. W. Gilman, Takashi Kashiwagi

National Institute of Standards and Technology
Gaithersburg, MD, USA

E. P. Giannelis and E. Manias

Cornell University
Ithaca, NY, USA

S. Lomakin

Guest Researcher at NIST
from the Russian Academy of Sciences, Moscow, Russia.

J. D. Lichtenhan, P. Jones

Air Force Research Laboratory, Edwards Air Force Base, CA, USA

In the pursuit of improved approaches to flame retarding polymers a wide variety of concerns must be addressed. The low cost of commodity polymers requires that the fire retardant (FR) approach be of low cost. This limits the solutions to the problem primarily to additive type approaches. These additives must be easily processed with the polymer, must not excessively degrade the other performance properties, and must not create environmental problems in terms of recycling or disposal. Currently, some of the commonly used flame retardant approaches for polymers can reduce the thermal and mechanical properties of the polymer[1-3]. Polymer-clay nanocomposites are hybrid organic polymer inorganic layered materials with unique properties when compared to conventional filled polymers. The mechanical properties for nylon-6 clay nanocomposite, with clay mass fraction of 5 %, show excellent improvement over those for the pure nylon-6. The nanocomposite exhibits a 40 % higher tensile strength, 68 % greater tensile modulus, 60 % higher flexural strength, 126 % increased flexural modulus, and comparable impact strengths. The heat distortion temperature (HDT) is increased from 65°C to 152°C[4]. Previously, we reported on the flammability properties of nylon-6 clay nanocomposites[5]. Here, we will briefly review these results, present the results of radiative gasification experiments and report on our initial studies of the flammability of *intercalated* polymer-clay nanocomposites prepared from polystyrene, PS, and polypropylene-graft-maleic anhydride, PP-g-MA.

1. EXPERIMENTAL

1.1. Cone Calorimeter

Evaluations of flammability were done using the Cone Calorimeter[6]. The tests were done at an incident heat flux of 35 kW/m^2 and 50 kW/m^2, using a cone shaped heater. A heat flux of 35 kW/m^2 represents a typical small-fire scenario[7]. Peak heat release rate, mass loss rate and specific extinction area data, measured at 35 kW/m^2, are reproducible to within ± 15 %. The carbon monoxide and heat of combustion data are reproducible to within ± 10 %. The uncertainties for the Cone calorimeter are based on the uncertainties observed while evaluating the thousands of samples combusted to date. Typically two samples were combusted in each case in the Cone, and the results averaged. Cone samples were prepared by compression molding the samples (20g to 50g) into rectangular or round plaques, 4mm to 15 mm thick, with a typical area of ~ 0.004 m^2, using a press with a heated mould.

1.2. Radiative Gasification

Figure 1 shows a gasification apparatus which is similar to the Cone calorimeter. The gasification apparatus allows pyrolysis, in a nitrogen atmosphere, of samples identical to those used in the Cone calorimeter, at heat fluxes like those experienced in a fire (30 kW/m^2 to 100 kW/m^2). It allows study of the condensed phase decomposition processes decoupled from the gas phase combustion and resulting heat feedback from the flame. In a typical experiment thermocouples are imbedded in the sample to monitor the temperature at which the pyrolysis and decomposition processes occur. A load cell gives mass loss rate data which can be compared to mass loss rate data from the Cone calorimeter experiments. A video camera records the pyrolysis and charring phenomena.

1.3. Nylon-6 clay nanocomposites

Nylon-6 clay nanocomposites (clay mass fraction of 2 % and 5 %) and nylon-6 were obtained from UBE industries and used as received.[8] The above nanocomposites will be referred to as nylon-6 clay nanocomposite (2%) and nylon-6 clay nanocomposite (5%), respectively.

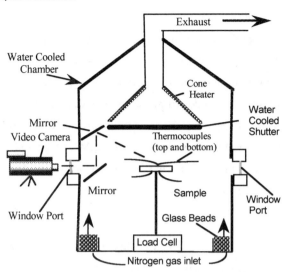

Figure 1. *A schematic of the radiative gasification apparatus (1 m diameter, 2 m height). The gasification apparatus allows pyrolysis, in a nitrogen atmosphere, of samples identical to those used in the Cone calorimeter.*

1.4. PS-clay-nanocomposite

Preparation of PS-clay-nanocomposite (clay mass fraction of 3 %) by melt blending PS with bis(dimethyl)bis(octadecyl)ammonium-exchanged montmorillonite, yields a nanocomposite with the *intercalated* structure (see Scheme 1). The inter-gallery spacing, by X-ray diffraction, XRD, is 3.1 nm ($2\theta = 2.7$). In the *intercalated* form and at this low clay concentration this nanocomposite is essentially a blend, with *intercalated*-PS-clay domains dispersed in pure PS. The *immiscible* PS-clay mix, where the clay is only mixed in at the primary-particle size scale (~5 μm), is prepared under the same melt blending conditions except the alkylammonium used to compatibilize the montmorillonite has only one octadecyl R group instead of two. This renders the ion exchanged montmorillonite slightly less organophilic and intercalation does not occur. These PS-clay combinations will be referred to as *intercalated*-PS-clay (3 %) nanocomposite and *immiscible* PS-clay (3 %) mixture. The PS used was Styron 6127 from Dow Chemical Co.

1.5. PP-g-MA-clay-nanocomposite

Preparation of PP-g-MA-clay-nanocomposite (clay mass fraction of 5 %) by melt blending was accomplished by pressing the PP-g-MA mixed with the bis(dimethyl)bis(tallow)ammonium-exchanged montmorillonite, Closite 15A, at 160°C for 30 minutes using a Carver press, followed by heating in a vacuum oven for several hours at 160°C. This yields a nanocomposite with the *intercalated* structure (see scheme 1). The inter-gallery spacing, by XRD analysis, is 3.6 nm. PP-g-MA (m.p.: 152°C) was purchased from Aldrich and contains a mass fraction of 0.6 % maleic anhydride. It has a melt index of 115 g / 600 s, a Mw of ~ 10K and Mn ~ 5K.

1.6. Characterisation

X-ray diffraction spectra were collected on a Phillips diffractometer using Cu Kα radiation, (λ= 0.1505945 nm). Powder samples were ground to a particle size of less than 40 μm. Solid polymer-clay monoliths were typically 14 mm by 14 mm with a 2 mm thickness.

Thermogravimetric analysis, TGA, was done on a Perkin-Elmer 7 Series TGA. Four runs of each sample type were typically run, the results averaged and the uncertainties calculated using standard methods. The samples were heated from 30°C to 600°C at a heating rate of 10°C/ minute in a nitrogen atmosphere. For the differential TGA plots (Figure 10) the uncertainty in the maximum of the mass loss rate ($d(m/m_0)/dT$ (°C^{-1})), in the normalised mass loss rate versus temperature plots, was found to be ± 20 % (± 1 standard deviation). The uncertainty in the temperature at the maximum, in the normalised mass loss rate versus temperature plots, was found to be ± 2 % (± 1 standard deviation).

TGA-FTIR was performed on a TA Instruments (Model TGA-951) TGA coupled to a FTIR gas analyser manufactured by Nicolet Inc. (Model 7-SX). Samples (5 mg - 10 mg) were first flushed with nitrogen at 100 cm^3/minute for 30 minutes and then heated at 10°C/minute from room temperature (25°C) to 1000°C under nitrogen. Evolved gases from the sample were swept through a heated (250°C) capillary transfer line to a gas analysis cell and then to the spectrometer sample compartment by nitrogen purge gas. FTIR spectra were recorded once every 6 s at resolution of 8 cm^{-1}.

Transmission electron microscopy (TEM), the char was broken into small pieces, embedded in an epoxy resin (Epofix), and cured overnight at room temperature. Ultra-thin sections were prepared with a 45° diamond knife at room temperature using a DuPont-Sorvall 6000 ultramicrotome. Thin sections (nominally 50 nm-70 nm) were floated onto water and mounted on 200-mesh carbon-coated copper grids. Bright-field TEM images were obtained with a Philips 400T microscope operating at 120 kV, utilising low-dose techniques.

2. BACKGROUND

The polymer-clay nanocomposites contain montmorillonite clay that has had the sodium ions removed by ion-exchange with various alkyl ammonium salts. This modification renders the usually hydrophilic clay organophilic. A molecular representation of the layered structure of sodium montmorillonite is shown in Figure 2.

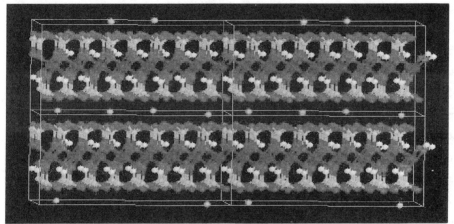

Figure 2. *Molecular representation of sodium montmorillonite, showing two aluminosilicate layers with the Na^+ cations in the interlayer gap or gallery (1.14 nm spacing between layers).*

The nylon-6 clay nanocomposites are synthesised by ring-opening polymerization of ε-caprolactam in the presence of cation exchanged montmorillonite clay[9]. This process creates a polymer layered silicate nanocomposite with either a *delaminated* structure or an *intercalated* structure (see Scheme 1), depending on the clay content. The *intercalated* structure, which forms when the mass fraction of clay is greater than 20 %, is characterized by a well ordered multilayer with spacing between the silicate layers (gallery spacing) of only a few nanometers. The *delaminated* structure, which forms when the mass fraction of clay is less than 20 %, contains the silicate layers individually dispersed in the polymer matrix. The *delaminated* structure is less ordered and the gallery spacing is greater, 10 nm to 100 nm.

3. FLAMMABILITY STUDIES

3.1 Nylon-6 clay nanocomposite

To evaluate the feasibility of controlling polymer flammability via a nanocomposite approach, we examined the flammability properties, using the Cone calorimeter, of nylon-6 clay nanocomposites and compared them to those for pure nylon-6. The Cone calorimeter data, shown in Figure 3, indicates that the peak heat release rate (HRR), an important parameter for predicting fire hazard[10], is reduced by 63 % in a nylon-6 clay-nanocomposite containing a clay mass fraction of only 5 %. These samples were exposed to a heat flux of 35 kW/m^2.

This reduction in HRR is comparable to that found for commercial flame retarded (FR) polymers, but at a lower mass fraction of "additive" than is typical (see Table 1). This system maintains effectiveness even at higher heat fluxes. The peak HRR is reduced by 68 % when the samples are exposed to a heat flux of 50 kW/m^2. From the Cone calorimeter data the nanocomposites were found to have the same heat of combustion as the pure nylon-6. Furthermore, the nanocomposites did not increase the rate of carbon monoxide or soot (measured by the specific extinction area, SEA) formation during the combustion, as some flame retardants do[11,12]. Figure 4 shows the mass loss rate data for nylon-6, and nylon-6 clay-nanocomposite (5 %). The two curves closely resemble the HRR curves, indicating that the reduction in HRR for the nanocomposites is primarily due to the reduced mass loss rate and the resulting lower fuel feed rate to the gas phase. This data indicates that the nano-dispersed clay modifies the condensed phase and not the gas phase processes of the polymer during the combustion.

We did not find the same behavior (i.e., reduced mass loss rate) when we studied the milligram scale thermal properties of these materials. We found that the nylon-6 nanocomposite had the same thermal decomposition behavior when the thermal stability was probed using thermogravimetric analysis combined with Fourier transform infrared spectroscopy, TGA-FTIR. The peak of the derivative of the mass loss versus time curves were both at 460°C ± 10°C. The FTIR data for both samples were identical and corresponded to the spectrum for ε-caprolactam with traces of CO and CO_2. The TGA-residue yields were 0.3 wt. % (± 0.1 wt. %) for nylon-6 and 5.5 mass % (± 1.0 wt. %) for the nanocomposite.

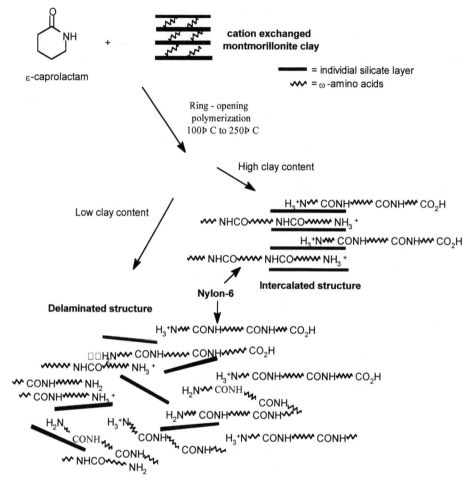

Scheme 1. *Diagram of the process used to prepare polymer layered silicate nanocomposites with either a delaminated structure or an intercalated structure.*

After accounting for the amount of clay present (5 wt. %) in the nanocomposite, these <u>carbonaceous</u> residue yields are essentially the same. This was somewhat surprising since other studies of the thermal reactions in layered organic-clay intercalates, at 400°C, reported formation of carbonaceous-clay residues and other condensation and crosslinking type reaction products[13]. These data indicate that the mechanism of flame retardancy is not via retention of a large fraction of carbonaceous char in the condensed phase.

Visual observations of the combustion experiments, in the Cone calorimeter, reveal different behavior for the nylon-6 clay-nanocomposites, compared to the pure nylon-6. A thin char layer forms, on the top of all the samples, in the first few minutes of exposure, prior to ignition. In the case of pure nylon-6, this char layer fractures into small pieces early in the combustion. The char does not fracture with the nylon-6 clay-nanocomposites. This tougher char layer survives and grows throughout the combustion, yielding a rigid multicellular char-brick with somewhat larger dimensions as compared to the original sample.

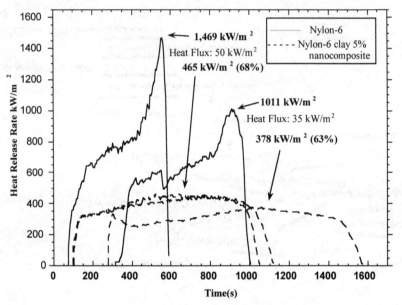

Figure 3. *Heat release rate versus time plot for nylon-6 clay-nanocomposite (5 mass %) and pure nylon-6. The data for the 35 kW/m² and 50 kW/m² flux exposures are shown. Two experiments at the 50 kW/m² flux exposure are included to show the typical reproducibility. The nanocomposite has a 63 % lower HRR at 35 kW/m² and a 68 % lower HRR at 50 kW/m².*

We proposed that the *delaminated* hybrid structure collapses as the nylon-6 decomposes. This forms a reinforced char layer which acts as an insulator and a mass transport barrier, slowing the escape of the volatile products (e.g. ε-caprolactam) generated as the nylon-6 decomposes. Indeed, transmission electron microscopy (TEM) of a section of the char-residue from the combustion of the nylon-6 clay-nanocomposite, shown in Figure 5, reveals a multilayered-silicate structure. X-ray diffraction, XRD, shown in Figure 6, and TEM give an interlayer spacing of 1.3 nm ($2\theta = 6.9°$).

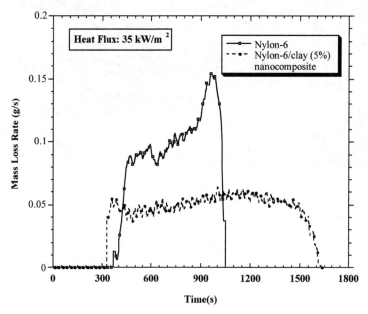

Figure 4. *The mass loss rate data for nylon-6, and nylon-6 clay-nanocomposite (5 %).*

XRD also shows a peak for the 0.33 nm spacing between layers in graphitic-carbon ($2\theta = 26.7°$). The carbon presumably occupies the interlayer space between the silicate layers. The nanocomposite structure of the <u>char</u> appears to enhance the performance of the char, just as the nanocomposite structure enhances the performance of the nylon-6. Since the nanocomposites have excellent barrier properties, an additional effect, due to the low permeability of liquids and gases through the <u>nylon-6</u> nanocomposite char-residue, may also be responsible for the slow transport of volatile fuel to the gas phase[14]. Recent molecular dynamics simulations of the thermal degradation of nano-confined polypropylene support this type of mechanistic hypothesis[15].

4 RADIATIVE GASIFICATION

4.1 Nylon-6 clay nanocomposite

Figure 1 shows the gasification apparatus. The gasification apparatus allows pyrolysis, in a

Figure 5. *TEM of a section of the combustion char from the nylon-6 clay-nanocomposite (5 %) showing the silicate (1 nm thick, dark bands) multilayered structure. This layer may act as an insulator and a mass transport barrier.*

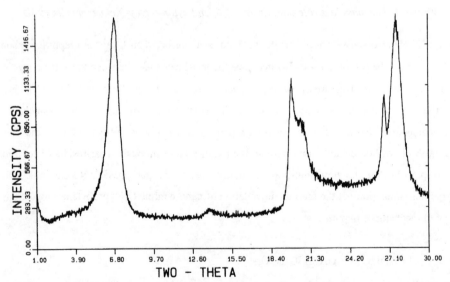

Figure 6. *X-ray diffraction pattern of the char residue from combustion of nylon-6-clay nanocomposite in the Cone calorimeter. This gives an interlayer spacing for the clay in the char residue of 1.3 nm (2θ = 6.9°).*

nitrogen atmosphere, of samples identical to those used in the Cone calorimeter. Figure 7 shows the mass loss rate curves for a set of experiments, carried out in the gasification apparatus, aimed at comparing the pyrolysis behavior of the nanocomposite to that for pure nylon-6. The slope of the mass loss curve for the nanocomposite significantly differs from that for the pure nylon-6 at ~ 180 seconds. The digitized video images, shown in Figure 8, of the pyrolysis experiments, reveal that the nanocomposite (center column) begins to char at the edges of the sample at 120 seconds. The thermocouple on the bottom of the sample shows that the temperature, at 240 seconds, is ~ 50°C to 75°C lower, under the nanocomposite sample, than under the pure nylon-6 sample. These images also show that the formation of a char layer on the top of the nanocomposite sample (center column) occurs at the same time, 180 seconds, that the mass loss rate of the sample slows. Presumably it is this char layer that is responsible for both the lower back-side temperature and the lower mass loss rate for the nylon-6 nanocomposite. Figure 8 also shows the effect of oxygen on the charring process for the nylon-6 nanocomposite. The images in the column on the right are of the pyrolysis of the nylon-6 nanocomposite in a nitrogen - oxygen atmosphere with an oxygen volume fraction of 7.5 %. It appears that the presence of oxygen causes the charring process to occur earlier in the pyrolysis. Even at 60 seconds the entire sample is darkened and at 120 seconds most of the surface is charred.

This is similar to the pre-ignition charring, mentioned above, for the nanocomposite in the Cone experiment. However the mass loss rate for the nylon-6 nanocomposite is not significantly different in the oxygenated pyrolysis atmosphere. The residue yields are also the same (5.0 wt. % ± 0.5 wt. %) within the experimental uncertainty. Comparison of the mass loss data from the Cone calorimeter experiment (Figure 4) to the mass loss data from the gasification experiment (Figure 7), shows that the nanocomposite has a lower mass loss rate in the Cone than in the gasification apparatus. In the Cone calorimeter, the mass loss rate for the nanocomposite, is 62 % lower than the mass loss rate for nylon-6. In the gasification apparatus, the mass loss rate for the nanocomposite, is only 35 % lower than the mass loss rate for nylon-6. Some of this difference is due to the fact that in the Cone a sample with a lower HRR experiences a reduced heat feed back (~ 10 kW/m^2) from the flame and hence is exposed to a lower net flux.

This further reduces the HRR and the mass loss rate. This does not occur in the gasification experiment. Each sample is exposed to the same flux throughout the pyrolysis.

Some of this effect may be due to the effect of oxygen on the mechanism of flame retardancy for the nanocomposite.

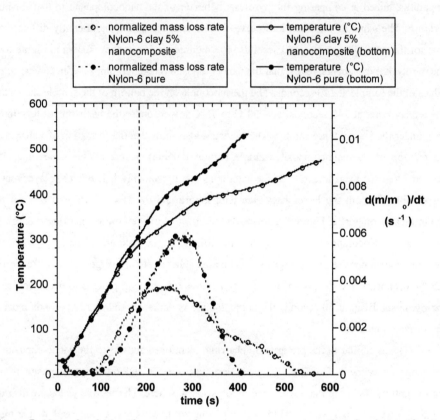

Figure 7. *Normalized Mass loss rate and temperature versus time plots for the gasification experiments for nylon-6 and nylon-6 clay (5%) nanocomposite with N_2 atmosphere. All samples were exposed to a flux of 40 kW/m² in a N_2 atmosphere. The mass loss rate curves begin to differ at 180 seconds when the surface of the nanocomposite sample is partially covered by char.*

Another difference was also observed for the nanocomposites between the Cone results and the gasification results. Although the residue yields for both experiments are the same (~5.5 wt. %), the volume of the residue from the Cone sample is <u>several times greater</u> than that from the gasification sample. Recall that above we mentioned the formation of a rigid multicellular char-brick with the same or somewhat larger dimensions as the original sample.

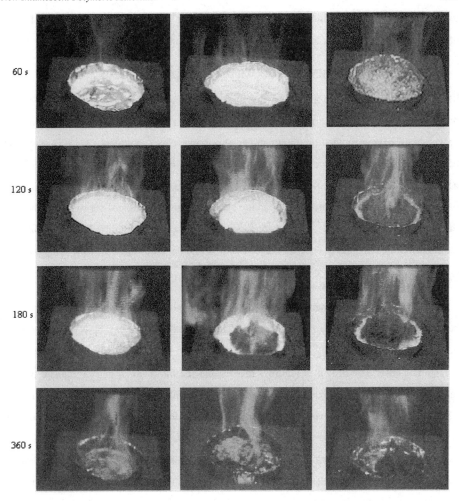

Figure 8. *Images from radiative gasification, at various times, of nylon-6 with N_2 atmosphere (left), nylon-6 clay (5 %) nanocomposite with N_2 atmosphere (center) and nylon-6 clay (5 %) nanocomposite in a N_2/O_2 atmosphere containing a volume fraction of 7.5 % O_2 (right). All samples were exposed to a flux of 40 kW/m^2.*

We have observed similar differences between Cone data and gasification data in another system, the results of which were published previously. In this case the reduction in mass loss rate, due to the presence of silica-additives, in the gasification experiment was only half of that observed in the Cone experiment. However, in this system the residue yield was 15 wt. %

in the Cone and only 9 wt. % in the gasification apparatus[16]. Further study of the possible role of oxygen in the mechanism of flame retardancy for the nanocomposite is underway.

5. VINYL POLYMER NANOCOMPOSITES FLAMMABILITY

Many other polymers have also been prepared as polymer-clay nanocomposites[17]. For the purpose of investigating the general effectiveness of this new approach to flame retarding polymers we have examined the flammability of polymer-clay nanocomposites prepared from polystyrene, PS, and polypropylene-graft-maleic anhydride, PP-g-MA. Like the nylon-6 nanocomposites these polymer-clay nanocomposites are prepared using organic-modified montmorillonite clay. In contrast to the nylon system, these nanocomposites were prepared by the melt blending process[17]. In this process the appropriately modified (compatibilized) montmorillonite clay and the polymer are combined in the melt to form the nanocomposite.

5.1. PS-clay nanocomposite

As described in the experimental section the PS-clay nanocomposite (mass fraction of 3 %) has an *intercalated* structure (see Scheme 1). The inter-gallery spacing, by XRD analysis, is 3.1 nm ($2\theta = 2.7°$).

The HRR data shown in Figure 9 reveals that the *intercalated* nanocomposite, with a mass fraction of only 3 %, has a 45 % lower peak HRR than the *immiscible* PS-clay mixture. The data in Table 1 also shows that the peak HRR for the *intercalated* PS-clay nanocomposite is ~ half of that for *pure* PS as well as for the *immiscible* PS-clay mixture. These data also indicate that, simple mixing the clay into the PS, so that the clay is dispersed like a conventional filler, only to the level of the particle size of the clay (5 μm), is not sufficient to modify the flammability of the PS. Since the HRR of the *immiscible* PS-clay mixture is essentially the same as the HRR for the *pure* PS. It appears that the *intercalated* nano-morphology is necessary for improved flammability. Like the results for nylon-6, the cone data shows that it is the reduced mass loss rate of the *intercalated* PS-clay nanocomposite that is responsible for the improved flammability. Comparison of this Cone data with that for PS flame retarded using decabromodiphenyl oxide, DBDPO, and antimony trioxide, Sb_2O_3, (total mass fraction of 30 %) (Table 1) shows that the *intercalated* nanocomposite results in a similar reduction in the peak HRR for PS, but without as much of an increase in the soot (measured using the specific extinction area, SEA) or CO yields.

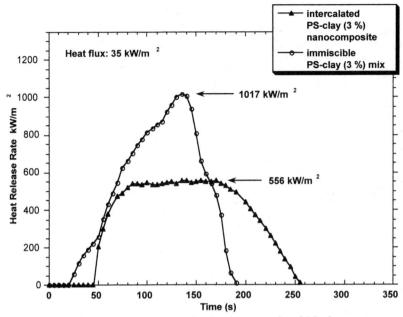

Figure 9. *Heat release rate versus time plot for the intercalated PS-clay nanocomposite and that for the immiscible PS-clay mixture.*

These results were surprising since, the *intercalated* form of the nanocomposite, at such a low clay mass fraction, is inhomogeneous, with *intercalated*-PS-clay domains dispersed in pure PS. The thermal stability of this system is seen in the TGA behavior shown in Figure 10. The TGA of the *intercalated*-PS-clay nanocomposite reveals almost a 50°C increase in the peak for the derivative of the TG plot (DTG) as compared to the peak of the DTG for the *immiscible* PS-clay mixture.

This is in contrast to the nylon-6 clay nanocomposite which showed <u>no</u> change in the peak DTG. Also apparent from the DTG data is the presence of two decomposition peaks for the *intercalated*-PS-clay nanocomposite. The first, at 440°C, coincides with the DTG peak for the *immiscible* PS-clay mixture, the second at 488°C, is most likely from the decomposition of the *intercalated*-PS. Studies, published previously, of the thermal decomposition behavior of several polymer-clay nanocomposites also showed that the *intercalated* form had the highest stability, even greater stability than the *delaminated* nanocomposite[18,19]. We determined that the ratio of the area under the DTG curves for these two decomposition processes is 1.0 to 1.5.

Table 1. *Cone Calorimeter Data*

Sample (structure)	Residue Yield (%) ±0.5	Peak HRR (Δ%) (kW/m²)	Mean HRR (Δ%) (kW/m²)	Mean H_c (MJ/kg)	Total Heat Released (MJ/m²)	Mean Specific Ext. Area (m²/kg)	Mean CO yield (kg/kg)
Nylon-6	1.0	1,011	603	27	413	197	0.01
Nylon-6 clay-nanocomposite 2% *delaminated*	3.0	686 (32%)	390 (35%)	27	406	271	0.01
Nylon-6 clay-nanocomposite 5% *delaminated*	5.7	378 (63%)	304 (50%)	27	397	296	0.02
PS	0	1,118	703	29	102	1,464	0.09
PS clay- mix 3% *immiscible*	3.2	1,080	715	29	96	1,836	0.09
PS clay-nanocomposite 3% *intercalated*	3.7	567 (48%)	444 (38%)	27	89	1,727	0.08
PS w/ DBDPO/Sb₂O₃ 30%	2.6	491 (56%)	318 (54%)	11	38	2,577	0.14
PP-g-MA	0	2,028	861	38	219	756	0.04
PP-g-MA clay nanocomposite 5% *intercalated*	8.0	922 (54%)	651 (24%)	37	179	994	0.05

Hc : Heat of combustion

This indicates that a mass fraction of <u>only 3 % clay</u> in the *intercalated*-PS nanocomposite increases the thermal stability of 60 % of the PS. This ratio was determined using the DTG curve for the *immiscible* PS-clay decomposition to approximate the decomposition contained in the shoulder of the *intercalated*-PS-clay DTG curve. We find that multiplication by the appropriate factor (0.38) gives a ratio of the area under the DTG curves for these two decomposition processes, *immiscible* and *intercalated* respectively, of 1.0 to 1.5. This is in agreement with studies on other polymer-clay nanocomposites where the mass fraction of polymer directly effected by the presence of the clay was found to be ~ 60 %[20].

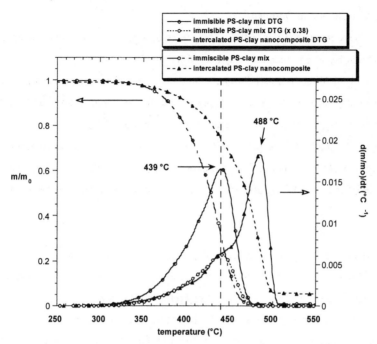

Figure 10. *Normalized TGA plots and DTG plots for immiscible PS-clay (3 %) mixture and intercalated PS-clay (3 %) nanocomposite. The intercalated-PS clay clearly has a greater thermal stability than the immiscible PS-clay mixture.*

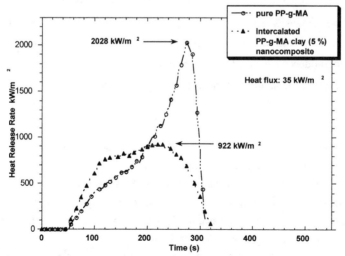

Figure 11. *Heat release rate versus time plot for the intercalated PP-g-MA-clay nanocomposite and that for the pure PP-g-MA.*

5.2. PP-g-MA-clay-nanocomposite

We have also examined the flammability properties of a polypropylene-clay nanocomposite. The PP-g-MA-clay-nanocomposite (clay mass fraction of 5 %) was prepared by melt blending PP-g-MA with bis(dimethyl)bis(tallow)ammonium-exchanged montmorillonite. In this case the melt blending also yielded a nanocomposite with the *intercalated* structure (see Scheme 1). The inter-gallery spacing, by XRD analysis, is 3.6 nm ($2\theta = 2.4°$).

Figure 11 shows that the peak HRR for the *intercalated* PP-g-MA-clay-nanocomposite is 54 % lower than that for pure PP-g-MA. Like the nylon-6 and PS clay nanocomposites, the reduction in HRR for the PP nanocomposite is a direct result of a reduced mass loss rate.

6 SUMMARY

The Cone data in Table 1 shows that the FR performance of the PS-clay nanocomposite and the PP-g-MA-clay nanocomposite is very similar to that for the nylon-6 clay nanocomposite. The Cone data also shows that the FR mechanism for the PS-clay nanocomposite and the PP-g-MA-clay nanocomposite is very similar to that for the nylon-6 clay nanocomposite. That is, the heat of combustion, and rates of soot and CO formation are unchanged, within the experimental uncertainty, by the presence of the clay in the nylon-6, PS and PP nanocomposites. Furthermore, the residue yields from combustion, for all the polymer, are not high enough to account for the reduced flammability due to retention of carbon in the condensed phase. A condensed phase mechanism where a reinforced char layer forms, which acts as an insulator and a mass transport barrier, slowing the escape of the volatile decomposition products generated as the polymer decomposes appears to be one likely explanation for the excellent FR performance of the clay-nanocomposites.

References

1. G. L. Nelson, *"Fire and Polymers II"*, *ACS Symposium Series* 599, ed. G. L. Nelson, American Chemical Society, Washington, DC, pp. 1-28 (1995).
2. S. Levchik, G. Camino, L. Costa, and G. Levchik, *Fire and Materials*, **19**, 1 (1995).
3. A. Hochberg, *"Proceedings of the Fall FRCA Meeting"*, Naples, FL., 159 (1996).
4. Y. Kojima, A. Usuki, M. Kawasumi, A. Okada, Y. Fukushima, T. Kurauchi, and O. Kamigaito, *J. Mater. Res.*, **8**, 1185 (1993).
5. J. W. Gilman, T. Kashiwagi, J. D. Lichtenhan, *SAMPE Journal*, **33(4)**, 40 (1997).
6. V. Babrauskas, R. Peacock, *Fire Safety Journal*, **18**, 255 (1992).

7. V. Babrauskas, *Fire and Materials*, **19**, 243 (1995).

8. Certain commercial equipment, instruments, materials, services or companies are identified in this paper in order to specify adequately the experimental procedure. This in no way implies endorsement or recommendation by NIST.

9. A. Usuki, Y. Kojima, M. Kawasumi, A. Okada, Y. Fukushima, T. Kurauchi, and O. Kamigaito, *J. Mater. Res.*, **8**, 1179 (1993).

10. V. Babrauskas, R. Peacock, *Fire Safety Journal*, **18**, 255 (1992).

11. A. Grand, *SAMPE Journal*, **33(4)**, 47 (1997).

12. R. Harris, Jr., V. Babrauskas, B. C. Levin and M. Paabo, NISTR 4649, 1991.

13. J. M. Thomas, *"Intercalation Chemistry,* Chapter 3"*, Academic Press Inc., London, 1982, p. 55.

14. E. Giannelis, and P. Messersmith, *J. Polym. Sci. A: Polym. Chem.*, **33**, 1047 (1995).

15. M. Nyden, J. W. Gilman, *"Computational and Theoretical Polymer Science"*, in press, 1998.

16. J. Gilman, S. Ritchie, T. Kashiwagi, and S. Lomakin, *Fire and Materials*, **21**, 23 (1997).

18. E. Giannelis, *Adv. Mater.*, **8**, 29 (1996).

19. J. Lee, T. Takekoshi and E. Gannelis, *Mat. Res. Soc. Symp. Proc.*, **457**, p. 513, (1997).

20. J. Lee and E. Giannelis, *Polymer Preprints*, **38**, 688, (1997).

21. S. Burnside, *Ph. D. Thesis*, Cornell University, 1997.

Acknowledgements

The authors would like to thank the Federal Aviation Administration for partial funding of this work, through Interagency Agreement DTFA0003-92-Z-0018. We would also like to thank Mr. Michael Smith for Cone Calorimeter analysis, Lori Brassel for TGA analysis, Dr. Catheryn Jackson and Dr. Henri Chanzy for TEM analysis of the char samples, Dr. Marc Nyden for the montmorillonite structure. We would also like to express our gratitude to Dow Chemical Co. for PS samples (STYRON 6127), Ube America Inc. for nylon-6 clay nanocomposite samples and Southern Clay Products for the organic-modified clays, Closite 15A and 3A.

SYNERGY IN INTUMESCENCE: OVERVIEW OF THE USE OF ZEOLITES

S. Bourbigot and M. Le Bras

Laboratoire de Physicochimie des Solides, E.N.S.C.L., U.S.T. Lille,
BP 108, F-59652 Villeneuve d'Ascq Cedex, France
E-mail: serge.bourbigot@ensc-lille.fr

1. INTRODUCTION

To intumesce was used by the Tragedian John Webster (1580-1624) in Elizabethan days with two meanings: "to grow and to increase in volume against the heat" or "to show a blowing up effect by bubbling". The definition of Webster allows the description of the behaviour of an intumescent coating or of an intumescent material which heated beyond a critical temperature, begins to swell and then blows up. The result of this process is a foamed cellular charred layer on the surface, which protects the underlying material from the action of the heat flux or the flame [1-2]. The proposed mechanism is based on a charred layer acting as a physical barrier which slows down heat and mass transfer between gas and condensed phase.

Intumescent technology has found its place in Polymer Science as a method of providing flame retardancy to polymeric materials[3-4]. The association of an ammonium polyphosphate *(APP)* and pentaerythritol *(PER)* has been shown to be an efficient fire-retardant (FR) intumescent system for paints and varnishes[2] and, more recently, for polyolefinic materials[5-6]. This provides good fire-retarding properties in FR polypropylene[5-7] and polyethylene-based formulations[8-9]. Nevertheless, we have recently shown that the combination of zeolites with the APP/PER or the diammonium pyrophosphate/PER systems leads to a high degree of improvement of the fire-retardant properties (Table 1) within different polymeric matrices[8].

This paper reviews the role played by zeolites in the classical intumescent APP/PER system in different polyolefines in the sense first, of flame performance using traditional fire-testing (LOI[10], UL-94[11] and oxygen consumption calorimetry using a cone calorimeter[12]). The thermal behaviour of some representative formulations was then modelled to understand the part of zeolites in the degradation of the materials. Spectroscopic tools for the characterisation of the carbonaceous materials are summarised. Finally, the different steps of the intumescence process are discussed in order to propose a mechanism of action of the zeolite in the enhancement of the performance of an intumescent material.

Table 1. *FR performance of intumescent formulations with and without zeolite.*

System	LOI (volume %)	UL-94 rating	System	LOI (volume %)	UL-94 rating
PP-APP/PER	30	V0	PP-APP/PER/13X	45	V0
LDPE-APP/PER	24	V0	LDPE-APP/PER/4A	26	V0
PP-PY/PER	32	V0	PP-PY/PER/13X	52	V0
PS-APP/PER	29	V0	PS-APP/PER/4A	43	V0
LRAM3.5-APP/PER	29	V0	LRAM3.5-APP/PER/4A	39	V0

The additives' loading is kept at a constant equalling 30 wt.-% and the synergistic effect exhibits a maximum at 1 or 1.5 wt.-% zeolite loading[13].
Glossary: PP: isotactic polypropylene, LDPE: low density polyethylene, PS: polystyrene and LRAM 3.5: ethylene-butylacrylate-maleic anhydride terpolymer (Lotader P3200 from Elf-Atochem).

2. EFFECT OF THE ZEOLITE

There are different kinds of zeolites and their effects on the fire performances have to be mentioned. Zeolites are tectosilicates characterised by a three-dimensional framework of AlO_4 and SiO_4 tetrahedra[14]. The framework contains channels and interconnected voids which are occupied by the cations and water molecules. The negative charges due to AlO_4 is balanced by exchangeable cations. The size of the voids or the channels is approximately the size of the usual organic molecules. The chemical ideal formula is $M_{x/n}\left[(AlO_2)_x,(SiO_2)_y\right]zH_2O$. The part in brackets is the framework of the zeolite with a ratio $y/x \geq 1$ and M^{n+} is the balancing cation. Study of the fire behaviour of the materials requires therefore consideration of the physical and chemical characteristics of the zeolites.

Comparison of several zeolites used in intumescent formulations has shown[15] that there is no relation between the type of exchangeable cation or the aperture size of the zeolites and the FR performances (Figure 1) and that very high Si/Al ratio leads to the decrease of the FR performances (Figure 2). Moreover, comparison of FR performances with systems containing other aluminosilicate compounds, such as clays or silicon-based systems, shows that the use of an aluminosilicate with a zeolite structure gives the best FR performance .

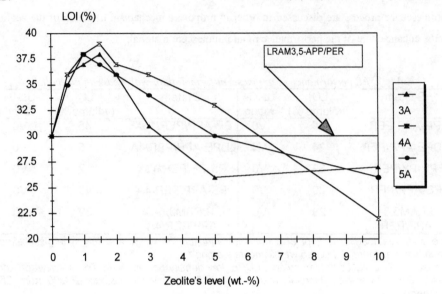

Figure 1. *LOI values vs. A zeolites' loading of formulations LRAM3.5-APP/PER-Zeolites (additives loading remaining constant equalling 30 wt.-%).*
Aperture sizes: 0.3 (3A), 0.35 (4A) and 0.42 nm (5A); Si/Al=1 at/at

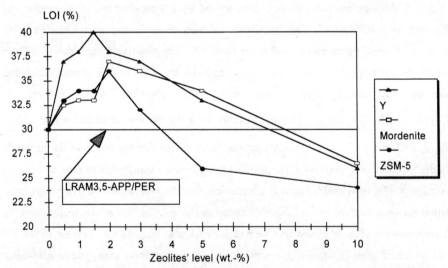

Figure 2. *LOI values vs. Y, mordenite and ZSM-5 zeolites' loading of formulations LRAM3.5-APP/PER-Zeolites (additives' loading remaining constant equalling 30 wt.-%).*
Aperture sizes: 1 (Y), 0.7 (Mordenite) and 0.55 nm (ZSM-5);
Si/Al=2.4 (Y), 5 (Mordenite) and 140 at/at (ZSM-5)

3. EFFECT OF THE POLYMERIC MATRIX

The zeolite 4A (NaA zeolite) added in the intumescent fire retardant (FR) APP/PER system leads to a very significant synergistic FR effect in polyethylene-based formulations[8]. Nevertheless the effect of the polymeric matrix on the fire-retarding performance, for a given formulation, can be very large (Figure 3 shows a typical example in a copolymer ethylene-vinyl acetate (EVA) versus the vinyl acetate amount in the polymer). It was shown in the case of some ethylenic copolymers and terpolymers that the performance depends on the chemical nature and the amount of the co-monomer in the polymer[8]. It was proposed that the possibility of acidity reinforcement by the functionalised co-monomers may explain the increase of the performance. This effect is highest when the thermal degradation of the polymer leads to the formation of carboxylic acidic species which may play a part in the condensed phase process. A typical matrix leading to this important FR effect is the ethylene-butylacrylate-maleic anhydride terpolymer (LRAM3.5). Note that, in the following, we will consider only LRAM3.5 with zeolite 4A because the synergistic effect is the highest.

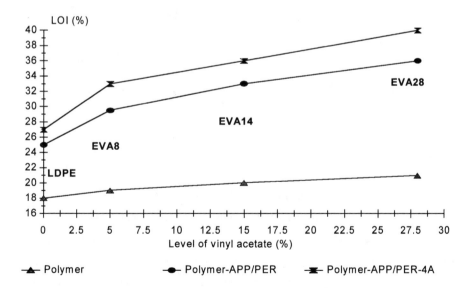

Figure 3. *LOI values versus the level of vinyl acetate and versus the formulation (EVAx are EVA polymers with x% vinyl acetate and LDPE is low density polyethylene).*

4. THERMAL BEHAVIOUR AND MODELLING

The burning process depends on chemical processes occurring in the bulk of the polymeric materials (condensed phase) and above the material surface (gas phase) and also on the physical process of heat transfer from the gas phase back to the condensed phase. All three of these depend primarily on the thermal degradation reactions in the condensed phase, which generate volatile products and which may also change the physical character of the surface of the sample. It is therefore necessary to study the thermal behaviour of the formulations.

TG analyses of the LRAM3.5-based formulations containing 1.5 wt.-% of the zeolite 4A (Figure 4) show that the intumescent process leads to the protection of the polymeric matrix within the temperature range 250-450°C and that the presence of the zeolite provides a residue comparatively stable at T>550°C.

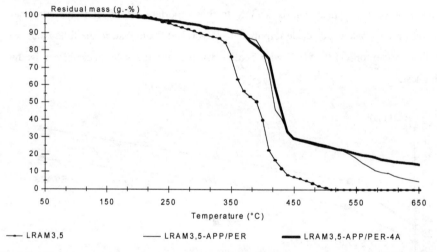

Figure 4. *TG curves of the LRAM3.5-based formulations(heating rate = 3°C/min, under air).*

In the 250-450°C range an intumescent material is formed which degrades in the 450-550°C range to provide a carbonaceous residue without any expanded character. This residue degrades at temperature higher than 550°C. In the case of the formulation with zeolite the residue forms in the 450-500°C range. Its comparatively high thermal stability at temperatures higher than 550°C may be related to the formation of a stable carbonaceous material and to oxides arising from the degradation of 4A.

The invariant kinetic parameter (IKP) method proposed by Lesnikovich et al.[16] and then applied to fire-retarded polymers in this laboratory[17] allows the modelling of the degradation velocity of the polymeric materials and therefore of the "fuel flow" feeding the flame. Figure 5 shows the degradation rates of the formulations versus the temperature and the conversion degree of the degradation reaction (α).

The degradation rate of the virgin polymer is very high compared to the intumescent ones and is maximum at the beginning of the degradation ($0<\alpha<0.1$). The degradation rates, in the case of the fire-retarded polymers, increase up to $\alpha=0.7$ (system without zeolite) or $\alpha=0.5$ (system with zeolite) and then decrease. Between $0<\alpha<0.5$, the curves are very close and at $\alpha>0.5$, the degradation rate of the formulation containing a zeolite becomes comparatively very low. This demonstrates therefore that the zeolite slows down the degradation process.

5. FIRE BEHAVIOUR

In order to explain the part played by the zeolite in the formation of the intumescent coating during the combustion process of the formulations, the oxygen consumption calorimetry (cone calorimeter) is a powerful tool[19].

The curves of the variations of the rate of heat release (RHR) versus time (typical examples in Figure 6) show a significant decrease of the RHR maximum values of the flame retardant polymers in comparison to the pure polymer. In the case of the intumescent formulations, the rate of heat release (RHR) decreases strongly after the ignition.

It is to be noted that the RHR behaviours of the flame retardant polymers are similar. The three maxima curves of the intumescent materials may be explained by three successive steps of burning :

- first, the formulations degrade and intumescent shields form,
- then, these coatings degrade and, consequently, the residual materials degrade and form a new intumescent coating,
- finally, the whole of the residual materials degrades.

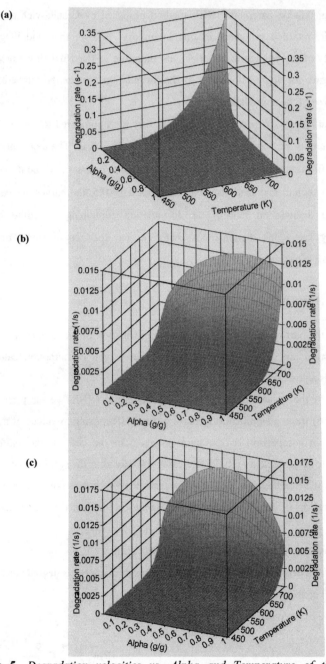

Figure 5. *Degradation velocities vs. Alpha and Temperature of the LRAM3.5-based formulations; (a) LRAM3.5, (b) LRAM3.5-APP/PER, (c) LRAM3.5-APP/PER-4A*[18].

The maximum RHR values are different. These differences become very significant for high time values which correspond to the highest temperature of the samples. As an example at t=600s, the RHR value is only 150 kW/m² for the system with zeolite whereas it is 300 kW/m² for the system without zeolite.

The toxicity of gaseous products evolving during combustion is an essential parameter which can be estimated using the cone calorimeter. The formulation with zeolite yields less smoke than the formulation without zeolite and than the virgin polymer during the first 250s (Figure 7) which is very important for safety.

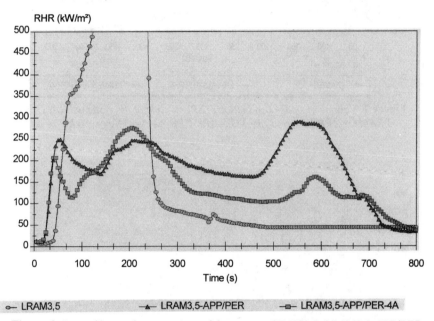

Figure 6. *Rate of heat release vs. time of the systems LRAM3.5, LRAM3.5-APP/PER and LRAM3.5-APP/PER-4A (heat flux = 50 kW/m²).*

Moreover, the zeolite reduces the evolved amount of CO and CO_2 during combustion (Figure 8 and 9) what reduce the toxicity of the materials and the oxygen depletion in fire. These results imply that the characteristics of the protective coatings formed are very different and confirm that the zeolite changes the degradation of the formulations in the high temperatures range. It may be proposed that the presence of the zeolite leads to a lower rate of the evolution of the resulting fuel and/or to a different composition of this fuel.

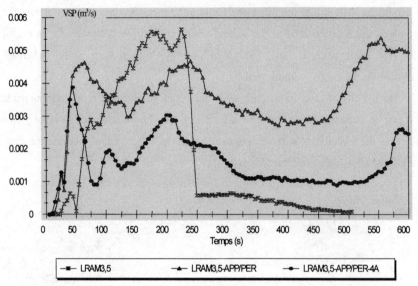

Figure 7. *Volume of smoke production (VSP) vs. time of the systems LRAM3.5, LRAM3.5-APP/PER and LRAM3.5-APP/PER-4A (heat flux = 50 kW/m²).*

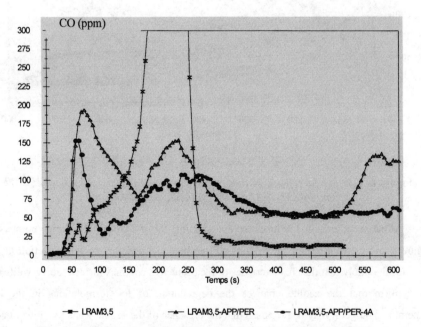

Figure 8. *CO vs. time of the systems LRAM3.5, LRAM3.5-APP/PER and LRAM3.5-APP/PER-4A (heat flux = 50 kW/m²).*

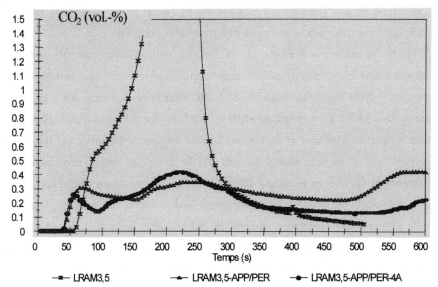

Figure 9. *CO₂ vs. time of the systems LRAM3.5, LRAM3.5-APP/PER and LRAM3.5-APP/PER-4A (heat flux = 50 kW/m²).*

6. CHEMISTRY OF THE INTUMESCENCE

The study of the thermal behaviour of the formulations has shown that the development of the intumescence is a process having several steps. In order to elucidate the action mechanisms of the zeolite, the spectroscopic characterisation of these steps is needed.

Our methodology consists of heat-treatment of the formulations at characteristic temperatures determined by TGA. The materials were then examined in the solid state rather than by solvent extraction. Indeed we consider that the extraction by a solvent may modify the sample. Further it is impossible to render completely soluble an intumescent material which has a carbonaceous structure and therefore, any resulting analyses would not be representative of the material. Nevertheless it is suggested by E. D. Weil et al.[20] the possibility to wash the char with cold water and then pH titration, to characterise the outer surface of the char.

To characterise the carbonaceous materials, several spectroscopic tools may be used. FTIR of black bodies allowed us to show the formation of a phosphocarbonaceous structure in the intumescent shield which is confirmed by NMR of phosphorus in the solid state[13,21]. Note that no P-C bonds was characterised as suggested by Lecher et al.[22] when they studied the thermolysis of some aromatic compounds with phosphoric anhydride.

NMR of carbon and proton shows that the carbonaceous materials consists of polyaromatic species. Moreover, micro-Raman spectroscopy reveals a local structuring of the "carbon"[13, 23]. Note that in the case of ^{13}C NMR, Gilman et al.[24] proposed a quantitative approach for studying the combustion chars to determine the role of non-protonated carbon present in the char. This kind of carbon may control the char performance and flammability since this type of carbons should be part of the crosslinks which may improve the performance of the char by increasing thermal stability and enhancing mechanical strength. Further XPS identifies the nitrogenated compounds and shows the formation of pyridinic and pyrrolic nitrogen and quaternary nitrogen in polyaromatic groups[25-26]. No P-N bonded structure was characterised in our case as previously proposed by Levchik et al.[27] in the case of the formulation APP-Polyamide-6. Finally, NMR of aluminium and silicon allows characterisation of the species formed from reactions between the zeolite and the phosphate[15,21].

The spectroscopic studies of additive systems and of polymer-additive formulations allow us to propose a reaction scheme of the carbonisation process[28]. When the temperature increases, the intumescent structure develops. At 280°C it is formed by stacks of polyaromatic species linked principally by phosphohydrocarbonated bridges. These bridges provide the dynamic properties of interest to the structure which is then able to accommodate the stresses. At this time, the structures developed from the formulations with and without zeolite are distinguished by the organisation of the carbon, the zeolite slowing down its organisation process.

The condensation of aromatic species and the decrease of phosphocarbonated species by scission of the P-O-C bindings are then observed when the temperature increases (T > 280°C). Consequently the enlargement of the size of the polyaromatic stacks increases drastically the viscosity of the material and so, leads to the loss of the properties of interest of the coating. The addition of the zeolite in the formulation keeps a large number of polyethylenic links in the structure stabilised by «organic aluminosilicophosphate complexes», and reduces the scission of the P-O-C bonds and therefore the increase of the size of the polyaromatic stacks[21]. Moreover, the pyridinic nitrogen is observed at all temperature in the case of zeolite and only up to 350°C in the case of the system without zeolite. The zeolite leads to the improvement of the «mechanical properties» and it was proposed that the delay of the condensation of the polyaromatic network allowing the retention of pyridinic-N may participate in the improvement of the mechanical properties of interest and therefore to the improvement of the fire proofing properties of the material[26].

7. GENERAL DISCUSSION

The intumescent coating synthesised from the intumescent formulations containing a zeolite has fire-retardant properties much higher than those of the system without zeolite. The study of the carbonaceous coatings obtained from intumescent formulations permits us to show clearly the part played by the zeolite in the structures. Intumescent systems develop a phosphocarbonaceous structure which is thermally stabilised by the zeolite. This latter allows to block polyethylenic links under the form of organic phosphates and/or aluminophosphates and limits therefore the depolymerisation, i.e., the evolution of small flammable molecules able to feed the fire in fuel.

Moreover, it is shown that the zeolite allows the formation of a more «coherent» structure and therefore orients differently the structure of the materials. The formation of a « coherent » macromolecular network and the participation of polyethylenic links seem therefore to be favourable for obtaining the fire-retardancy of interest. Indeed the development in the intumescent shield of a structure consisting in polyaromatic species creates a rigid material (case of the LRAM3.5-APP/PER formulation). On the other hand, the stabilisation of polyethylenic links in the intumescent structure able to bridge polyaromatic species via alumino- and/or silicophosphate groups and/or because of the paramagnetic character of the intumescent structure may provide the mechanical properties of interest of the intumescent coating. They provide flexibility to the carbonaceous shield which allows this shield under the conditions of a fire, a retarded creation and propagation of cracks in which oxygen diffuse to the polymeric matrix and through which small molecules may be released as fuel.

8. CONCLUSION

The adduct of zeolite improves strongly the fire retardant properties (performance and smoke toxicity) of the materials. The comparison of several zeolites show that very high Si/Al ratio leads to the decrease of the FR performance. Moreover, the efficiency of the zeolite depends strongly on the polymeric matrix. The influence of the chemical nature and the amount of the comonomer on the fire retardant properties is clearly demonstrated.

Kinetic study using the invariant kinetic parameter method, shows that the addition of zeolites slows down the degradation rates of the intumescent formulations.

The chemical and structural characterisation of the carbonaceous materials arising from intumescent formulations demonstrates that interactions occur between the polymer and the additives. In the case of the classical APP/PER system, the protection arises principally from the additives which develop a thermal barrier between the flame and the material. The zeolite plays an essential role because it allows the formation of structures stabilising the polymer which participates to the formation of the intumescent shield and therefore, to its own protection. Moreover, the properties of the protective material obtained from the formulations depend on the « quality » of the coating obtained from the additives.

The very large improvement of the fire proofing properties in polymeric materials containing the zeolite is explained: the key to form an very efficient intumescent shield is the ability of the additives to synthesise a structure which can trap the polymer in order to decrease first, the fuel rate feeding the flame and second to provide the dynamic properties of interest.

References

1. H.L. Tramm, US 2. 106938, assigned to Ruhrchemie Aktiengesellschaft, 1938.
2. H.L. Vandersall, *J. Fire & Flammability,* 1971, **2**, 97.
3. G. Montaudo, E. Scamporino and D. Vitalini, *J. Polym. Sci., Polym. Chem.*, 1983, **21**, 3361.
4. G. Camino, « *Actes du 1er Colloque Francophone sur l'Ignifugation des Polymères* », J. Martel ed., IUT Pub., Saint-Denis (France), 1985, pp. 36.
5. G. Camino, L. Costa and L. Trossarelli, *Polym. Deg. & Stab.*, 1985, **12**, 213.
6. R. Delobel, M. Le Bras, N. Ouassou and F. Alistiqsa, *J. Fire Sci.,* 1990, **8**(2), 85.
7. S. Bourbigot, M. Le Bras and R. Delobel, *J. Fire Sci.*, 1995, **13**(1-2),3.
8. S. Bourbigot, M. Le Bras, R. Delobel, P. Bréant and J-M. Trémillon, *Polym. Deg. & Stab.*, 1996, **54**, 275.
9 M. Le Bras, S. Bourbigot, Y. Le Tallec and J. Laureyns, *Polym. Deg. & Stab.*, 1997, **56**, 11.

10. *"Standard test method for measuring the minimum oxygen concentration to support candle-like combustion of plastics."*, ASTM D2863/77, Philadelphia (1977).

11. *"Tests for flammability of plastic materials for part in devices and appliances"*, Underwriters Laboratories, ANSI/ASTM D635-77, Northbrook, (1977).

12. V. Babrauskas, *"Development of cone calorimeter – A bench scale rate of heat release based on oxygen consumption"*, NBS-IR 82-2611, US Natl. Bur. Stand. (1982).

13. S. Bourbigot, M. Le Bras, R. Delobel, R. Décressain and J.-P. Amoureux, *J. Chem. Soc., Faraday Trans.*, 1996, **92**(1),149.

14. J. V. Smith, *"Origin and Structure of Zeolites"*, in: *"Zeolite Chemistry and Catalysis"*, J. A. Rabo Ed., ACS Monograph 171, Washington DC (1976).

15. S. Bourbigot, M. Le Bras, P. Bréant, J.-M. Trémillon and R. Delobel, *Fire & Materials*, 1996, **20**, 145.

16. A. I. Lesnikovich, S. V. Levchik and V. G. Guslev, *Thermochim. Acta*, 1984, **77**, 357.

17. S. Bourbigot, R. Delobel, M. Le Bras and D. Normand, *J. Chim. Phys.*, 1993, **90**, 909.

18. S. Bourbigot, M. Le Bras, P. Bréant, J.-M. Trémillon and R. Delobel, in: *"Recent Advances in Flame Retardancy of Polymeric Materials"*,. M. Lewin Ed., BCC Pub., Norwalk (1996).

19. V. Babrauskas and S.J. Grayson, in : *"Heat Release in Fires"*, S. J. Grayson Ed., Chapman & Hall Pub., London (1995).

20. W. Zhu, E. D. Weil and S. Mukhopadhyay, *J. Appl. Polym. Sci.*, 1996, **62**, 2267.

21. S. Bourbigot, M. Le Bras, R. Delobel and J-M. Trémillon, *J. Chem. Soc., Faraday Trans.*, 1996, **92**(18), 3435.

22. H. Z. Lecher, T. H. Chao, K. C. Whitehouse and R. A. Greenwood, *J. Am. Chem. Soc.*, 1954, **76**, 1045.

23. S. Bourbigot, R. Delobel and M. Le Bras, *Carbon*, 1993, **31**(8), 1219.

24. J. W. Gilman, S. Lomakin, T. Kashiwagi, D. VanderHart and V. Nagy, in *"Abstract of San Francisco, California Meeting – Polymer Preprints ACS Symposium Series"*, Volume 38(1), American Chemical Society, Washington DC (1997).

25. S. Bourbigot, M. Le Bras, L. Gengembre and R. Delobel, *Appl. Surf. Sci.*, 1994, **81**, 283.

26. S. Bourbigot, M. Le Bras, L. Gengembre and R. Delobel, *Appl. Surf. Sci.*, 1997, **120**, 15.

27. S.V. Levchik, G. Camino, L. Costa and G.F. Levchik, *Fire and Mater.*, 1995, **19**, 1.

28. S. Bourbigot, *Doctoral Dissertation*, University of Lille, 1993.

Acknowledgement

The Authors thank Professor E. D. Weil (Brooklyn Polytechnic Institute) for helpful discussions and interesting comments about this paper.

INFLUENCE OF HIGH ENERGY RADIATION ON THE THERMAL STABILITY OF POLYAMIDE-6

A. I. Balabanovich, W. Schnabel,

Hahn-Meitner-Institut,
Glienicker Strasse 100, D14109 Berlin-Wansee, Germany

G. F. Levchik, S. V. Levchik,

Institute for Physical Chemical Problems, Byelorussian State University,
Leningradskaya 14, 220080 Minsk, Belarus

C. A. Wilkie

Department of Chemistry, Marquette University,
P. O. Box 1881, Milwaukee, WI 53201 USA

1. INTRODUCTION

The thermal decomposition of polyamide-6 (PA-6) at temperatures exceeding 350°C, results in volatile products that burn after ignition. The decomposition products consist mainly of monomer, caprolactam, and a lesser amount of cyclic oligomers[1]. Upon heating PA-6 is totally converted to volatiles provided the temperature is sufficiently high and the heating time is sufficiently long.

Much work has been devoted to improving the thermal stability of aliphatic polyamides and to make them flame-resistant. Additives such as melamine[2] (MA), phospham[3] (PH), and ammonium polyphosphate[4] (APP), have been used; significant improvement is only attainable at a rather high additive content on the order of 20 wt. %. Therefore, it appeared interesting to determine if chemical modification of PA-6 could bring about improvement; an elegant procedure for the chemical modification of polymers is the utilisation of high energy radiation. High energy radiation has the opportunity to lead to the formation of unsaturated groups and intermolecular cross-links which, in turn, can have an effect on char generation. It is known[5] that the formation of char is frequently accompanied by a reduction in both the yield and the rate of formation of volatile products[6-7]. This has been ascribed to the char acting as a physical barrier against heat transmission and oxygen diffusion[8]; char can also act as scavenger of free radicals that may otherwise initiate chain reactions yielding volatile products[9].

The potential of high energy radiation-induced chemical modification to alter the flammability of polymers has been little examined to date. Nyden et al.[10-11] have reported that irradiation of polyethylene with γ-rays causes a significant increase in the time to ignition and this implies an improvement in the flame retardancy of the polymer. Irradiation of polyethylene affects char formation upon burning; no char is formed in the case of unirradiated polyethylene but at an absorbed dose of 1 MGy, the char yield is 2.7 %. Similarly chemically cross-linked poly(methyl methacrylate), PMMA, forms char while linear PMMA gives no char[10,12]. These findings and molecular dynamic modelling prompted Nyden et al., to postulate that cross-linked polymers tend to cross-link further when burned thus eventually forming a thermally stable char[10]. Based upon these results it appeared worthwhile to perform similar studies with aliphatic polyamides which have been reported to undergo intermolecular cross-linking and become partially insoluble upon irradiation with high energy radiation, such as γ-rays or electron beam (EB) radiation[13-14].

PA-6, upon irradiation with ^{60}Co-γ-rays under nitrogen, loses crystallinity and undergoes simultaneous cross-linking and main chain scission with $G(X) = 0.67$ and $G(S) = 0.68$, respectively[14]; additional radiolysis products are -CON=CH- groups and gaseous compounds (H_2, CO, CO_2, CH_4)[13]. In the work reported herein, samples of virgin PA-6 as well as PA-6 containing various additives have been irradiated with ^{60}Co-γ-rays or EB radiation and the products have been characterized for the extent of cross-linking and thermal properties.

2. EXPERIMENTAL

2.1. Materials

Two commercial polyamide-6 samples were used: PA-6a, a product of Khimvolokno, Grodno ($M_n = 3.5 \times 10^4$ g mol^{-1}) and PA-6b a product of Du Pont (Zytel, RV=49.5, according to ASTM D 789). The following compounds were used as additives: melamine, MA (Chemie Linz), phospham, PH, synthesised according to a prescription reported in the literature[15], ammonium polyphosphate, APP (Exolit 422, Hoechst), triallyl cyanurate, TAC (Aldrich), melamine cyanurate, MA-CY (DMS, Netherlands) and melamine pyrophosphate, MA-PR (DMS, Netherlands). The structures of these various additives are shown below.

2.2. Sample Preparation and Irradiation.

Polyamide/additive mixtures were prepared under nitrogen in a closed mixer operating at 60 rpm at 240°C. Specimens (10.0x1.0x0.3 cm) were cut from slabs formed from the melt. The specimens were irradiated in vacuum or under air either with γ-rays emitted by a ^{60}Co-source at an absorbed dose rate Dr_{abs} = 3 kGy/h or with 16 MeV electrons generated by a linear accelerator at Dr_{abs} = 0.15 MGy/min.

TAC	MA	MA-CY

MA-PR	APP PH

2.3. Characterisation of Irradiated Samples.

The gel content of the irradiated samples was determined by extraction for 10-12 hours in a Soxhlet apparatus with m-cresol boiling at reduced pressure at about 100°C. The gel content is the ratio of the mass of the sample after extraction to the mass before extraction. The sample was thoroughly dried to remove all solvent before the mass was determined. The swelling ratio of the irradiated sample was determined from the ratio of the mass of the swollen sample obtained immediately after extraction to that of the thoroughly dried sample. Thermogravimetric analysis and differential scanning calorimetry were carried out on the samples; both TGA and DSC were performed using a Mettler 3000 Thermoanalyzer at a heating rate of 10°C per minute under both an inert atmosphere and an air atmosphere. The oxygen indices of the samples were determined using a Stanton-Redcroft apparatus following the standard ASTM procedure, ASTM 2863-77. Infrared spectra were obtained on a Perkin-Elmer 2000 apparatus.

3. RESULTS AND DISCUSSION

3.1. Cross-linking.

Electron beam irradiation of the polyamide samples is expected to lead to intermolecular cross-linking of the samples, eventually resulting in the formation of an insoluble gel provided that the absorbed dose is larger than the minimum which is required. The gel content is a measure of the total amount of material which has cross-linked while the swelling ratio is a measure of the number of cross-links which have been produced. Since polyamide-6 undergoes simultaneous intermolecular cross-linking and main chain scission[14], the gel content is expected to increase with increasing absorbed dose and to level off at a limiting value which is significantly less than 100 %[16]. One may view a non-cross-linked polymer as having an essentially infinite swelling ratio because the polymer dissolves. As the extent of cross-linking increases, the possibility of swelling must decrease so a decrease in the swelling ratio will indicate an increased number of cross-links formed. The data for gel content and swelling ratio as a function of applied dose is shown in Table 1.

Table 1. *Gel Content and Swelling Ratio of γ-Ray Cross-linked Polyamide-6a Samples.*

Additive		Absorbed Dose,	Swelling Ratio	Gel Content
Name	Content, wt. %	MGy		wt. %
none	-	0.15, vacuum	1830	70
none	-	4.5, vacuum	1160	73
none	-	4.5, air	1240	79
TAC	2.5	0.15	1260	88
TAC	2.5	1.45	1100	83
TAC	5.0	0.15	880	88
TAC	5.0	1.45	780	86

For the three irradiations which were conducted with no additive added, one sees that the gel content is constant regardless of the absorbed dose. One can see from the gel content that approximately 70 – 80 % of the sample is cross-linked. Irradiations in air or under vacuum give similar gel contents and swelling ratios. Triallyl cyanurate, TAC, is a material which has been reported to enhance the radiation-induced cross-linking of polyamide-6[17].

In the presence of TAC the gel content increases a little and the swelling ratio decreases a little, indicating that TAC promotes cross-linking.

The effect of all additives which have been studied on the extent of cross-linking is reported in terms of the minimum dose required to achieve gel formation; this is shown in Table 2. It is interesting to note that phospham, a material not known as a cross-linking enhancer, effects cross-linking at a lower dose than does triallyl cyanurate which is a cross-linking enhancer. Melamine appears to have little effect while ammonium polyphosphate appears to inhibit cross-linking.

Table 2. *Gel Dose of PA-6a in the Presence and Absence of Additives.*

Additive		Dose Required for Gel Formation,
Name	**Content, wt. %**	kGy
none		200
TAC	2.5	<150
PH	20	80
MA	20	175
APP	20	>300

3.2. Thermal Analysis

3.2.1. Polyamide-6a without Additives.

Both thermogravimetric analysis and differential scanning calorimetry data are available for all of the samples and these may be compared in a wide variety of ways. The regions of interest in the TGA curves are the onset temperature of degradation and the amount of non-volatile residue which remains at 600°C. The onset temperature of degradation may be defined in a variety of ways. Measurements have been made of the temperature at which 1 %, 3 %, and 10 % mass loss has occurred. A 1 % mass loss seems to be too small because this may be due to simply sorbed moisture so normally the temperature at which 3 % or 10 % mass loss occurs will be used to compare the various samples. The TGA curves in nitrogen for unirradiated and irradiated polyamide-6a are shown in Figure 1 and the results for TGA in nitrogen and in air are summarised in Table 3.

The fact that a non-volatile residue is obtained in nitrogen shows that irradiation does have an effect on the char forming tendency of the polymer. The temperature at which 3 % mass loss occurs falls dramatically as the irradiation dosage increases but there is a much smaller effect on the temperature for a 10 % mass loss. It is likely that irradiation causes the cleavage of many bonds and some of these do not form cross-links but rather participate in other reactions, such as hydrogen abstraction, and produce small and volatile fragments which can easily escape to the gas phase as the temperature is increased.

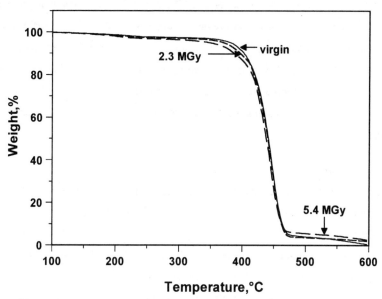

Figure 1. *TGA curves for virgin and irradiated polyamide-6 in nitrogen at a scan rate of 10 ℃/min recorded at various absorbed doses.*

Table 3. *Temperatures for 3 % and 10 % Mass Loss and Amount of Non-volatile Residue at 600 ℃ for Pure Polyamide-6a*

Atmosphere	Dose, MGy	Temperature, °C, for 3 % mass loss	Temperature, °C, for 10 % mass loss	Non-volatile residue at 600°C, wt. %
Nitrogen	0	350	407	0
	2.3	317	397	4
	5.4	246	387	4
Air	0	345	396	0
	2.3	300	394	0
	5.4	288	393	0

In the DSC curve, shown in Figure 2, two endothermic transitions are observed: in unirradiated PA-6a these occur at 220°C and at 455°C (broad). The first of these is the melting point while the second must be degradation. In the irradiated polymer the melting transition occurs at 200°C, irradiation probably changes the crystallinity of the sample which causes this change in the melting endotherm; the degradation endotherm is unchanged.

The results are quite similar when the experiment is performed under an air atmosphere. The temperature at which a 10 % mass loss occurs is essentially unchanged from that obtained under an inert atmosphere; the one difference between air and nitrogen is that no non-volatile residue is observed for an irradiated sample when the TGA is performed in air. Irradiation increases the amount of transitory residue by 5 % at 475°C. In an air atmosphere the DSC shows the same change in the melting endotherm and the higher temperature degradation endotherm observed under nitrogen is replaced by an exotherm at about the same temperature. It appears that irradiation has little effect on the DSC under either inert or an air atmosphere.

3.2.2. Polyamide-6a with 20 % Additives.

The presence of additives has a large effect on the degradation of PA-6a. This can be seen in both the amount of non-volatile residue as well as in the temperatures at which mass loss occurs and the results are shown in Table 4.

One scheme for flame retardation is to promote the modification of the polymer in question at lower temperature so that the reaction scheme is changed into one which can produce char. This appears to be applicable in these cases.

The TGA curves for PA-6a with phospham are shown in Figure 3 and these show that at all temperatures the degradation occurs more facilely in the presence of phospham than in its absence. There is a very slight indication of a second step in the degradation curve; the initial degradation step commences at a little below 300°C while the second step begins at about 350°C.

Table 4. *Temperatures for 3 % and 10 % Mass Loss and Amount of Non-volatile Residue at 600°C for Polyamide-6a with 20 % Additive*

Additive	Dose, MGy	Temperature, °C, for 3 %mass loss	Temperature, °C, for 10 % mass loss	Non-volatile residue at 600°C, wt %
PH	0	299	319	16
PH	2.3	286	322	19
PH	4.9	281	321	21
MA	0	268	327	3
MA	2.3	263	311	5
MA	5.4	252	304	7
APP	0	293	316	12
APP	5.4	245	309	18

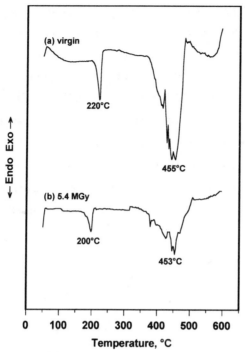

Figure 2. *DSC curves for virgin and irradiated polyamide-6 in nitrogen at a scan rate of 10 ℃/min recorded at various absorbed doses.*

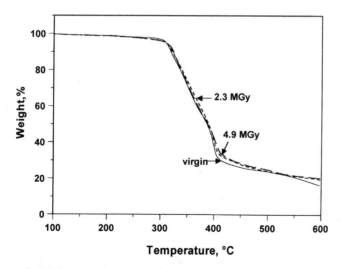

Figure 3. *TGA curves for polyamide-6 containing 20 % phospham in nitrogen at a scan rate of 10 ℃/min recorded at various absorbed doses.*

Irradiation has little or no effect on the temperatures of degradation or on the amount of non-volatile residue. It is likely that whatever process is promoting char formation in the virgin polymer is also active in the presence of this, and as will be seen shortly, the other additives. The DSC curves show three endotherms under a nitrogen atmosphere. Again the melting endotherm is observed at 220°C which falls to 200°C after irradiation. For virgin PA-6a the degradation endotherm is seen near 450°C; in the presence of phospham this is replaced by two endotherms one at 350°C and the other at 400°C. The first of these is moved to 325°C upon irradiation while the second is not changed by irradiation. This is in agreement with the TGA data; degradation occurs more easily in the presence of phospham by a two step process and irradiation lowers the initial temperature at which degradation occurs.

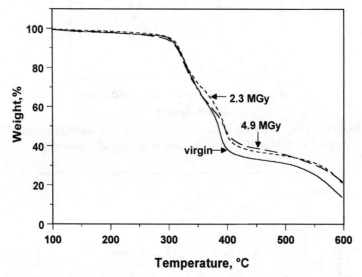

Figure 4. *TGA curves for polyamide-6 containing 20 % phospham in air at a scan rate of 10 °C/min recorded at various absorbed doses.*

The TGA curve for the combination of PA-6a and phospham under an air atmosphere is shown in Figure 4. In both nitrogen and in air the initial degradation step is ended by 400°C; the amount which remains is higher in air than nitrogen. It should also be noted that more material remains after irradiation than in the virgin sample. Irradiation also increases the amount of material which is non-volatile at 600°C. Finally, extensive foaming is seen with irradiated samples of PA-6a with phospham in air; such intumescence is not observed in nitrogen or in air with unirradiated samples. Under nitrogen the unirradiated polyamide/phospham mixture degrades at 400°C while in air this degradation occurs at 474°C.

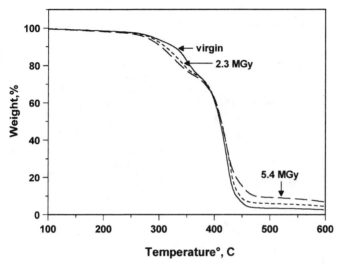

Figure 5. *TGA curve for polyamide-6 containing 20 % melamine in nitrogen at a scan rate of 10 °C/min recorded at various absorbed doses.*

The TGA curves for the melamine - PA-6a combination are shown in Figure 5. The degradation clearly shows two steps; the first begins at about 260°C while the second commences at about 400°C. Upon irradiation the initial step becomes more pronounced but the temperatures are unchanged. Melamine itself undergoes volatilisation in this region. The most important effect is that irradiation facilitates the production of char. The curve for the degradation in air is similar except there is some material which is formed at about 420°C which has transitory stability and is absent in nitrogen; this must indicate an extra step in the degradation pathway in air. This is also seen in the DSC. Under nitrogen the unirradiated sample with melamine degrades endothermically at 340°C and 420°C; the melting transition is also observed at 220°C. Upon irradiation the melting transition is again observed at 200°C and the degradation endotherms are at 290°C and 425°C. Again irradiation facilitates the degradation of the system and the DSC results complement those from TGA. In air the endothermic degradation of the unirradiated sample at 340°C is absent but that at 420°C is still present; the exothermic oxidation reactions at higher temperatures, 430°C and 480°C, dominate the DSC curve. In the case of irradiated sample there is still some small amount of endothermic degradation at 290°C as well as a small endothermic peak near 420°C and the higher temperature exotherms have moved to even higher temperatures, 450°C and 530°C.

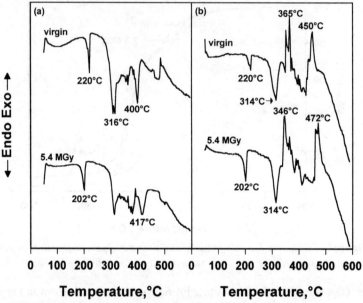

Figure 6. *DSC curves for unirradiated and irradiated polyamide-6 containing 20 % APP at a scan rate of 10 °C/min in nitrogen (a) and in air (b).*

Both ammonium polyphosphate and phospham give much higher char yields when combined with PA-6a than does melamine; since melamine completely thermally degrades below 400°C while the other two additives are more thermally stable, this is not surprising. For APP the amount is greater than would be expected from APP alone and this indicates that there is an interaction between APP and PA-6a. Under both a nitrogen and an air atmosphere, there is a step in the degradation at about 400°C; the amount of material which has not volatilised at this point is greater for the irradiated sample than for the unirradiated sample and it is greater in air than under nitrogen. This greater residue in air is attributable to transitory char which disappears at higher temperatures. The DSC curves confirm this; in nitrogen one observes three endothermic steps, melting at 220°C and degradation at 316°C and 400°C. In the irradiated sample the melting point is again depressed and the degradation temperature is raised. The same is true when the experiment is performed in air. The DSC curves are shown in Figure 6.

3.2.3. Combinations of PA-6b with TAC and Flame Retardant Additives.

Several combinations of triallyl cyanurate, a cross-linking enhancer, with the same flame retardants noted above have been examined by thermogravimetric analysis.

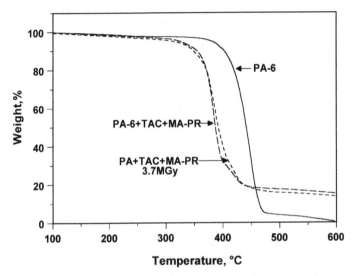

Figure 7. *TGA curve for polyamide-6, polyamide-6 with TAC and melamine pyrophosphate, and an irradiated sample of polyamide-6, with TAC and melamine pyrophosphate at a scan rate of 10 °C/min in nitrogen.*

The addition of 5 % TAC to PA-6b causes an increase in the amount of material which is non-volatile at 600°C from 2 % to about 8 %. The shape of the TGA curve indicates that this material will volatilise at slightly higher temperatures. The onset temperature of the degradation is little changed by the presence of this additive. Both melamine cyanurate, MA-CY, and melamine pyrophosphate, MA-PR, have been used together with TAC in combination with the polyamide. For both the onset temperature of degradation is significantly depressed, by perhaps 50°C; for MA-CY a minimal amount of non-volatile is seen at 600°C and this will volatilise at slightly higher temperature. MA-PR gives a significant yield of non-volatile, about 20 %, both with and without irradiation. The shape of the TGA curve indicates that this is unlikely to be transitory char. The TGA curves for the combination of PA-6b, 5 % TAC, and 20 % MA-PR are shown in Figure 7. When the TGA experiment is performed in air the general shape of the curves is about the same.

3.2.4 Oxygen Index Measurements.

The measurement of the oxygen index gives an indication of the relative flammability of a material. There is essentially no difference in the oxygen index of virgin PA-6a and that of the irradiated material, all values are 25 ± 1. When samples containing TAC are irradiated the oxygen index initially drops by about 4 % if the sample is irradiated at 0.15 MGy but increases slightly at higher dosage. The experiments described here were designed to test some ideas concerning the mode of flame retardant action of certain additives. Several mechanisms have been proposed for the action of melamine. These include promotion of dripping[2] or cooling the flame due to its endothermic evaporation[18]. Cross-linked polymers cannot drip, so one can distinguish between these modes of action. If dripping is the important phenomenon, then the oxygen index will decrease while if cooling due to endothermic evaporation or similar processes are operative, the oxygen index will not change. The oxygen index data for all additives are presented in Table 6.

Table 6. *Effect of Absorbed Dose on Oxygen Index for Polyamide-6a Containing 10 Weight % of Additive.*

Additive	Absorbed Dose, MGy	Oxygen Index
MA	0	31
MA	5.3, vacuum	27
MA	5.3, air	24
PH	0	29
PH	5.3, vacuum	27
PH	5.3, air	23
APP	0	26
APP	5.3, vacuum	28
APP	5.3, air	25

In the presence of 10 % melamine the oxygen index decreases upon irradiation in vacuum and a greater decrease is observed in air. This supports the dripping mechanism as the likely mode of action for melamine. At higher concentrations of melamine irradiation in either air or nitrogen has little effect and this may suggest that both mechanisms are important. Both phospham and APP should cause char formation and this is expected to lead to an increase in oxygen index. At 10 % phospham the oxygen index decreases upon irradiation, indicating that irradiation degrades the flame retardant effect of phospham. Irradiation of PA-6a combinations with ammonium polyphosphate causes a small increase in the oxygen index so it is likely that there is a small flame retardant effect.

Oxygen index measurements on those PA-6b samples which contain both TAC and the additives are presented in Table 7. This set of experiments was also designed to examine the effect of dripping. In the presence of glass fibers there is no dripping while virgin PA-6b shows low dripping and in the presence of melamine cyanurate there is a high degree of dripping. Melamine pyrophosphate does not affect dripping but does promote char formation. The presence of the flame retardant additive is seen to marginally increase the oxygen index relative to those in which only TAC is present but it is a marginal increase.

Table 7. *The Oxygen Index as a Function of Absorbed Dose for γ-Irradiated Polyamide-6b Containing 5 Weight % TAC and 20 Weight % of Another Additive.*

Absorbed Dose, MGy	PA-6b	PA-6b + MA-CY	PA-6b + MA-PR	PA-6b + glass fibers
0	22	23	28	22
0.5	21	22	26	23
1.0	20	22	25	24
2.0	20	24	26	22
2.5	20	24	26	24
3.7	21	24	27	24

It is apparent that irradiation and dripping have little or no effect on the oxygen index while the presence of a char forming additive has a significant effect.

3.2.5. Mechanical and optical properties.

These properties were not measured systematically, but visual and manual inspection revealed the following: all specimens have lost their elasticity and became hard and brittle upon irradiation both in the presence and absence of air. These alterations become quite significant when the absorbed dose exceeds about 1 MGy. The surface of samples irradiated in air is hygroscopic and tacky. In all cases the samples became discoloured upon irradiation; the samples have a yellowish/green colour.

3.2.6. Infrared Spectroscopy of the Polyamide-6a Samples.

The infrared spectra of the unirradiated samples are identical with those of the irradiated samples; no new bands are observed which means that no new functional groups are produced. A possible interpretation is that irradiation effects the breaking of bonds to give radicals and these then react to cause the cross-linking. The important conclusion from this spectroscopic study is that the same types of bonds are present after irradiation as are present before irradiation and the presence of the additives does not change this observation.

3.2.7. Mode of Action.

The data confirm that cross-linking occurs upon irradiation of polyamide-6. It is also seen that the gel content does not depend upon the absorbed dose in the dose range studied in this work whereas the density of cross-linking increases with increasing absorbed dose. Previous workers have shown that irradiation of polyamide samples causes simultaneous cross-linking and main-chain degradation.[14] Thus one may expect that the degraded portion of the polyamide will volatilise earlier, so an earlier onset of degradation will be observed. On the other hand, cross-linking will enhance the char forming tendency of the polymer. Irradiation of the samples, either alone or in the presence of additives, decreases the onset temperature of the degradation and increases the amount of residue which is non-volatile at 600°C. The infrared spectra show that no new functional groups are produced so irradiation rearranges the aliphatic structure of the polymer into a cross-linked aliphatic structure. The density of cross-links does have an effect on the fraction which is non-volatile at high temperatures but it is likely that the type of the cross-links is more important than the number of cross-links.

4. CONCLUSION

It is quite possible to "insolubilize" polyamide-6 by the use of high energy radiation. This is due to intermolecular cross-linking which is accompanied by main-chain scission. The presence of a low molar mass fraction, generated by main-chain scission, is presumable responsible for the observation that the onset temperature for the mass loss is lower for the irradiated than for the unirradiated polyamide-6. Upon heating polyamide-6 to 600°C, a non-trivial portion of the polymer forms char in the presence of all additives studied in this work and irradiation to rather high absorbed doses leads to only a small increase in this portion. Infrared measurements showed that the chemical nature of the bonds is not altered, *i.e.*, no completely new bonds are formed by irradiation. This explains why irradiated polyamide-6 is not noticeably less flammable than the virgin polymer. If one wishes to render a polymeric material less or non-flammable by cross-link formation, a method should be used which cross-links the polymer without side reactions, especially main-chain scission. Moreover the treatment should result in the formation of new bonds which have a higher thermal stability than those which are lost. It seems that the quality of the chemical bonds which are produced by cross-linking is likely to be more important than the density of cross-links.

References

1. S. V. Levchik, L. Costa and G. Camino, *Polym. Degrad. Stab.*, 1992, **36**, 229.
2. S. V. Levchik, A. I. Balabanovich, G. F. Levchik, and L. Costa, *Fire and Materials.*, 1997, **21**, 75.
3. E. D. Weil, and N. G. Patel, *Fire and Materials*, 1994, **18**, 1.
4. S. V. Levchik, G. F. Levchik, A. I. Balabanovich, G. Camino, and L. Costa, *Polym. Degrad. Stab.*, 1996, **54**, 217.
5. G. Camino and L. Costa, *Rev. Inorg. Chem.*, 1986, **8**, 69.
6. S. K. Brauman, *J. Fire Retard. Chem.*, 1979, **6**, 249.
7. T. Kashiwagi, *Polym. Combust. Flammability.*, 1994, **25**, 1423.
8. P. Carty and S. White, *Fire and Materials*, 1994, **18**, 151.
9. P. Carty and S. White, *Polymer*, 1994, **35**, 1213.
10. M. R. Nyden, J. E. Brown and S. M. Lomakin, *Mat. Res. Soc. Symp. Proc.*, 1992, **278**, 47.
11. M. R. Nyden, G. P. Forney and J. E. Brown, *Macromolecules*, 1992, **25**, 1658.
12. S. M. Lomakin, and G. E. Zaikov, *Int. J. Polym. Mater.*, 1996, **33**, 133.
13. J. Zimmermann,., in *The Radiation Chemistry of Macromolecules,* ed. Dole, M., Academic Press, New York, 1973, p. 121.
14. B. J. Lyons, and L. C. Glover, *Radiat. Phys. Chem.*, 1990, **35**, 139.
15. R. A. Shaw and T. Ogawa, *J. Polym. Sci., Part A,* 1965, **3**, 3343.
16. A. Charlesby, *"Atomic Radiation and Polymers"*, Pergamon Press, Oxford, 1960, p. 170.
17. U. Stenglin, *Plastverarbeiter* 1991, **42**, 56.
18. E. D. Weil and V. Choudhary, *J. Fire. Sci.*, 1995, **13**, 104.

Acknowledgements

The authors gratefully acknowledge the help of Professor G. Camino (University of Torino, Italy) for the discussion of this paper. This work was financially supported by a NATO Linkage Grant, OUTR.LG960310, which is gratefully acknowledged.

THE USE OF ORGANOPHOSPHORUS FUNCTIONALITY TO MODIFY THE THERMAL AND FIRE RETARDANT BEHAVIOUR OF POLYAMIDES

J. W. Wheeler, Y. Zhang and J. C. Tebby

School of Sciences, Staffordshire University,
College Road, Stoke-on-Trent ST4 2DE, UK

1. INTRODUCTION

1.1. Organophosphorus Fire Retardant Additives

Due to the great use of and dependence upon synthetic materials placed by today's society, it is necessary to make commercial polymers both non-flammable and fire retardant. One approach by which this may be achieved is the incorporation by blending of suitable additives within the bulk of a given polymer or of coating upon its surface.

The fire retardant properties of small molecule organophosphorus compounds have been recognised and developed over a number of years and have thus found use as additives in a variety of polymeric materials[1,2]. Many additives are now available commercially, examples of which are given in Figure 1. Ammonium polyphosphate (Exolit 422, Hoechst) is used as an intumescent fire retardant for thermoplastics such as polypropylene, as is the Great Lakes Chemical product Char-Guard CN329. Akzo Nobel's Fyrol FR-2 has been incorporated into flexible polyurethane foams.

Generally, organophosphorus fire retardants work in the condensed or solid phase, promoting and yielding the formation of large amounts of char. This acts to both insulate the material underneath from the heat of the fire and thus reduce its thermal degradation into volatile, combustible products and also to isolate it from the source of ignition. This family of fire retardants is found to work well for oxygenated polymers, such as polyesters and polyurethanes, but alone are not as effective for hydrocarbon-like materials, which tend to melt, drip and depolymerise upon combustion[3].

Also, many organophosphorus additives have low molecular weights and are therefore somewhat volatile, leading to the possibility of their loss from the polymer during either high temperature processing or in the early stages of combustion. Additionally, their transition to the gaseous phase can cause the smoke from the burning material to become loaded with toxic phosphorus-containing compounds. Clearly, there is a need to increase the permanence of the additive within the polymer in order to maximise the fire retardancy and to reduce the evolution of toxic species. Therefore, the incorporation of organophosphorus functionality within the polymeric structure is a logical progression of this field and may be expected to accomplish the two previously stated aims.

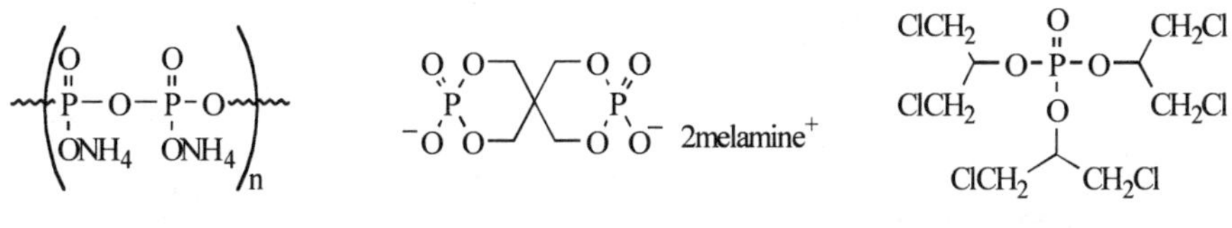

Exolit 422	*Char-Guard CN329*	*Fyrol FR-2*
(ammonium polyphosphate)	(Great Lakes Chemical)	(Akzo Nobel)
(Hoechst)		

Figure 1. *Examples of commercially available organophosphorus fire retardant additives.*

1.2. Polymers Incorporating Organophosphorus Functionality

The incorporation of organophosphorus functionality within the structure of polymeric materials, either as a constituent of the backbone or as a group appended to it, has been an area of interest for a number of years[4-6]. Recent reports concerning phosphorus-containing polymers have highlighted many applications, including fire retardant behaviour. Several examples are given in Figure 2.

Connell has described a series of polyimides[7], *e.g.* (**1**), that demonstrate high char yields and oxygen plasma resistance. The appending of phosphate groups to poly(hydroxystyrene)s, leading to materials such as (**2**), was found to greatly improve the fire retardancy of the original polymer[8]. Polyphosphonates, *e.g.* (**3**), have been reported to be effective fire retardant additives for rubbers[9] and poly(methylmethacrylate)s with pendant bicyclophosphonates, *e.g.* (**4**), were resistant to combustion but do not however, undergo intumescent decomposition[10]. Although many useful materials have been developed there is still great potential for the production of materials with improved fire retardancy and additionally, more favourable physical and mechanical properties.

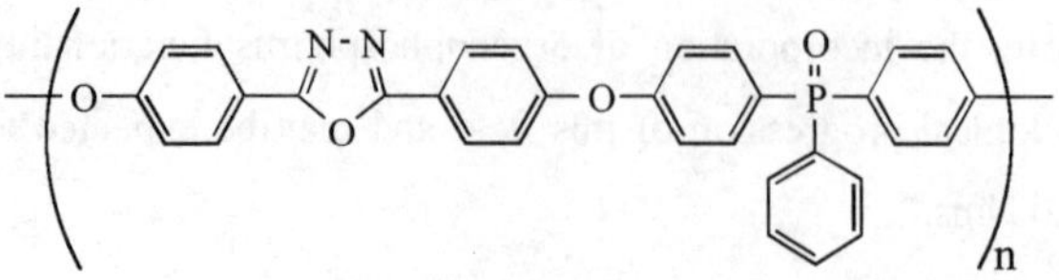

(1)

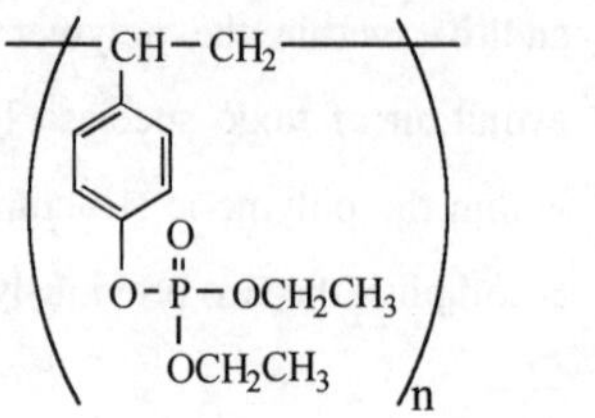

(2)

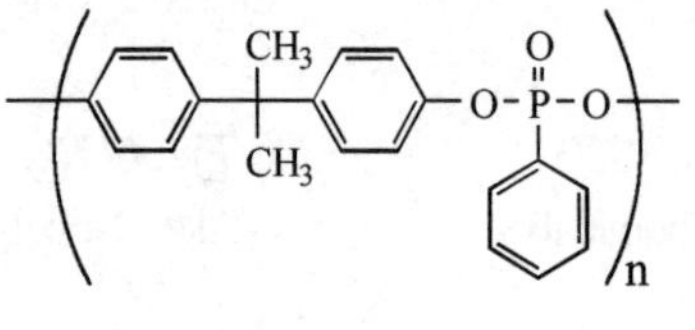

(3)

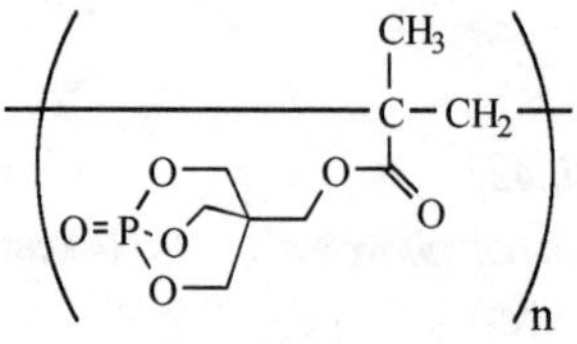

(4)

Figure 2. *Examples of polymers incorporating organophosphorus functionality that display fire retardant behaviour.*

1.3. Aromatic Polyamides

Aromatic polyamides (Figure 3) were of interest to us for a number of reasons. They possess many desirable characteristics, such as excellent mechanical properties, good thermal stability and high solvent resistance, and have accordingly found use in a number of applications, in particular as high temperature fibres[11].

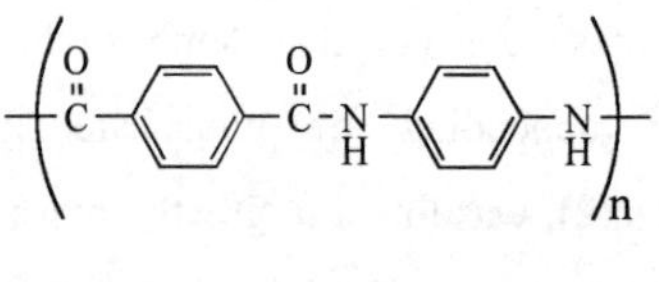

(5)

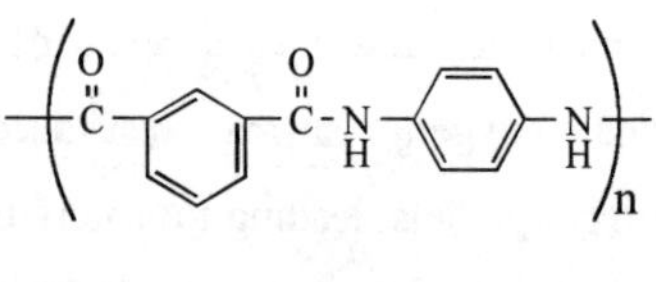

(6)

Figure 3. *Aromatic polyamides*

However, they can be somewhat difficult to process due to their limited solubilities, especially in common and environmentally-friendly solvents, and often have extremely high glass transition (Tg) or melt temperatures. Additionally, their fire retardant nature is often quite variable, being highly influenced by the precise polymeric structure. Clearly, if the backbone of aromatic polyamides can be made more flexible then the resultant polymers should demonstrate lowered *Tg* values and enhanced solubilities. Also, the covalent attachment of fire retarding groups can be expected to improve the resistance to ignition and combustion.

2. POLYAMIDES INCORPORATING ORGANOPHOSPHORUS GROUPS WITHIN THE POLYMER BACKBONE

2.1. Aims of this Study

The aims of our work were to undertake a systematic study of the synthesis of a range of aromatic and aromatic-aliphatic polyamide structures and to observe the effects on their thermal and physical properties upon the introduction of organophosphorus groups within the polymer backbone. We postulated that it would be possible to produce materials with greater and more permanent fire retardant behaviour, since the organophosphorus functionality will be covalently bound to the polymeric structure and clearly will not be lost during processing. Also, the incorporation of organophosphorus groups will provide points of flexibility within the polymer backbone which should result in lower *Tg* values and enhanced solubilities compared to traditional polyamides, thus improving the processibility of these materials.

Triarylphosphine oxide groups appealed to us as candidates for inclusion into the polyamide backbone due to the chemical stability of their P–C bonds, thus endowing them with good thermal and hydrolytic stabilities. Their ease of synthesis and good solubilities in a wide range of organic solvents were additional advantageous factors which led us to commence our quest towards the preparation of polyamides incorporating phosphine oxide groups. These materials will have potential for use as processible, fire retardant polymers and additionally, may find application as high molecular weight fire retardant additives for commercial polymeric materials.

2.2. Polymerisation Strategy

Polyamides may be synthesised by the reaction of stoichiometric quantities of dicarboxylic acids with diamines under the appropriate conditions. This presented us with two possible strategies by which phosphine oxide containing polyamides could be prepared. Firstly, dicarboxylic acid monomers that incorporate the required organophosphorus functionality can be reacted with a series of diamino bridging groups (Scheme 1). A number of dicarboxylic acid phosphine oxides, such as bis(4-carboxyphenyl)phenylphosphine oxide (7)[12] , have been reported previously and may be synthesised easily from readily available starting materials.

Scheme 1

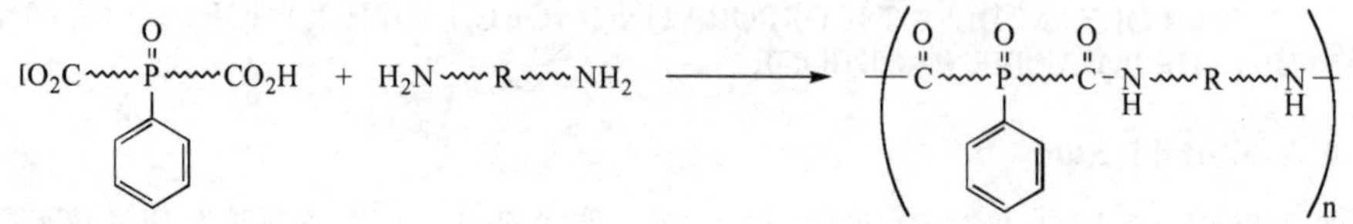

| Phosphine oxide dicarboxylic acid | Diamine | Phosphine oxide Polyamide |

Secondly, diamino monomers containing phosphine oxide units could be synthesised and reacted subsequently with commercially available dicarboxylic acids (Scheme 2). However, far fewer numbers of suitable diamines could be identified by literature searches and so as an additional part of this study we designed and synthesised a new polymerisation monomer, bis[4-(2-aminoethyl)aminophenyl]phenylphosphine oxide[13], in order to be able to utilise the second of these two strategies.

Scheme 2

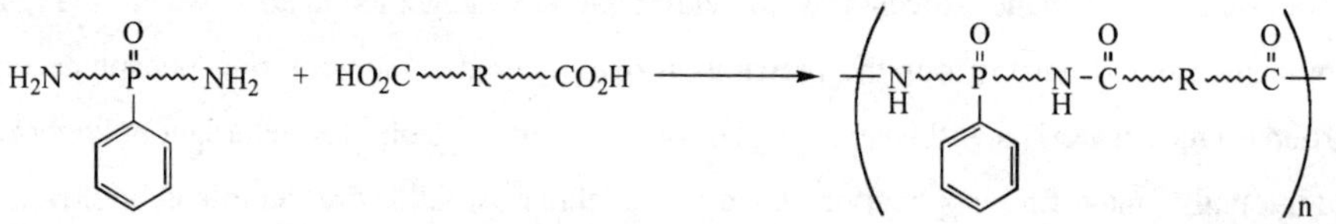

| Phosphine oxide diamine | Dicarboxylic acid | Phosphine oxide Polyamide |

2.3. Polymerisation Reactions

The method employed for the polymerisation synthesis of the polyamides was dependent upon the basicity of the diamino monomer being used. For monomers containing weakly basic aromatic amino functional groups the phosphorylation polymerisation method of Yamazaki[14] was utilised. When the reactions involved aliphatic primary amines this method was inappropriate since the stronger basic nature of the amine meant that ammonium carboxylate salt formation was more favourable than amide bond synthesis. Therefore, in such cases a melt polycondensation method was used.

2.3.1. Method A - Phosphorylation Polycondensation.

Polycondensation of equimolar quantities of dicarboxylic acid and aromatic diamine was undertaken successfully in the presence of lithium chloride, triphenyl phosphite and pyridine in N-methylpyrrolidinone as solvent[15,16]. Reaction times of 3 hours and temperatures of 100°C gave polyamides in yields of 85 – 100 % and high molecular weights, as determined by viscosity measurements. For example, polyamide (**8**) was prepared in 100 % yield by the reaction of bis(4-carboxyphenyl)phenylphosphine oxide (**7**) with 1,4-diaminobenzene (Scheme 3). Yields and viscosity values for polyamides synthesised by Method A are given in Tables 1 and 2.

Scheme 3

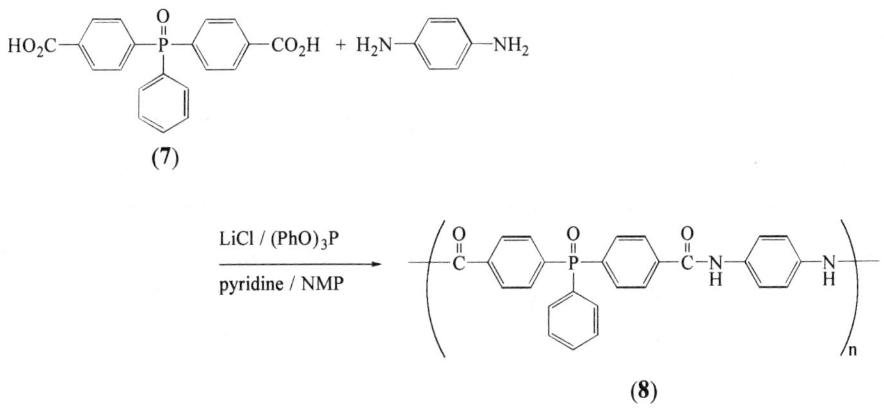

(**7**)

(**8**)

2.3.2. Method B - Melt Polycondensation.

Polycondensation of aliphatic-like diamines with dicarboxylic acids was undertaken in two stages. Firstly, in order to ensure a 1:1 ratio of the two reactants prior to reaction, the ammonium carboxylate salt was prepared at room temperature and isolated. The salt was subsequently heated at 200 - 300°C under nitrogen and high vacuum for 4 hours to give essentially quantitative yields of polyamides[13,17]. For example, polyamide (**19**) was prepared in 90 % yield by the reaction of bis(4-carboxyphenyl)phenylphosphine oxide (**7**) with

1,2-diaminoethane (Scheme 4). Yields and viscosity values for polyamides synthesised by Method B are given in Tables 3 and 4.

Scheme 4

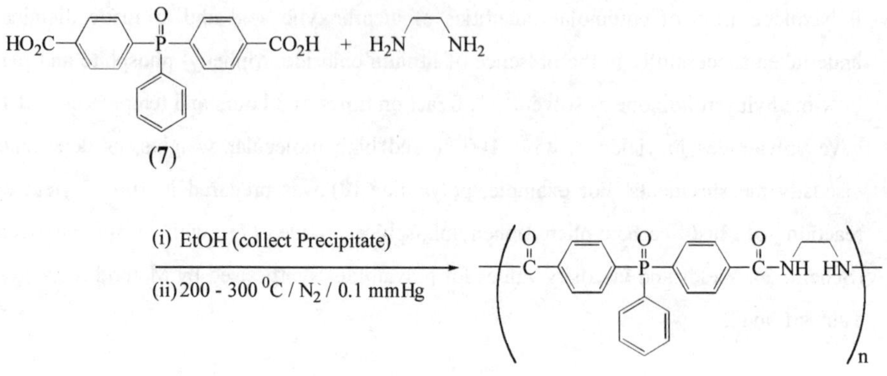

(7)

$$\xrightarrow[\text{(ii) } 200 - 300\,^{0}\text{C / N}_2\text{ / 0.1 mmHg}]{\text{(i) EtOH (collect Precipitate)}}$$

(19)

2.4. Thermal Behaviour and Potential For Fire Retardancy

The thermal behaviour of all new polyamides was investigated by differential scanning calorimetry (DSC) and thermogravimetric analysis (TGA). Generally, a wide endotherm was observed between 50 and 150°C, or so, in the first run of DSC. This was attributed to the loss of water associated with these materials and therefore, two runs were required to obtain consistent DSC traces. All of the polyamides studied were found to be amorphous and no crystalline temperatures were observed. Data and discussions for representative examples from each of the series of polymers that have been prepared and studied are given in the following sections.

2.4.1. Wholly Aromatic Phosphine Oxide Polyamides.

Polyamides incorporating phosphine oxide groups within a wholly aromatic backbone (Figure 4) were synthesised in good yields and high molecular weight by synthetic Method A (Table 1) [15]. Glass transition temperatures (*Tg*) within the 225-254°C range were recorded. These values demonstrate a good correlation with other related systems[16] and are significantly lower than wholly aromatic polyamides lacking a phosphine oxide bridging group, for example, polyamide (5)[11]. Polyamides such as (10), incorporating two phosphine oxide groups within each repeating unit of the polymer, had the lowest *Tg* values, probably as a result of the increased flexibility of the backbone.

Table 1. *Characterisation and thermal data for wholly aromatic phosphine oxide polyamides.*

Polymer	Synthetic Method	Yield / %	Viscosity [η] / dLg^{-1}	Tg / °C (DSC)	Temp of 5 % weight loss	Char Yield at 700 °C
(8)	A	100	0.471	252	425	24
(9)	A	98	0.435	254	430	50
(10)	A	93	0.332	225	435	38
(5)	ref. 11			>375	513 (10 % wt. loss)	61
(6)	ref. 11			250	512 (10 % wt. loss)	64

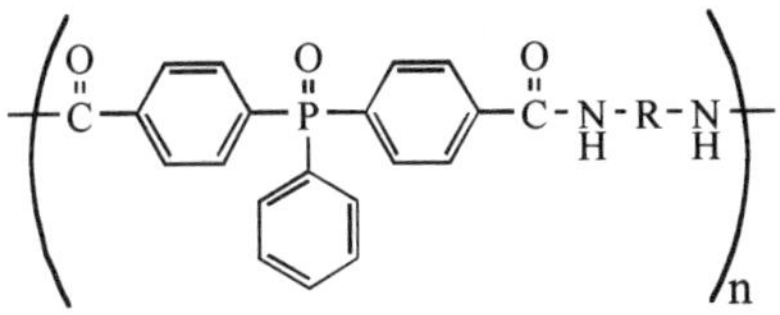

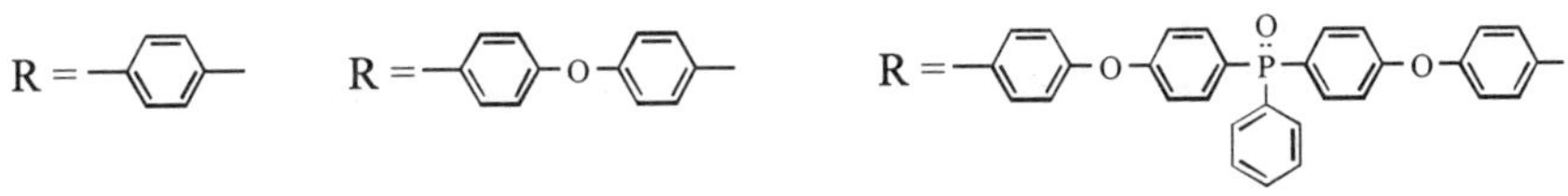

Figure 4. *Wholly aromatic phosphine oxide polyamides*

Results of the TGA measurements indicated that the presence of the phosphine oxide moiety clearly results in excellent thermo-oxidative stability at high temperatures. Figures for 5 % weight loss were observed at greater than 425°C and, although these fall a little short of those for polyamides such as (5) and (6), are comparable to those of other reported phosphine oxide containing polymers[16]. Extensive degradation was only observed in excess of 500°C, with polyamides having two phosphine oxides in each polymeric repeating unit showing higher stabilities. All wholly aromatic polyamides gave substantial amounts of char upon prolonged heating at elevated temperatures which augers well for their use as fire retardant materials.

2.4.2. Aromatic Phosphine Oxide Sulphone Polyamides.

The preparation of polyamides (11) - (14) incorporating diarylsulphone units in addition to triarylphosphine oxides (Figure 5) in good yields and high molecular weights was undertaken by Method A. The DSC studies revealed *Tg* values at 170 - 200°C for this series of materials (Table 2). The additional inclusion of sulphonyl groups increased the flexibility of the polymer chain and thus brought about an average decrease of 40 - 50°C in the *Tg* value compared to those materials which consist only of phosphine oxide moieties within a wholly aromatic backbone, for example polymers (8) - (10). The observation that polyamides (11) and (14) had the lowest *Tg* values within this series was attributed to having only *m*- or *p*-phenylene, respectively, in the backbone, the presence of which decreases the extent of polar interaction between neighbouring chains.

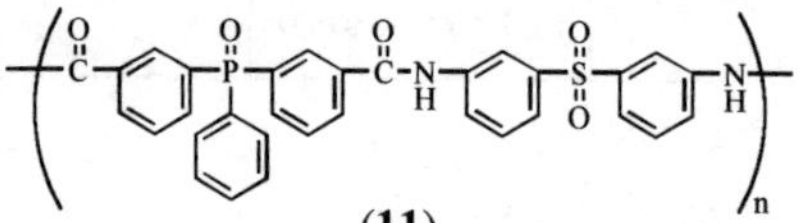

(11) 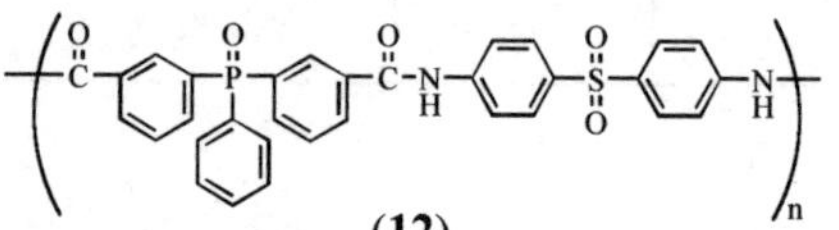   (12)

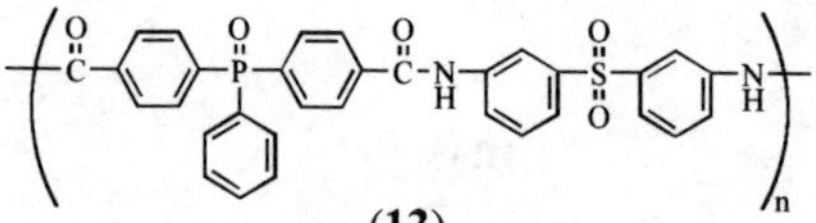

(13) 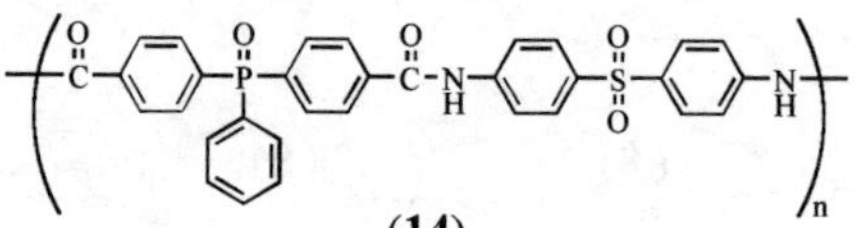   (14)

Figure 5. *Aromatic phosphine oxide sulphone polyamides.*

Table 2 *Characterisation and thermal data*
for aromatic phosphine oxide sulphone polyamides.

Polymer	Synthetic Method	Yield / %	Viscosity $[\eta]$ / dLg^{-1}	Tg / °C (DSC)	Temp of 5 % weight loss	Char Yield at 800 °C
(11)	A	88	0.795	170	415	27
(12)	A	85	0.866	198	440	34
(13)	A	90	0.912	200	410	20
(14)	A	87	0.819	186	415	28

The results of the TGA measurements confirm that the presence of both phosphine oxide and sulphonyl moieties results in materials with excellent thermo-oxidative stabilities, with the onset of thermal degradation showing no significant variance from aromatic polyamides possessing only the phosphine oxide unit. All of this series of polyamides gave substantial amounts of char upon heating at 800 °C under air flow, again the figures being comparable to those for materials lacking the sulphonyl groups.

2.4.3. Aromatic - Aliphatic Phosphine Oxide Polyamides.

During our studies it became evident that increasing the flexibility of the polymer chain resulted in a decrease in the glass transition temperature of the polyamide. Therefore, in order to exploit this trend further we designed the new polymerisation monomer bis[4-(2-aminoethyl)aminophenyl]phenylphosphine oxide and undertook melt condensation reactions (Table 3) with a series of dicarboxylic acids to yield aromatic - aliphatic phosphine oxide polyamides (Figure 6) [13].

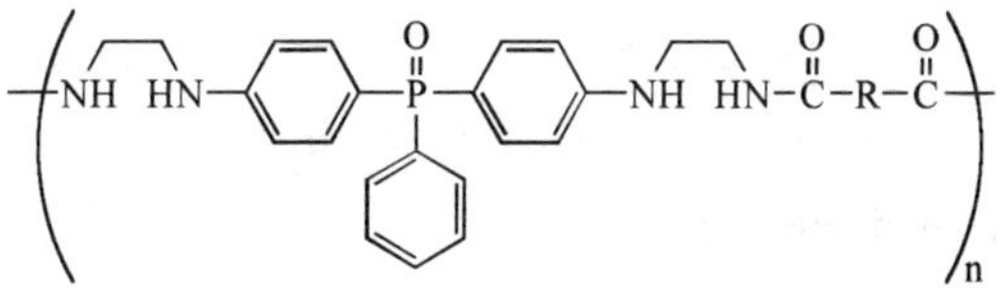

(**15**) R = ⟨benzene ring⟩   (**16**) R = –CH$_2$–   (**17**) R = –CH$_2$CH$_2$–   (**18**) R = –CH=CH–

Figure 6. *Aromatic-aliphatic phosphine oxide polyamides synthesised from bis[4-(2- aminoethyl)aminophenyl]phenylphosphine oxide.*

Table 3. *Characterisation and thermal data for aromatic-aliphatic phosphine oxide polyamides synthesised from bis[4-(2-aminoethyl)aminophenyl]phenylphosphine oxide.*

Polymer	Synthetic Method	Yield / %	Viscosity [η] / dLg^{-1}	Tg / °C (DSC)	Temp of 5 % weight loss	Char Yield at 650 °C
(**15**)	B	99.5	0.901	208	385	33
(**16**)	B	98.5	1.126	180	360	38
(**17**)	B	98.5	1.050	190	345	34
(**18**)	B	100	1.257	200	370	26

The materials thus formed demonstrated *Tg* values in the 180 - 208°C range, these values being up to 74°C lower than those for wholly aromatic polyamides such as (8) - (10) incorporating phosphine oxide groups. It was also observed that the *Tg* values of polyamides (15) - (18) were comparable to those of aromatic polymers containing the additional sulphonyl functionality, (11) - (14), with the lowest values corresponding to polyamides (16) and (17) that posses the high degree of aliphaticity and hence, the most flexible parent chain.

The aromatic - aliphatic materials (15) - (18) began to lose weight above 340°C in air, followed by rapid thermal decomposition on further heating. In comparison to wholly aromatic polyamides, such as (8) - (10), the introduction of aliphatic bridging groups into the parent chain leads to a general decrease in thermo-oxidative stability, reflected by a drop of up to 90°C in the 5 % weight loss figures. However, all polymers of this series gave substantial amounts of char upon prolonged heating under air flow at 650°C, the yields being comparable to all other phosphine oxide containing polyamides that we have synthesised in this study. Therefore, although the increased aliphatic nature of these materials led to lower thermo-oxidative stabilities, they may be processed at reduced temperatures due to having reduced *Tg* values and additionally retain the potential for fire retardant behaviour as demonstrated by our related polymer systems.

Table 4. *Characterisation and thermal data for aromatic-aliphatic phosphine oxide polyamides synthesised from bis(4-carboxyphenyl)phenylphosphine oxide (7).*

Polymer	Synthetic Method	Yield / %	Viscosity [η] / dLg⁻¹	Tg / °C (DSC)	Temp. of 5 % weight loss	Char Yield at 700 °C
(19)	B	90	0.74	205	390	38
(20)	B	90	1.43	191	375	35
(21)	B	92	1.56	184	370	34
(22)	ref. 4			150		

A further series of aromatic - aliphatic polyamides incorporating phosphine oxide groups (Figure 7) was synthesised by Method B by the polymerisation of bis(4-carboxyphenyl) phenylphosphine oxide (7) with a range of aliphatic diamines[18]. The *Tg* values (Table 4) were found to occur in a similar region to those of the previously prepared aromatic - aliphatic polyamides **(15)** - **(18)** and demonstrated a steady decrease as the proportion of aliphaticity within the polymer backbone was increased. This data corresponded well to that reported by Korshak[4] for analogous materials incorporating aliphatic chains, including polyamide **(22)** which had a softening point of 150°C.

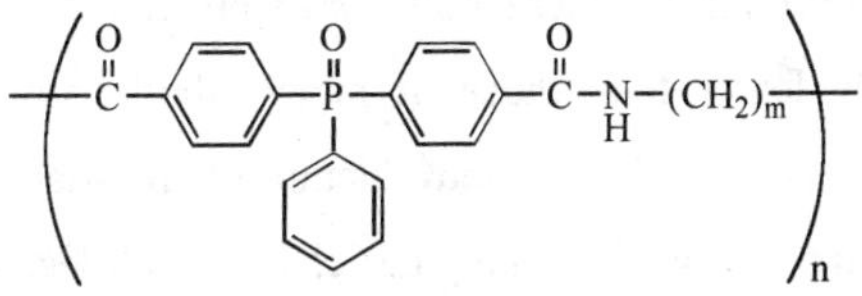

(19)	m = 2
(20)	m = 4
(21)	m = 6
(22)	m = 10

Figure 7 *Aromatic-aliphatic phosphine oxide polyamides synthesised from bis(4-carboxyphenyl)phenylphosphine oxide.*

In comparison to wholly aromatic phosphine oxide polyamides, the inclusion of aliphatic chains within the backbone of polyamides **(19)** - **(22)** brought about an understandable reduction in thermo-oxidative stability that was found to correlate closely with that of our other aromatic - aliphatic polymers containing phosphine oxide groups. However, the char yields obtained from prolonged heating at 700°C under air flow (Table 4) were again encouraging and found to be similar to those of our previously studied polyamides.

2.5. Solubility

Generally, all of the polyamides incorporating phosphine oxide groups under study demonstrated good ambient temperature solubility (> 1 %) in aprotic polar solvents, such as N-methylpyrrolidinone, dimethyl sulphoxide and dimethylformamide. Some selected partial solubility (0.1 - 1 %) in common organic solvents was also observed, for example **(9)**, **(20)** and **(21)** showed some solubility in solvents such as chloroform, tetrahydrofuran, acetone and methanol. The solubility appeared to be related to the chain flexibility, although no direct correlation was discovered. None of the materials prepared were found to be at all soluble in water. The most interesting solubility behaviour was displayed by the aromatic phosphine oxide sulphone polyamides **(11)** - **(14)**, which were completely insoluble in common organic solvents but became significantly more soluble upon the addition of small amounts of water to the solvent. The solubility in 5 % water / tetrahydrofuran was especially good. We proposed that this behaviour was due to the water molecules aiding solvating of the polar functionalities of the polymeric structures and thus increasing the solubility as a whole[17].

3. CONCLUSIONS

During the present study we have synthesised new polymerisation monomers and have utilised these and others, that have been reported previously, to successfully synthesise a wide range of polyamides incorporating phosphine oxide groups by following two synthetic strategies. Glass transition temperatures for these polymers were found to be in the 170 - 254°C range, being significantly lower than those of many wholly aromatic polyamides lacking the phosphine oxide moiety. Clearly, the introduction of the triarylphosphine oxide into the backbone increases the flexibility of the polymer chain which leads to a decrease in *Tg* value, the lowest of which were observed for polyamides with the most flexible polymer backbone.

The thermo-oxidative stabilities of the whole series of materials were found to be excellent, if a little lower than traditional wholly aromatic polyamides. Values for 5 % weight loss were obtained in the 345 - 445°C range, with the lower temperatures demonstrated by the polymers incorporating the less thermally stable aliphatic chains. Substantial amounts of char (20 - 50 %) were obtained for all polyamides upon prolonged heating at elevated temperatures, which appears to be especially good since they contain only 5.5 - 7.9 % phosphorus by mass. Good solubility was observed in aprotic polar solvents with some polyamides demonstrating some solubility in common organic solvents.

Overall, the new materials that we have studied have a wide processibility range due to their reduced *Tg* values and good solubility properties. In addition to this, their thermal behaviour suggests that they have potential for use as fire retardant materials and as high molecular weight fire retardant additives for commercially-available polymers. Our current work is focused upon undertaking thorough ignition and flammability tests on the polyamides we have prepared thus far and further developing polymer structures and methods for their synthesis.

References

1. A. M. Aaronson in, '*Phosphorus Chemistry: Developments in American Science*', E. N. Walsh, E. J. Griffith, R. W. Parry and L. D. Quin eds, American Chemical Society, 1992, 218-228.
2. J. Green, *J. Fire Sci.*, 1992, **10**, 470-487.
3. S. V. Levchik, G. F. Levchik, G. Camino and L. Costa, *J. Fire Sci.*, 1995, **13**, 43-58.

4. V. V. Korshak, *J. Polym. Sci.*, 1958 **31**, 319-326.

5. J. Pellon and W. G. Carpenter, *J. Polym. Sci., Part A,* 1963, **1**, 863-876.

6. S. Maiti, S. Banerjee and S. K. Palit, *Prog. Polym. Sci.*, 1993, **18**, 227-261.

7. J. W. Connell, J. G. Smith Jr and P. M. Hergenrother, *Polymer*, 1995, **36**, 5-11.

8. Y.-L. Liu, G.-H. Hsiue, Y.-S. Chiu, R.-J. Jeng and C. Ma, *J. Appl. Polym. Sci.*, 1996, **59**, 1619-1625.

9. S. Ghosh and S. Maiti, *J. Polym. Mater.*, 1994, **11**, 49-56.

10. D. W. Allen, E. C. Anderton, C. Bradley and L. E. Shiel, *Polym. Degrad. Stab.*, 1995, **47**, 67-72.

11. H. H. Yang, in *'Aromatic High-Strength Fibers'* (Chapter 5), Wiley, New York, 1989, 191.

12. P. W. Morgan and B. C. Herr, *J. Am. Chem. Soc.*, 1952, **74**, 4526-4529.

13. Y. Zhang, J. C. Tebby and J. W. Wheeler, *J. Polym. Sci. Part A: Polym. Chem.*, 1997, **35**, 2865-2870.

14. N. Yamazaki, F. Higashi and J. Kawabata, *J. Polym. Sci. Part A: Polym. Chem.*, 1974, **12**, 2149-2154.

15. Y. Zhang, J. C. Tebby and J. W. Wheeler, *J. Polym. Sci. Part A: Polym. Chem.*, 1996, **34**, 1561-1566.

16. Y. Delaviz, A. Gungor, J. E. McGrath and H. W. Gibson, *Polymer*, 1993, **34**, 210-213.

17. Y. Zhang, J. C. Tebby and J. W. Wheeler, *J. Polym. Sci. Part A: Polym. Chem.*, 1997, **35**, 493-497.

18. Y. Zhang, J. C. Tebby and J. W. Wheeler, *J. Polym. Sci. Part A: Polym. Chem.*, manuscript submitted for publication.

COMPREHENSIVE STUDY OF PROTECTION OF POLYMERS BY INTUMESCENCE - APPLICATION TO ETHYLENE VINYL ACETATE COPOLYMER FORMULATIONS.

M. Le Bras, S. Bourbigot

Laboratoire de Physico-Chimie des Solides, Ecole Nationale Supérieure de Chimie de Lille, U.S.T.L., BP 108, 59652 Villeneuve d'Ascq Cedex, France.
E.-Mail: michel.le-bras@ensc-lille.fr

C. Siat and R. Delobel

Centre de Recherche et d'Etude sur les Procédés d'Ignifugation des Matériaux (C.R.E.P.I.M.), Zone Initia, 62700 Bruay-la-Buissière, France.

1.INTRODUCTION

Recently, we have developed new additives mixtures using polymers as carbonisation agents. In particular, polyamide-6 (PA-6) / ammonium polyphosphate (APP) additive mixture, blended with thermoplastics, leads to FR properties of interest, by developing an intumescent shield, which may at least partially, limit mass and heat transfers[1-2]. The intumescent additive mixtures have been developed for use in ethylene propylene rubber, ethylene - vinyl acetate copolymers[3] and polystyrene. In these blends, the polyamide plays both the role of a polymeric matrix and of a carbonisation agent[4-5].

This study deals with the association of ammonium polyphosphate as acid source and polyamide-6 as carbonisation agent used directly as fire retardant intumescent additive in an ethylene vinyl acetate 8 % copolymer (EVA8) matrix. Fire performance (Limiting Oxygen Index[6] (LOI) and UL-94 rating[7]) are reported in Figure 1. Addition of APP in EVA8 leads to a blowing up and a weak carbonisation (LOI: 24 vol. %, UL94: no rating, with 40 wt. % loading). Fire retardancy of interest are obtained (LOI: 31 vol. %, UL94: V-0 rating) in the formulation EVA8/APP/PA-6 (ratio APP/PA-6 = 5:1 wt/wt and 40 wt. % of APP/PA-6 loading in the EVA8 matrix).

The protective effect brought by the blowing up will be compared in EVA8/APP to the intumescent process with formation of a carbonaceous coating in EVA8/APP/PA-6. In a first part, this work uses oxygen consumption calorimetry to assess the effectiveness of the additives APP and APP/PA-6 in EVA8 using a cone calorimeter. The burning process depends mainly on the thermal degradation reactions in the condensed phase. Therefore the kinetic of the thermo-oxidative degradations, using the invariant kinetic parameters IKP method [8] is investigated.

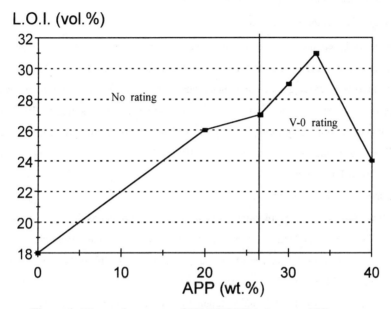

Figure 1. *Fire performances of EVA8/APP/PA-6 versus APP content.*

Finally, the part played by mass and heat transfers will be discussed.

2. EXPERIMENTAL

2.1 Materials

Raw materials were EVA8 (Lactene 1005 VN3, melt index: 0.4 g/10 mn, as pellets supplied by Elf Atochem), PA-6 (as pellets supplied by Rhône-Poulenc), and APP $((NH_4PO_3)_n$, n=700, Exolit 422, soluble fraction in H_2O < 1 wt. %, supplied by Hoechst).

Materials were mixed at 230°C using the Brabender Laboratory Mixer measuring head (type 350/EH, roller blades, checking of the mixing conditions using the data processing torque rheometer system Brabender Plasticorder PL2000, constant shear rate: 50 rpm). Sheets were then obtained using a Darragon pressing machine in standard conditions (220°C, 10^6 Pa).

2.2 Tests methods for evaluation of FR polymers

2.2.1. UL94 and LOI

The tests were carried out using normalised procedures[6-7].

2.2.2. Oxygen Consumption Calorimetry

The method is based on the principle of oxygen consumption calorimetry[9]. The Stanton Redcroft cone calorimeter was used according to the procedure ASTM 1356-90 (the standard procedure involves exposing specimens measuring 100 mm x 100 mm x 3 mm in horizontal orientation to an external heat flux of 50 kW/m² representing generalised fire[10-11]). The conventional data : time to ignition [TTI, s.], rate of heat release [RHR, kW/m²], peak of RHR [PkRHR, kW/m²] and total heat evolved [THE, kJ] were computed using a software developed in our Laboratory.

2.2.3. Temperature measurement

Temperature measurements were carried out using five Thermocoax K type thermocouples (0.25 mm diameter) embedded in 20 mm thick horizontal sheets. Thermocouples were set at 2, 5, 8, 11, 14 mm from the surface of the samples (Figure 2). The surfaces of the polymer sheets were about 85 mm x 85 mm. Temperatures versus time and position from the surface were computed every two seconds under standard conditions in cone calorimeter.

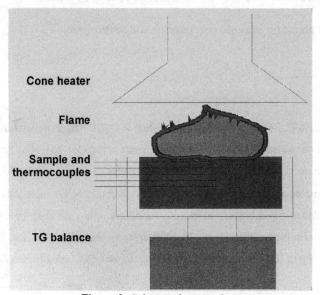

Figure 2. *Scheme of cone calorimeter experiment.*

2.2.4. Thermogravimetric analyses

Thermogravimetric analyses were carried out at 5 heating rates β_v (2.5, 4, 8, 11 and 13.5°C/min) under synthetic air flow (Air Liquide grade, flow rate: 5.10^{-7} Nm^3/s) using a Setaram MTB 10-8 thermobalance. Samples were about 10^{-5} kg and were positioned in open vitreous silica pans. Precision on temperature measurements is ±1.5°C in the whole range.

Curves of weight difference between experimental and calculated TG curves allow to point out increases and decreases of the thermal stability of the specimens. These curves are computed as follows:

$$m_{diff}(T) = m_{exp}(T) - m_{calc}(T),$$

where $m_{calc}(T)$ are computed by linear combinations of the experimental TG curves of the sole components. For example, m_{calc} (EVA8/APP) = 0.6 m_{exp} (EVA8) + 0.4 m_{exp} (APP).

From TG data and using the invariant kinetic parameters (IKP) method, the invariant kinetic parameters (activation energy E_{INV} and pre-exponential factor A_{INV}) were computed as well as the most probable degradation functions $f(\alpha)$ (with α the conversion degree) among 18 kinetic functions tested[12]. The main advantage of this method -in comparison with classical methods- is first not to put forward the assumption on the form of the kinetic degradation function. Moreover, the values E_{INV} and A_{INV} obtained are no longer apparent but real characteristics of the blend. From these parameters, the degradation rate V can be modelled as following:

$$V = d\alpha/dt = k \cdot f(\alpha) = A_{INV} \exp(-E_{INV}/RT).f(\alpha).$$

3. QUANTIFYING HAZARD IN FIRE

Heat release has long been recognised as the major fire reaction parameter because it defines fire size[11]. Illustrative measurements of RHR of the EVA8 and the EVA8/APP and EVA8/APP/PA-6 systems are shown in Figure 3.

Associated data for EVA8 are: TTI = 36 s, PkRHR = 1760 kW/m² (145 - 165 s.), THE at 300 s. = 1120 kJ. The RHR curve versus time presents in our experimental conditions an intermediate plateau between 105 and 120 s. (RHR = 750 kW/m²). This phenomenon may be explained by the elimination of acetic acid issued from the degradation of vinyl acetate links, leaving a polyethylenic backbone which degrades later[13].

Addition of APP in EVA8 decreases strongly the RHR values compared to the virgin polymer and leads to delay the time to ignition. Associated data for the EVA8/APP system are : TTI = 45 s, PkRHR = 660 kW/m² (174 s), THE at 300 s = 700 kJ. The RHR curves reveal two main steps : the first one (between 55 and 140 s) may be interpreted as ignition, flame spread on the surface of the material and formation of a protective coating resulting from the blowing up of the specimen. The resulting shield leads to the protection of the material which is efficient up to 140 s (RHR = 450 kW/m²). Beyond 140 s, the second step results from the destruction of this coating and thus of the residual material.

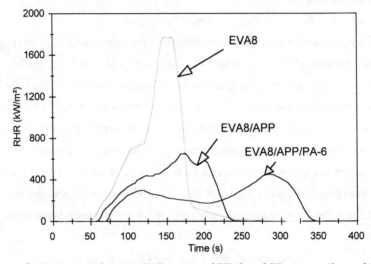

Figure 3. *Comparison between RHR curves of EVA8 and FR systems (3 mm thick).*

Addition of PA-6 in the EVA8/APP system delays the ignition of the specimen (TTI = 57 s) and leads to a significant decrease of the peak of rate of heat release (PkRHR = 450 kW/m², t = 285 s). Associated data for EVA8/APP/PA-6 system are: THE at 400 s = 700 kJ. Ignition and flame propagation are followed by the formation of an expanded carbonaceous shield (70 – 115 s) which limits the heat transfer to the substrate and the oxygen consumption (200 < RHR < 300 kW/m²). The heat resistance of the shield is efficient up to 215 s. Further, the destruction of the protective shield leads to the degradation of the residual polymer. Thus, the addition of PA-6 in the EVA8/APP system has the advantage to delay the ignition of the specimen and to slow down its degradation by increasing the stability of the coating.

4. THERMO-OXIDATIVE DEGRADATION

The first stage of the thermo-oxidative degradation of EVA8 (Figure 4) occurs under 400°C (about 21 wt. %) and corresponds to the complete elimination of acetic acid leading to the formation of a polyethylene crosslinked network[13]. At temperature above 400°C, a major weight loss (79 wt. %) is observed with formation of a carbonaceous residue which is stable up to 520°C and then degrades.

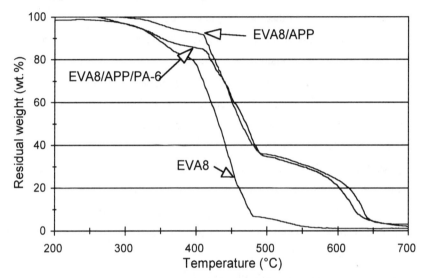

Figure 4 *TG curves of EVA8 and the FR formulations (heating rate : 7.5°C/min, under air flow).*

In the presence of APP, the thermo-oxidative degradation of EVA8 starts at a temperature 50°C higher than for the virgin polymer (Figure 4). If there was no interaction between the degradation products of EVA8 and APP, a weight loss of 17 wt. % (corresponding to the single acetic acid elimination and the beginning of the degradation of APP) would be observed at 410°C. The experimental weight loss, i.e. 8 wt. %, proves therefore that APP allows the stabilisation of the polymer. Afterwards, decomposition occurs via a two step mechanism leading to the formation of a charred material (about 35 wt. % at 490°C), which decomposes then slowly up to 600°C.

Addition of polyamide-6 in the EVA8/APP blend does not introduce a significant modification in the thermal behaviour of the material (Figure 4): the three steps process is observed once again. Nevertheless, considering the three components of the mixture separately, the acetic acid out-gassing and a partial degradation of APP and PA-6 at a temperature of 410°C, a mass loss of 14.5 wt. % should be observed. In fact, a real mass loss of 15 wt. % is noticed in the EVA8/APP/PA-6 blend. As a result, addition of the additives APP and PA-6 does not lead to a significant modification of the virgin material. On the other hand, when considering the stabilisation effect observed with EVA8/APP system, and then the addition of polyamide-6, a mass loss of 10 wt. % would be expected instead of the 15 wt. % observed. It can be concluded that the addition of PA-6 in the EVA8/APP system leads to an initial lower stability of the material. The ablative phenomenon – in comparison with the EVA8/APP mixture – can be explained by the exothermic reaction which occurs at 230°C between PA-6 and APP[2-3] and leads to the carbonisation phenomenon[2]. Accordingly, the exothermic reaction and the presence of free radicals lead to modifications in the EVA8 chain with acetic acid elimination and crosslinking[13]. The carbonaceous material formed (about 35 wt. % at 490°C) degrades slowly up to 600°C.

Figures 5 and 6 show the weight difference curves m_{diff} (T) of the FR systems. Addition of APP in EVA8 leads to a preservation of the polymeric material in the 290 – 490°C range (Figure 5). The maximum of stability is reached for a temperature of 410°C. The resulting residue is stable in the 490 – 540°C range and then begins to degrade slowly. The comparatively low stability of the residue observed between 540 and 670°C may be explained by the degradation of phosphocarbonaceous species leading to the formation of phosphorus oxides which evolve.

In the case of the EVA8/APP/PA-6 system (Figure 6), the same phenomenon as above is observed in the 290 – 430°C range. The addition of APP/PA-6 mixture allows the setting up of a first coating with a maximum of protection at 410°C. Above this temperature, the shield looses partially its efficiency. Above 430°C, a second protective process is observed: the presence of PA-6 allows the formation of a second shield which acts as a relay of the first one. The maximum of efficiency of this coating is reached at 470°C. This can be related to the results obtained in the cone calorimeter experiments. It is observed that the addition of PA-6 increased the heat resistance and the stability of the coating, delaying the evolution of degradation gases. Finally, the comparatively low stability observed in the 580 – 680°C range may be assigned to the degradation of the phosphocarbonaceous species.

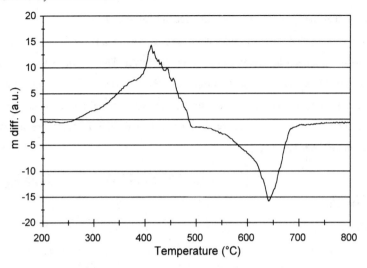

Figure 5. *Difference between experimental and calculated TG curves of the EVA8/APP system.*

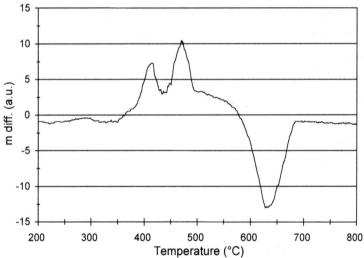

Figure 6. *Difference between experimental and calculated TG curves of the EVA8/APP/PA-6 system.*

The IKP method allows the computation of the invariant kinetic parameters and the modelling of the thermal degradation of the materials. In this part, we assume that the main step of the degradation of the three systems is determining for the synthesis of fuel in fire conditions and that the resulting flammability depends on this step.

Table 1 shows that the invariant kinetic parameters of the two FR systems are quite the same. The values obtained for the virgin polymer are lower. This confirms that the protection occurs in the condensed phase by a stabilisation of the material. Moreover, in a kinetic point of view, there is no significant difference between blowing up and intumescence. As a result, kinetic parameters may not be considered alone to explain the different fire behaviours of EVA8/APP and EVA8/APP/PA-6.

	EVA8	**EVA8/APP**	**EVA8/APP/PA-6**
E_{inv} (KJ/MOL)	89 ± 10	200 ± 20	220 ± 20
LOG A_{inv} (A_{inv} : s^{-1})	3.7 ± 0.1	11.6 ± 0.2	12.9 ± 0.2

Table 1. *Invariant activation energies and pre-exponential factors of the thermo-oxidative degradation of the materials.*

The computation of the probability distribution in association with the kinetic functions leads to the most probable shape of the degradation of the specimens[14]. The mechanism of degradation is very complex and cannot be represented by a single function but by a set of functions. The number of different overlapped reactions which occur during the degradation can explain this.

Using the invariant kinetic parameters and the kinetic functions $f(\alpha)$, the modelling of the degradation rate V shows that (Figure 7):

- for low values of temperature (T < 640 K): $V_{EVA8} > V_{EVA8/APP/PA-6} > V_{EVA8/APP}$. APP plays the role of a stabiliser,

- for higher temperatures (640 < T < about 735 K): $V_{EVA8/APP/PA-6} > V_{EVA8} > V_{EVA8/APP}$. APP plays the role of a stabiliser. On the other hand, the classical phenomenon of ablation characteristic of intumescence[4] is observed for the FR system EVA8/APP/PA-6: the material degrades quickly in order to protect itself. It leads to the formation of a thermally stable carbonaceous material,

- finally, for temperature higher than 735 K : $V_{EVA8/APP/PA-6} > V_{EVA8/APP} > V_{EVA8}$. The formation of the carbon or phosphate glass-based materials allows the protection of the FR specimens.

Thus, the addition of APP in EVA8 leads to a stabilisation and to the formation of an expanded coating which allows the encapsulation of the gaseous degradation products.

On the other hand, the addition of APP/PA-6 increases the stability of the material in the higher temperature range, the protection is provided by material resulting from the low temperature ablative phenomenon.

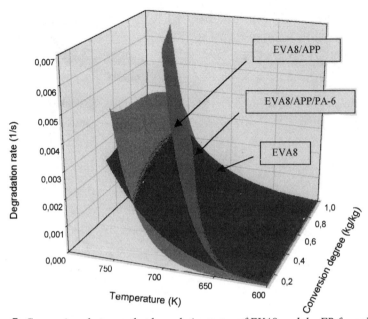

Figure 7. *Comparison between the degradation rates of EVA8 and the FR formulations.*

4. HEAT TRANSFER

A comparison of temperatures in EVA8 and in the FR specimens is presented in Figure 8 in combustion conditions. Two centimetres thick sheets have been used here to study the temperature changes in the materials. The combustion of the 2 cm specimens occurs more slowly than in the case of the 3 mm thick sheets (for example, the complete combustion of a 2 cm EVA sheet is reached within 1000 s). The temperature is strongly reduced in the case of the FR polymers when time > 100 s, especially when PA-6 was added. As an example, at t = 700 s, the temperature reaches 650°C in the case of EVA and lies only between 150 and 375°C depending on the distance in the case of EVA-APP/PA-6. The temperatures of the intumescent material are always below the temperature where the degradation rate of EVA becomes fast (about 380°C). This temperature profile demonstrates also that the intumescent shield developed from EVA/APP/PA-6 reduces strongly the heat transfer into the structure between the material surface and core ($\Delta T = 250°C$ at t = 700 s between 2 and 14 mm).

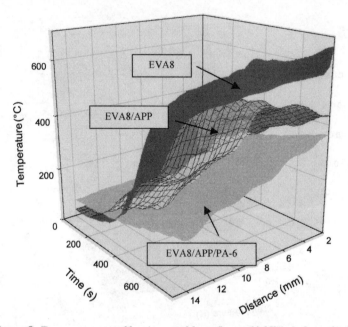

Figure 8. *Temperature profiles (external heat flux = 50 kW/m²; 2 cm thick sheets) of EVA8 and of the FR formulations.*

Using invariant kinetic parameters and results obtained with the cone calorimeter experiments, it is possible to calculate the temperature of the degradation front T_d, namely the place where the material degrades evolving gases and, therefore, the place where there is a real weight loss[8,14]. Figure 9 presents the evolution of T_d in the case of 2 cm thick sheets. The degradation front moves from the surface of the material - at the beginning of the combustion - to the inside of the sheet (Table 2).

Samples	T_d (°C) (500 s)	Time to reach 2 mm depth (s)	Time to reach 14 mm depth (s)
EVA8	≈ 450	250	700
EVA8/APP	440	340	1000
EVA8/APP/PA-6	380	890	>1750

Table 2 *Value and position of T_d in the different specimens (2 cm thick sheets).*

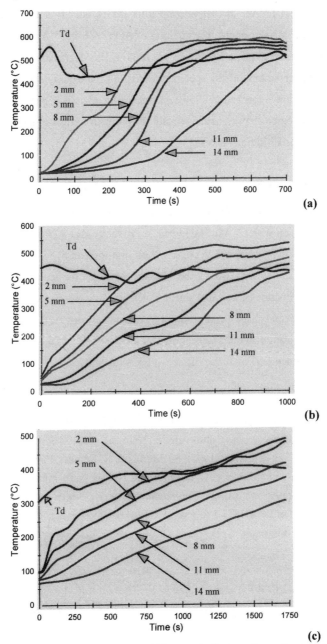

Figure 9 *Temperature profile and Td versus time and distance from the initial surface for EVA8 (a), EVA8/APP (b), EVA8/APP/PA-6 (c)(cone calorimeter conditions, heat flux of 50 kW/m², 2 cm thick sheets).*

In the case of EVA8, the degradation front moves quickly down. Addition of APP slows down its displacement. The addition of APP/PA-6 in EVA8 leads only to a comparatively superficial phenomenon up to 1000 s; the degradation front is thus located on the upper volume of the specimen during all the exposure to the heat flux (only a depth of 9 mm is reached in 1750 s). Comparison of times requested to reach a same degradation degree shows unambiguously the protective effect of swelling and of both the swelling/carbonisation process.

4. CONCLUSION

This work investigated the fire behaviour of the EVA8/APP and EVA8/APP/PA-6 systems in comparison with EVA8. The kinetic study shows that APP stabilises the specimen up to 430°C. Addition of PA-6 in EVA8/APP leads to a comparative improvement of the fire retardancy related to the increase of the heat-resistance of the shield. This resistance is explained by the formation of a second shield which acts as a relay of the shield formed from EVA8/APP. This second shield is the result of a reaction between APP and PA-6 leading to a protective ablation phenomenon. The heat transfer study shows that this protection is efficient enough to limit the degradation to the surface of EVA8/APP/PA-6.

References

1. P. Nathiez, *"Amélioration du comportement au feu de l'orgalloy R6000 par un système intumescent"*, Rapport au Conservatoire National des Arts et Métiers, available L.C.A.P.S, Villeneuve d'Ascq (1991); S. Bourbigot, R. Delobel, P. Nathiez and P. Bréant, *"Compositions ignifugeantes pour mélanges de résines thermoplastiques contenant une zéolithe et mélanges de résines thermoplastiques renfermant lesdites compositions"*, European patent n° 94.401317.8 (assigned to ELF Atochem S.A.), 1994.
2. M. Le Bras, S. Bourbigot, C. Delporte, C. Siat and Y. Le Tallec, *Fire & Materials*, **20**, 191-203 (1996).
3. C. Siat, S. Bourbigot and M. Le Bras, *"Structural study of the polymer phases in intumescent PA-6/EVA/FR additive blends"* in *"Recent Advances in Flame Retardancy of Polymeric Materials"* (Volume 7), M. Lewin Ed., BCC. Pub., Norwalk, 318-326 (1997).
4. S.V. Levchik, L. Costa and G. Camino, *Polym. Deg. & Stab.*, **36**, 229-237 (1992).
5. C. Siat, M. Le Bras and S. Bourbigot, *Polym. Deg. & Stab.*, **58**, 303-313 (1997).

6. *Standard Test Method for Measuring the Minimum Oxygen Concentration to Support Candle-like Combustion of Plastics*, ASTM D2863/77, Philadelphia (1977).

7. *Tests for Flammability of Plastic Materials for Part in Devices and Appliances*, ANSI/ASTM D635-77, Underwriters laboratories, Northbrook (1977).

8. S. Bourbigot, R. Delobel, M. Le Bras and D. Normand, *J. Chim. Phys.*, **90**, 1909 (1993).

9. C. Hugget, J. Fire and Flammability, **12** (1980).

10. V. Babrauskas, *"Development of Cone Calorimeter – A bench scale rate of heat release based on oxygen consumption"*, NBS-IR 82-2611, US NBS, Gaithersburg (1982).

11. S.J. Grayson, *Heat Release in Fire*, V. Babrauskas and S.J. Grayson ed., Elsevier Appl. Sci., London, (1992) 1-5.

12. N. Rose, M. Le Bras, S. Bourbigot and R. Delobel, *Polym. Deg. & Stab.*, **45**, 387-397 (1994).

13. J.D. Nam and J.C. Seferis, J. Polym. Sci., Part B, *Polymer Physics*, **29**, 601-608 (1991).

14. S. Bourbigot, M. Le Bras and R. Delobel, *J. Fire Sci.*, **13**, 3-22 (1995).

Acknowledgements

The authors gratefully acknowledge the Conseil Régional du Nord - Pas de Calais and the Conseil Général du Pas de Calais for their financial support.

POLYAMIDE-6 FIRE RETARDED WITH INORGANIC PHOSPHATES

G. F. Levchik, S. V. Levchik, A. F. Selevich, , A. I. Lesnikovich, A. V. Lutsko
Research Institute for Physical Chemical Problems
Byelorussian University, Leningradskaya 14, 220050 Minsk, Belarus

L. Costa
Dipartimento di Chimica Inorganica, Chimica Fisica e Chimica dei Materiali,
Universita di Torino, via P. Giuria 7, Torino 10125, Italy

1. INTRODUCTION

Ammonium polyphosphate (APP) is an effective fire retardant additive for polyamide-6 (PA-6)[1, 2]. Because APP involves PA-6 into the charring, an intumescent char can be produced without any additional charrable material. Practical usage of APP in aliphatic polyamides is limited because the temperature of the beginning of APP thermal decomposition is close to the temperature of the injection molding of polyamides.

Recently we showed[3, 4] that some inorganic pigments can trap acidic species evolved at the thermal decomposition of APP and therefore these pigments stabilise polyamide when it is compounded with APP. Furthermore, some inorganic pigments improve fire retardant behavior of APP[3-5]. Mechanistic studies of interaction between APP and pigments showed[6, 7] that binary metal - ammonium (BMAPs) are formed. These phosphates are probably to be responsible for the improving of fire retardant action of APP.

In this work we prepared various binary metal - ammonium phosphates and tested their fire retardant efficiency in polyamide-6. Thermal decomposition behavior of BMAPs and their formulations with PA-6 was studied by the methods of thermal analysis.

2. EXPERIMENTAL

It was developed a new method for synthesis of binary metal-ammonium phosphates (BMAPs)[8, 9]. APP (Exolit 422, Hoechst) and corresponding inorganic oxides or salts of volatile acids were mixed with the equivalent ratios 2:10 and heated in an oven at 200-400°C for 12-72 hours. Since the obtained BMAPs are mostly insoluble in water, the excess of polyphosphoric acid was removed by water or by acetone - water mixture. BMAPs were characterized by X-ray diffraction. Commercial polyamide-6 (PA-6) from Khimvolokno (Belarus, Grodno) was used for the preparation of formulations. PA-6 was mixed with 20 wt. % of BMAPs in a closed mixer at ca. 240 - 250°C for 5 minutes. Plates for combustion tests were moulded from the mixture at the same temperature.

Combustion of the formulations was studied either by the limiting oxygen index (LOI; following the ASTM D 2863 standard) or in air in horizontal configuration where the average time to extinguishing was calculated from the series of five burned specimens (Russian Standard GOST 28157-89).

Thermal analysis of the fire retarded polyamide-6 was carried out using Mettler TA 3000 thermal analysers supplied by TGA and DSC cells. Experiments were carried out in argon or air flow atmosphere (60 cm^3/min) at heating rate 10°C/min. For better separation of weight loss steps some experiments were carried out by using the high resolution thermogravimetry (HRTG) of Du Pont 2950 which decreases heating rate when the rate of weight loss increases.

Solid residues of the formulations with BMAPs were collected at different steps of thermal decomposition in inert atmosphere and characterized by infrared using the Specord 75 spectrometer.

3. RESULTS AND DISCUSSION

3.1. Thermal stability of BMAPs

Thermogravimetry curves for some newly prepared BMAPs are shown in Figure1. Apart from LaNH$_4$(PO$_3$)$_4$ all other products have thermal stability superior to APP. Solid residues from BMAPs at 600°C are usually higher than the residue from APP. This shows that BMAPs can be compounded to polyamide 6 with less impact than APP does.

3.2. Combustion

Pure PA-6 shows LOI=25 % (Table 1). It drips and does not self-extinguish at combustion in air. 20 wt. % of APP helps to increase LOI to 27.3 % but specimens do not extinguish in air. Fire retardant performance of BMAPs depends on their chemical composition (Table 1).

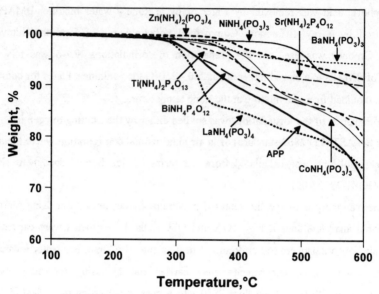

Figure 1. *Thermogravimetry of APP and some BMAPs.*
(Heating rate 10 °C/min. Argon flow 60 cm³/min).

K and Ba binary phosphates decrease LOI. The formulations with these BMAPs do not extinguish in air. They behave as inert fillers which usually increase flammability of polyamides if added at low concentration.

Sr, Zn, Bi and Ti binary phosphates are effective fire retardants, since they increase LOI and provide self-extinguishing at combustion in air. $Ti(NH_4)_2P_4O_{13}$ has LOI higher than APP. Mn and Ni binary phosphates show good LOI but they are not effective at combustion in air.

Co binary phosphates do not show correlation between LOI and combustion in air. $CoNH_4(PO_3)_3$ having high LOI performs badly in air, whereas $Co(NH_4)_2(PO_3)_4$ with lower LOI shows a good result in air. La binary phosphate provides both good LOI and good performance in air, however it is too expensive to be used in PA-6.

As it was shown in Table 1, $BiNH_4(PO_3)_4$ is very active at combustion in air. Therefore we synthesised a series of different bismuth phosphates and tested their efficiency in polyamide-6. Table 2 shows fire retardancy performance of these phosphates. Some phosphates (formulations 2-6) have LOI similar to pure PA-6, whereas bismuth-ammonium trihydrophosphate and bismuth ultraphosphate (formulations 7 and 8) increase LOI to 27.9 and 29.2 % respectively.

Table 1. *Combustion performance of polyamide-6 fire retarded by 20 wt. % of BMAPs.*

No	Additive	LOI Vol. %	Average time to extinguishing (s)
1	-	25.0	burns
2	APP	27.3	burns
3	$KNH_4(PO_3)_2$	23.8	burns
4	$BaNH_4(PO_3)_3$	23.8	burns
5	$Sr(NH_4)_2P_4O_{12}$	26.6	15
6	$Zn(NH_4)_2(PO_3)_4$	26.6	25
7	$BiNH_4(PO_3)_4$	25.9	2.5
8	$Ti(NH_4)_2P_4O_{13}$	30.5	7.0
9	$MnNH_4(PO_3)_4$	26.5	burns
10	$NiNH_4(PO_3)_3$	26.0	burns
11	$CoNH_4(PO_3)_3$	29.8	burns
12	$Co(NH_4)_2(PO_3)_4$	27.3	10
13	$LaNH_4(PO_3)_4$	29.8	20

Table 2. *Combustion performance of polyamide-6 fire retarded by 20 wt. % of bismuth phosphates.*

No	Additive	LOI Vol. %	Average time to extinguishing, s.
1	-	25.0	burns
2	$BiPO_4$	25.2	burns
3	$Bi_2P_4O_{13}$	25.9	5.0
4	$BiH(PO_3)_4$	25.2	4.0
5	$BiNH_4(PO_3)_4$	25.9	2.5
6	$BiNH_4P_4O_{12}$	25.1	2.5
7	$BiNH_4HP_3O_{10}$	27.9	2.0
8	BiP_5O_{14}	29.2	0

Bismuth phosphates in general perform better in air than in LOI test. The only formulation with $BiPO_4$ burns like pure PA-6, whereas all other bismuth phosphates provide self-extinguishing. The average combustion time goes down from the formulation 3 to 8. Bismuth ultraphosphate, BiP_5O_{14} makes PA-6 non combustible in air. However, the bars for combustion testing with BiP_5O_{14} are brittle which means that this phosphate deteriorates PA-6 upon compounding.

3.3. Thermal analyses

At linear heating (10°C/min) in inert atmosphere the formulations with BMAPs usually decompose in one step at temperatures lower than pure PA-6 but higher than the formulation with APP. In high resolution thermogravimetry where heating rate slows down if weight loss

occurs, thermal stability of some formulations with BMAPs is similar to that with APP
(Figure 2). High resolution thermogravimetry distinguishes three steps of weight loss in the
250 - 360°C temperature range for some formulations with BMAPs whereas the formulation
with APP shows only two steps of weight loss.

The formulation with $CoNH_4(PO_3)_3$ behaves differently from many BMAPs. It mostly
degrades above 300°C, showing a second step at 390 - 410°C.

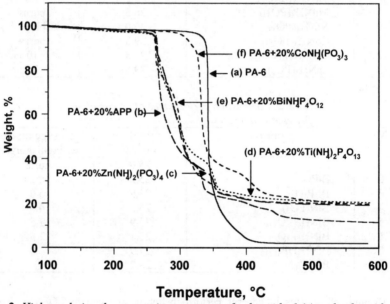

Figure 2. *High resolution thermogravimetry curves of polyamide-6 (a) and polyamide-6 fire
retarded by 20 wt. % of (b) APP, (c) $Zn(NH_4)_2(PO_3)_4$, (d) $Ti(NH_4)_2P_4O_{13}$, (e) $BiNH_4P_4O_{12}$ or
(f) $CoNH_4(PO_3)_3$. (Nitrogen flow 60 cm^3/min, heating rate 10°C/min, resolution factor +5).*

DSC gives further evidence for the interaction between PA-6 and BMAPs. Two
endotherms are observed for pure PA-6 in nitrogen: melting at 220°C and thermal degradation
at 431°C (Figure 3). $BiNH_4P_4O_{12}$ melts and decomposes simultaneously at 379°C and then
shows another endothermic peak of decomposition at 456°C. The formulation with
$BiNH_4P_4O_{12}$ begins to decompose at ca. 270°C which is lower than degradation temperature
ranges for pure PA-6 or $BiNH_4P_4O_{12}$. There are three endothermic processes which are likely

to correspond to the three steps of weight loss in Figure 2. Thus, DSC shows that both PA-6 and $BiNH_4P_4O_{12}$ mutually destabilise each other.

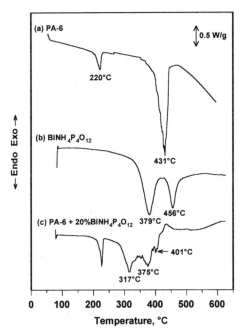

Figure 3. *DSC curves of: (a) polyamide-6, (b) $BiNH_4P_4O_{12}$ and (c) polyamide-6 fire retarded by 20 wt. % of $BiNH_4P_4O_{12}$ (Nitrogen flow 60 cm^3/min, heating rate 10 °C/min).*

Thermal decomposition behavior of Ny-6 with APP and Ny-6 with various BMAPs is compared in Figure 4 in terms of DSC. Apart from melting of Ny-6 the formulation with APP shows three steps of degradation. The formulations with $Zn(NH_4)_2(PO_3)_4$, $Ti(NH_4)_2P_4O_{13}$ and $BiNH_4P_4O_{12}$ also have three steps of thermal decomposition which proves similarity of the reactions of these BMAPs and APP with Ny-6 The formulation with $CoNH_4(PO_3)_3$ decomposes differently showing the endothermic peak at 412°C and two exothermic above 450°C.

3.4. Infrared characterisation of the solid residues

Solid residues obtained at different steps of the thermal decomposition of the formulations with BMAPs were analysed by infrared. All formulations showed similar behavior, therefore the infrared spectra of the formulation with $Ti(NH_4)_2P_4O_{13}$ are shown as an example (Figure 5).

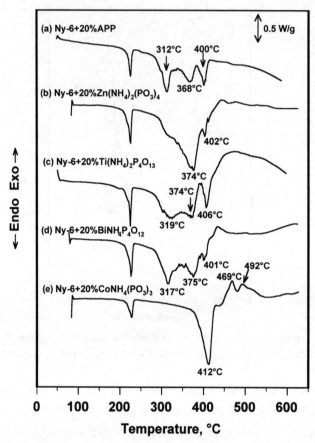

Figure 4. *DSC curves of polyamide-6 fire retarded by 20 wt. % of: (a) APP, (b) $Zn(NH_4)_2(PO_3)_4$, (c) $Ti(NH_4)_2P_4O_{13}$, (d) $BiNH_4P_4O_{12}$ or (e) $CoNH_4(PO_3)_3$. (Nitrogen flow 60 cm^3/min, heating rate 10 °C/min).*

In the initial spectrum, characteristic absorptions of polyamide-6[10] at 3307, 3090, 1642 and 1542 cm^{-1} and absorptions of $Ti(NH_4)_2P_4O_{13}$[11] at 1070, 976 and 909 cm^{-1} are observed. The bands at 1430 and 1262 cm^{-1} are likely to be superposition of the absorptions from PA-6 and the additive.

At the first step of weight loss the main changes in the spectrum are observed in the region of $Ti(NH_4)_2P_4O_{13}$ absorptions at 1300 - 800 cm^{-1} which indicate that the inorganic salt decomposes first. However, small band at 2247 cm^{-1} attributed to nitrile chain ends[12] proves that PA-6 also undergoes decomposition at this step. It is likely that $Ti(NH_4)_2P_4O_{13}$ transforms to phosphate with evolution of ammonia and polyphosphoric acid. The acid catalyses degradation of PA-6.

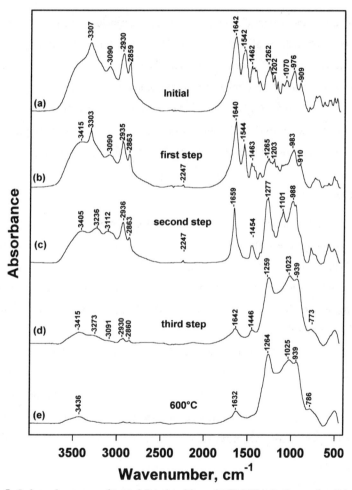

Figure 5. *Infrared spectra of initial PA-6 + 20 wt. % Ti(NH$_4$)$_2$P$_4$O$_{13}$ and solid residues collected in thermogravimetry in inert atmosphere at different steps of weight loss (Pellets in KBr).*

At the second step of weight loss, drastic changes in the infrared pattern of PA-6 are observed. New absorptions at 3500 - 1500 cm^{-1} are typical for aliphatic primary amides[12] whereas absorptions at 1300 - 900 cm^{-1} are likely to be a superposition of strong bands from titanium phosphate[11] and phosphoric esters[13]. The pattern of this spectrum is similar to the IR pattern of 5-amidopentyl polyphosphate[2] which is a product of reaction of polyphosphoric acid with polyamide-6 upon heating. This proves that BAPMs react with polyamide 6 similar to APP.

Thermal decomposition of 5-amidopentyl polyphosphate is observed in the third step. Very small concentration of aliphatics is still detected at this temperature (absorptions at 2930, 2980 and 1446 cm^{-1}). The band at 1642 cm^{-1} is likely to belong to aromatic char[14]. This stage is responsible for the formation of the intumescent char.

Absorptions of titanium phosphates dominate in the spectrum of solid residue at 600°C. The weak band at 1632 cm^{-1} seems to be due to aromatic char, however in the same region might contribute also absorbed water, the second band of which is seen at 3436 cm^{-1}.

4. CONCLUSIONS

Some binary metal - ammonium phosphates (BMAPs) are more thermally stable than ammonium polyphosphate (APP). BMAPs of transitory metal are often more effective in polyamide-6 (PA-6) than APP, whereas BMAPs of alkali or earth alkali metals are usually less effective. In the series of various bismuth phosphates fire retardant efficiency increases in the following order: $BiPO_4$ < $Bi_2P_4O_{13}$ < $BiH(PO_3)_4$ < $BiNH_4(PO_3)_4$ < $BiNH_4P_4O_{12}$ < $BiNH_4HP_3O_{10}$ < BiP_5O_{14}.

As it is seen from the thermal analysis, BMAPs destabilise PA-6. Similarly, BMAPs begin to degrade at lower temperature when they are mixed with PA-6. Polyphosphoric acid evolved from BMAPs catalyses thermal decomposition of the polymer. In spite of differences in combustion and thermal decomposition performances, BMAPs show chemical reactions (similar to APP) with PA-6 which provide intumescent char on the surface of the burning polymer.

References

1. S. V. Levchik, L. Costa and G. Camino, *Polym. Degrad. Stab.*, 1992, **36**, 229.

2. S. V. Levchik, G. Camino, L. Costa and G.F. Levchik, *Fire Mater.*, 1995, **19**, 1.

3. G. F. Levchik, A. F. Selevich, S. V. Levchik and A. I. Lesnikovich, Byelorussian Patent No. 00076-01, (to Byelorussian University), 1993.

4. S. V. Levchik, G. F. Levchik, G. Camino and L. Costa, *J. Fire Sci.*, 1995, **13**, 43.

5. S. V. Levchik, G. F. Levchik, G. Camino, L. Costa and A. I. Lesnikovich, *Fire Mater.*, 1996, **20**, 183.

6. G. F. Levchik, A. F. Selevich, S. V. Levchik and A. I. Lesnikovich, *Thermochim. Acta*, 1994, **239**, 41.

7. G. F. Levchik, S.V. Levchik, P. D. Sachok, A. F. Selevich, A. S. Lyakhov and A. I. Lesnikovich, *Thermochim. Acta*, 1995, **257**, 117.

8. A. F. Selevich, G. F. Levchik, A. I. Lesnikovich and S. V. Levchik, Byelorussian Patent No. 00260-01, (to Byelorussian University), 1993.

9. A. F. Selevich, S. V. Levchik, A. S. Lyakhov, G. F. Levchik, A.I . Lesnikovich and J.-M. Catala, *J. Solid State Chem.*, 1996, **125**, 43.

10. C. G. Cannon, *Spectrochim. Acta*, 1960, **16**, 302.

11. R. Ya. Melnikova, V. V. Pechkovskii and E. D. Dzyuba, «*The Atlas of Infrared Spectra of Phosphates. Condensed Phosphates*», Nauka, Moscow, 1985 (in Russian).

12. D. Lin-Vien, N. B. Colthup, W. G. Faterley and J. G. Grasselli, «*The Handbook of Infrared and Raman Characteristic Frequencies of Organic Molecules*», Academic Press, Boston, 1991.

13. L. C. Thomas, «*Interpretation of the Infrared Spectra of Organophosphorus Compounds*», Heyden, London, 1974

14. Zawadskii, in *"Chemistry and Physics of Carbon. A Series of Advances"*, Vol. 21, P.A. Thrower ed., Marcel Decker, New York, 1990, 147.

Acknowledgements

This research was sponsored by the International Association for the Promotion of the Cooperation with Scientists from the Independent States of the Former Soviet Union under the INTAS project 93-1846.

MICROENCAPSULATED FIRE RETARDANTS IN POLYMERS

A. Antonov, E. Potapova, T. Rudakova, I. Reshetnikov,

Polymer Burning Laboratory, Moscow Institute of Synthetic Polymeric Materials
70, Profsoyuznaya street., Moscow 117393, Russia

N. Zubkova, M. Tuganova,

Department of Technology of Chemical Fibres, Moscow State Textile Academy
1 Malaya Kaluzhskaya, Moscow, 117918, Russia

N. Khalturinskij

Semenov Institute of Chemical Physics
4, Kosygina street, Moscow, 117977, Russia

1. INTRODUCTION

A major shortcoming of engineering thermoplastics is their burning characteristics. The flammability of the some polymers is higher that than of a wood and the natural fibres. The calorific values of the common polymers such as polyethylene, polypropylene, polystyrene, polymethylmethacrylate are 46000 - 27000 kJ/kg, whereas this value of a wood is 19000 kJ/kg[1]. In addition, the combustion of some polymer materials is accompanied by smoke and soot formation, droplets and the isolation of highly toxic products. Thus, the wide application of polymer materials gives rise to the desirability to develop fire retarded polymer materials. An important technique to create improved flame retarded polymers consists in the addition of fire retardants (FR). It is the usual pathway for the development of fire retarded polymeric materials. On the other hand, this approach has a number of problems.

A major problem consists in replacement of low molecular weight containing FR, such as polybrominated diphenylethers. Using these additives gives rise to a number of problems such as sharply increased smoke formation, corrosion of equipment during processing and isolation of highly toxic brominated dibenzofurans and dibenzodioxins upon combustion or thermal degradation[2, 3].A good solution for these problems is replacement of the halogen containing low molecular FR by use of phosphorus or metal containing FR's, especially intumescent systems.

We have previously shown that the some compounds based on methylphosphonic acid can be applied as fire retardants for polyolefines, such as the product of the condensation of melamine-formaldehyde resin with methylphosphonic acid (MFMPA) and the diamide of methylphosphonic acid (DAPA)[4]. However, it was found out that DAPA reacts with polymers during processing[5]. In this connection, DAPA can not be used for the development fire retarded polymer materials based on polycaproamide (PCA) and polyethyleneterephthalate (PET).

The potential solution for this problem related to the use of some fire retardants consists in microencapsulation. Microencapsulated coatings prevent the interaction between the polymer and the fire retardants at processing temperatures as well as the sublimation and the exudation of fire retardants from the fire retarded polymer. Our work deals with the investigation of the efficiency of microencapsulated fire retardants containing DAPA in polymers.

2. EXPERIMENTAL

2.1. Materials

Polypropylene (PP; Russian grade 210020; molecular weight: 120,000; MFI: 1.8 g/min; density: 910.0 kg.m^{-3}), polyethylene low density (LDPE; Russian grade 10802-020; molecular weight: 28,000; MFI: 2.0 g/min; density: 918.5 kg.m^{-3}), polycaproamide (PCA; Russian grade 6-06-C9-83; molecular weight: 25,000; melting temperature: 215°C; density: 1130.0 kg.m^{-3}) and polyethyleneterephthalate (PET; Russian grade 6-05-830-83210020; molecular weight: 22,000; melting temperature: 260°C; density: 1320.0 kg.m^{-3}) were used to prepare the plastic compositions for this study.

The diamide of methylphosphonic acid (DAPA) was used as a fire retardant for the polymers[6]. The low molecular weight polyethylene (PE), the aromatic polyamide based on terephthalic acid and meta-phenylenediamine (PMFIA) and polyvinyltriethoxysilane (PVTS) were used as the microcapsulated shell. The technology for the development of microencapsulated fire retardants was previously described in detail[7].

Blends of the plastics containing fire retardants were prepared on a "Brabender PLE-330" plastograph at 160-260°C (depending on the plastic) during a 15 min period,, followed by molding at 15 Mpa and 170-270°C.

2.2. Limiting Oxygen Index Test

The limiting oxygen index (LOI) was measured using a Stanton Redcroft instrument on sheets (length 120 mm, width 60 mm, thickness 3 mm) according to the standard "oxygen index" test (ASTM D2863/77).

2.3. Thermogravimetric analysis

Thermogravimetric analysis (TGA) were carried out in air at a heating rate of 10°C/min using a derivatograph-C (system of F. Paulik, J. Paulik, L. Erdey (Hungary)).In each case samples weighed 20 mg.

2.4. Thermal properties of polymer blends

The heats and temperatures of melting of polymer blends were determined by differential scanning calorimetry (DSC) using DSM-3 device with a computer processing system. The measurements were carried out at a heating rate of 16°C per minute. The mass of the samples were 6-10 mg.

2.5. Physical properties of foamed char

The physical properties of the foamed char formed during the combustion of the fire retarded polymer compositions have been investigated as described previously[4].

2.6. Composition of pyrolysis products of polymer blends

The composition of the pyrolysis products of the polymer blends was studied by Stepwise Pyrolysis Gas Chromatography[8]. Chromatograph "Chrome 5" was used. The mass of the samples was 2-5 mg. Pyrolysis proceeded in the pyrolytic cell and face values was registered for each 50°C up to 1000°C. During the identification of the degraded products (CO_2, CO, H_2O), a standard ammonium oxalate was used.

2.7. High-temperature pyrolysis of polymer blends

The high-temperature pyrolysis of the polymer blends was investigated using the method as previously described[9].

2.8. Polymer "melt" viscosity at high temperature

The polymer "melt" viscosity of polymer compositions at high temperature during thermal degradation were measured by rotary viscosimetry[10].

2.9. Physico-mechanical data of the polymer blends

Physico-mechanical data of the polymer blends were characterised by their tensile strength and relative elongation (deformation speed 100 mm/min according to GOST 16272-89).

3. RESULTS AND DISCUSSION

The synthesised microencapsulated compounds based on DAPA have been investigated as fire retardants for PE, PP, PCA and PET. The LOI data for polymer blends containing the microencapsulated fire retardants based on DAPA are shown in Table 1.

Table 1. *The LOI data for polymer's blends containing microencapsulated fire retardants based on DAPA.*

N	Polymer	Microcapsulated shell	Content wt. %	LOI Volume %
1	PE	-	25	25,4
2	PE	PE	25	25,0
3	PE	PVTS	25	28,0
4	PP	-	25	24,8
5	PP	PE	25	24,6
6	PP	PVTS	25	27,0
7	PCA	-	30	24,8
8	PCA	PVTS	30	29,2
9	PCA	PMFIA	30	27,6
10	PET	-	10	27,5
11	PET	PVTS	10	27,7
12	PET	PMFIA	10	29,4

As can be seen from Table 1, in most cases the using of microencapsulated shells results in increasing the efficiency of DAPA, especially for the- PVTS and PMFIA shells used for PCA. It should be noted that addition of 10 % of PMFIA to PCA results in an increased LOI of PCA, 20 and 22,5 %, respectively.

To explain the action mechanism for microencapsulated FR for polymer blends, the thermal and thermo-physical properties of some tested polymer blends have been investigated. The thermal degradation of microencapsulated fire retardants based on DAPA and polymer blends containing FR have been investigated by means of TGA and DSC. There is no additional step in the TGA and DSC curves for DAPA with microencapsulating shells. There is not interaction between their microencapsulating shells and DAPA during thermal degradation. On the other hand, it should be noted that the rate of decomposition (W) and the maximum temperature of degradation (T_D) of DAPA decreased in the presence of the polymer shell (Table 2).

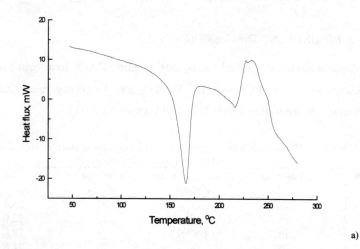

a)

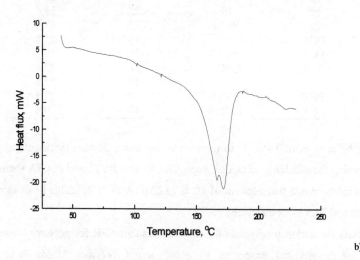

b)

Figure 1. *DSC data of PP (a) and PP+PVTS (b).*

Table 2. *The effect of polymer's coating on the thermal properties of DAPA.*

Polymer's shell	W, mg/min	T_D, °C
none	8,4	348
PE	5,3	348
PVTS	6,0	350
PMFIA	5,0	358

The thermal degradation of the blends based on PE, PP, PET and PCA with microencapsulating shells have been investigated. It was found that the microencapsulated shells don't affect the thermal properties of PE, the PE microencapsulated shells don't affect the thermal properties of all the tested polymers. DSC shows that PVTS protects the oxidation of PP (Figure 1). On the other hand the PVTS shell does not effect the thermal properties of PCA and PET. The glass transition temperatures for the blends of PCA and PET with PMFIA have been investigated. It has been established that the blends of PET + PMFIA and PCA + PMFIA are compatible (criteria of compatibility was a single transition temperature, as an example: PMFIA increases Tg of the matrixes; thus both blends are partially compatible) when the content of PMFIA up to 2 %. The addition of 2 wt. % of PMFIA to PET and PCA results in increasing their glass transition temperatures: 25 and 23°C, respectively.

Table 3. *The effect of the shell on the thermal properties of polymer composition containing DAPA.*

N	Polymer	Microcapsulated shell	W, mg/min	T_D, °C
1	PE	-	13,0(17,0)	472(490)
2	PE	PE	11,0	465
3	PE	PVTS	9,5	462
4	PP	-	8,5(12,0)	459(462)
5	PP	PE	9,0	456
6	PP	PVTS	7,6	455
7	PCA	-	12,0(17,0)	390(440)
8	PCA	PVTS	5,0	373
9	PCA	PMFIA	10,0	363
10	PET	-	15,0(24,0)	423(440)
11	PET	PVTS	6,5	430
12	PET	PMFIA	11,0	430

Numbers in brackets denote values for thermal degradation of original polymer.

The TGA curves of the binary mixture shows that DAPA containing PVTS microencapsulated shells do not affect the initial interval for PE decomposition. On the other hand, the FR decreases the thermal stability of PP (Figure 2), PET and PCA. It should be noted that the rate of decomposition (W) and the maximum degradation temperature T_D of polymer's composition containing DAPA decreases in the presence of shell (Table 3).

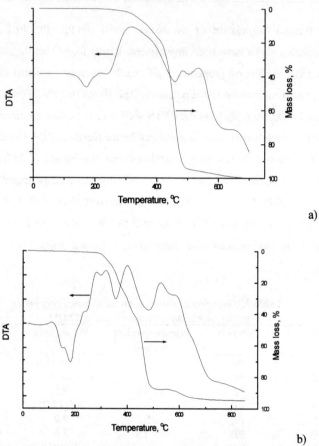

a)

b)

Figure 2. *TG and DTA data of PP (a) and PP + DAPA(PVTS) (b).*

Moreover, the char yield of the tested polymers increases in the presence of DAPA containing microencapsulated shell of PVTS increase. Thus, DAPA containing microencapsulated PVTS shell intensifies the polymer char formation.

The DTA curves (Figure 3) of polypropylene and polypropylene containing DAPA and DAPA with PVTS shell show that the thermal degradation of PP was accompanied with a large exothermic effect between 250 and 465 °C with the total heat effect of 3300 kJ/kg. The addition of DAPA and DAPA containing PVTS results in decreased total heat of the decomposition of PP, 2477 and 2563 kJ/kg, respectively. It should be noted that the small increase in the total heat effect for the decomposition of PP in the presence of DAPA containing PVTS in comparison with that of DAPA was related to oxidation reactions. The oxidation reactions result in the formation of crosslinked structures.

In order to conduct a more detailed analysis we investigated the effect of the microencapsulating shell on the rates of evolution (W) of some gases (CO, CO_2, H_2O) during the decomposition of some polymer blends. Data on the stepwise pyrolysis gas chromatography are presented (table 4). As can be seen, the amount of CO and CO_2 for compositions with DAPA and DAPA in the shell is less than that for the pure polymers. Additionally, the processes of oxidation of carbon monoxide to dioxide proceed to a lower extent. The increased amount of water in the presence of the FR points to of the dehydration reaction processes. Thus, microcapsulated DAPA inhibits the process not only in the gas phase but also in the condensed phase which results in increased char formation.

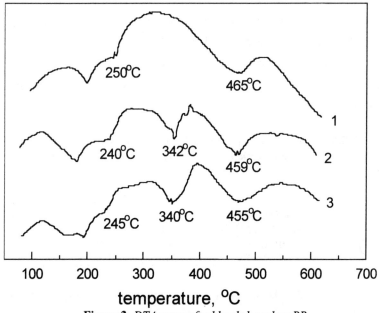

Figure 3. *DTA curves for blends based on PP
(1 - PP, 2 - PP+DAPA, 3 - PP+DAPA(PVTES)].*

Table 4. *The composition of pyrolysis products (%) of fire retarded plastics obtained at thermal decomposition conditions.*

N	Polymer	Microcapsulated shell	W, mg/min		
			CO	CO_2	H_2O
1	PE	without FR	164	210	3
2	PE	-	118	135	43
3	PE	PVTS	120	141	47
4	PP	without FR	184	230	3
5	PP	-	120	108	33
6	PP	PVTS	115	106	37
7	PCA	without FR	225	169	96
8	PCA	-	139	99	247
9	PCA	PMFIA	110	85	191
10	PET	without FR	261	187	25
11	PET	-	230	163	129
12	PET	PMFIA	201	140	105

It is known that the surface heating rates under conditions of burning may be greater than $4 \cdot 10^6$ J/s.m which would be equivalent to about 300 K/min. For this and other reasons, therefore, the result of studies of the controlled thermal decomposition of mixtures based on the organic polymers may not be directly relevant to the reactions taking place during the "decomposition stage" of polymer combustion. In this connection, we investigated the thermal degradation of some polymer blends using high-temperature pyrolysis[9]. From high temperature pyrolysis curves, the temperature corresponding to the maximum rate of decomposition T_D, the time of degradation (t) and the mass loss during the decomposition (%m) are given in Table 5.

Table 5. *High temperature pyrolysis data for fire retarded plastics.*

N	Composition	T_D °C	time s	Residual weight wt. %
1	PE	550	3	97
2	PP	540	4	98
3	PE+DAPA	520	8	80
4	PE+DAPA(PVTS)	650	16	75
5	PP+DAPA	560	10	90
6	PP+DAPA(PVTS)	510	6	70

The table shows that the high temperature pyrolysis data of polypropylene-based fire retarded plastics are in agreement with the TGA data. The decomposition proceeds in two stages: first, the degradation of PP (Figure 4) and the formation of crosslinked structures; second, the decomposition of formed char

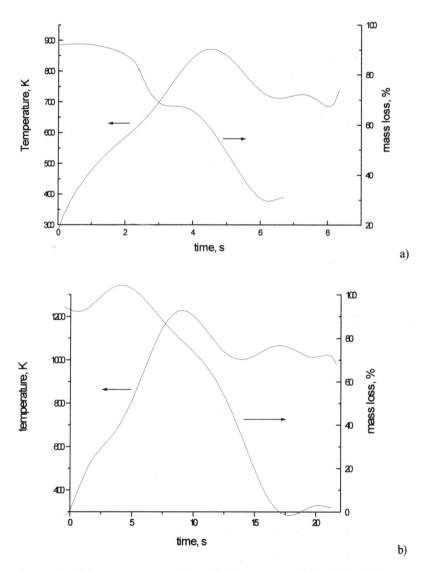

Figure 4. *High temperature pyrolysis of compositions with DAPA+PVTES: a)- in PP matrix, b) - in PE matrix.*

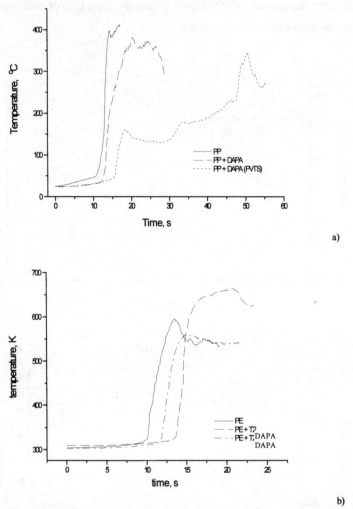

a)

b)

Figure 5. *Dependence of the temperature on the time for polymeric containing DAPA materials based on PP(a) and PE(b).*

The data for the high temperature pyrolysis of PE + DAPA + PVTS (Figure 4b) are not in agreement the with TGA data. As can seen from Table 2, the time for pyrolysis of PE + DAPA substantially increases in the presence of PVTS.

Physical properties of foamed char formed during the combustion of fire retarded polymer compositions have been investigated. The curves of the dependence of the temperature on the time for polymer materials based on PP containing DAPA are given in

Figure 5a. The comparison of the curves indicates that there is a great difference in the TPP of tested materials. The curves are very complex, but every interval of curve corresponds to specific system. The sharp rise of the temperature at the initial moment of time corresponds to the heating of materials up to the beginning of the formation of foamed char. The temperature following it, decreases for polymer compositions containing fire retardants and corresponds to the beginning of the formation of foamed char and its growth. There are bends in this moment for some polymer's blends. Thus, the coke formation for its composition doesn't occur.

The smooth interval following the recession of temperature correspond to the process of foaming of the entire sample volume (connected with the heat uptake during of vigorous decomposition of the entire volume of the polymer). The small rise of the temperature up to the constant value indicate that the forming of foamed char finished and the entire volume was transformed to foamed char.

The curves of the PP-based compositions indicate that the fire retardant effect is related to the formation of foamed char. The formation of coke during thermal degradation of PP doesn't take place. On the other hand, the foamed chars formed during degradation of PP containing DAPA have some TPP. Since the smooth interval is very small we can suppose that the amount of char is not enough in order to protect polymer for a long time. On the other hand, the comparison of the curves for PP containing DAPA and PP with DAPA in the PVTS shell indicates that the formation of char occurs faster and that foamed char have a large TPP in the presence of microcapsulated fire retardants. Thus, in the action of microencapsulated DAPA as fire retardants for polyolefines, apart from the factor which is based on the change in the direction of the process of thermal degradation of PP, leading to a decrease in amount of volatile flammable products of degradation[6], a role is also played by the intensification of coke formation, leading to the formation of the char having larger TPP.

The curves of the PE containing DAPA with PVTS shell indicate that the foamed char formed in the presence of PVTS (figure 5b) have a very low TPP. Thus, the action of DAPA with the PVTS shell as flame retardant based on the change in the direction of the process of thermal degradation of PE, leads to a decrease in the amount of volatile flammable degradation products. Additionally, DAPA with PVTS shell intensified the char formation of polymer materials during flash heating.

It is found out that the microencapsulated shell didn't affect the TPP of polymer blends based on PCA and PET. The TPP of PCA or PET with DAPA containing coating are the same as polymer blends without microcapsulated coating.

Figure 6 gives the effect of the polymer "melt" viscosity against the temperature for the PP, PP containing DAPA and PP containing DAPA with PVTS shell. The sharp fall of the melt of polymer materials initially corresponds to the heating of the materials up to the beginning of the formation of char. The bend of the curves correspond to the beginning of the char formation. As can be seen from Figure 1, the thermal degradation of PP passes without coke formation. The addition of DAPA results in the formation of coke during the decomposition of PP. The smooth interval for the melt degraded polymeric material corresponds to the process of charring of the entire volume of the sample. The sharp fall of the melt of coke indicates that the foamed char is degraded.

Thus, the experimental data indicates that the composition of the intumescent system for PP's blends affects the process for coke formation during their degradation. Probably, the fire retardant system based on DAPA containing microcapsulated PVTS have higher melt properties and formed cokes have better thermo-physical protected properties than that based on original DAPA.

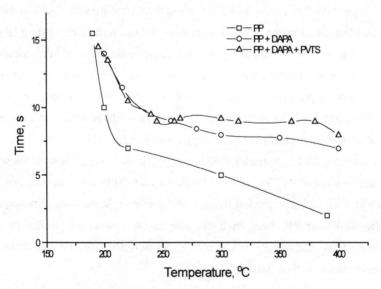

Figure 6. *Viscosity of the PP melt with additives.*

The Table 6 shows the mechanical properties of polyolefines containing DAPA. As can be seen from Table 6, the addition of DAPA containing microencapsulated shell does not affect the strength properties of polyolefines. Thus, the developed polymer materials based on polyolefines have not only reduced flammability but retain their high strength properties.

Table 6. *The strength properties polyolefines containing DAPA.*

N	Polymer	Microcapsulate d shell	Tensile strength, MPa	Relative elongation, %
1	PE	without FR	15	90
2	PE	PE	16	61
3	PE	PVTS	19	60
4	PP	without FR	23	280
5	PP	PE	21	246
6	PP	PVTS	24	250

References

1. Kopylov, B. B., Novikov, S. N. and Oksent'evich, L. A., *"Polymeric Materials with Low Flammability"*, Khimiya, Moscow, 1986.
2. Troitzsch, J., *Makromol. Chem., Macromol. Symp.*, **74** (1993) 125.
3. Luijk, R., Pureveen, J., Commandeur, J., Boon, J. *Makromol. Chem., Macromol. Symp.*, **74** (1993) 235.
4. Reshetnikov, I., Antonov, A., Rudakova, T., Aleksjuk, G. and Khalturinskij, N., *Polym. Degrad. Stab.*, **54** (1996) 137.
5. Vilesova, M., *J. of Appl. Chem. (Russia)*, **67** (1985) 79.
6. Zubkova, N., Tuganova, M., Bosenko, M., in *"First Int. Meeting on Flame Retarded Polymer Materials."*, Volgograd, (1995) 65.
7. Patent USSR N 759545, 1976.
8. Heinsoo, E., Kogerman, A., Kirret, O., *J. Anal. Appl. Pyrolysis*, **2** (1980) 131.
9. Galchenko, A., Khalturinskij, N., Berlin A., *Vysokomol. Soed.*, **22** (1980) 16.
10. Kochubej, A., Khalturinskij, N., Rachmangylova, N., *Vysokomol. Soed.*, **31** (1989) 659.

Acknowledgements

This work was carried out with the financial support of the INTAS Project No 93-1846.

FIRE RETARDANT ACTION OF RED PHOSPHORUS IN NYLON 6

G. F. Levchik, S. V. Levchik

Institute for Physical Chemical Problems, Byelorussian State University,
Leningradskaya 14, 220080 Minsk, Belarus

G. Camino

Dipartimento di Chimica Inorganica, Chimica Fisica e Chimica dei Materiali,
Universita di Torino, via P. Giuria 7, Torino 10125, Italy

and E. D. Weil

Polymer Research Institute, Polytechnic University,
Six Metrotech Center, Brooklyn, New York 11201, USA

1. INTRODUCTION

Red phosphorus is an effective fire retardant for pure and glass fiber reinforced nylons[1]. It was reported that char enhancers, e.g. phenolic resins[2-4] show a synergistic effect with red phosphorus preventing the polymer dripping. However, red phosphorus is potentially very flammable. It might generate highly toxic phosphine. Furthermore, it is difficult to mask the intense red colour.

The mechanism of fire retardant action of red phosphorus was extensively studied in various polymers[2, 5-11] but not in nylons. There is disagreement in the literature on the mode of action of red phosphorus. In some cases purely condensed phase mechanism is suggested[2, 5], whereas in other the contribution of gas phase mechanism is also proposed[6, 10, 11]. However, it is commonly accepted that red phosphorus is mostly effective in oxygen or nitrogen containing polymers but not in polyolefins or styrenics.

2. EXPERIMENTAL

Commercial nylons 6 either from Atochem (France, Orgamide 6, Ny-6o) or from Khimvolokno (Belarus, Grodno, Ny-6k) were used. Fire retardant grade stabilised red phosphorus (PR, Safest) was kindly supplied by Italmatch (Italy). In some experiments PR and various co-additives of the chemically pure grade (Reakhim, Russia) were used.

Fire retardant tests were performed either by the oxygen index (limiting oxygen index, LOI, following the ASTM D 2863 standard) or by the UL 94 standard on 1/8 inch specimens in the vertical configuration.

The char formed on specimens burned at a few units below the LOI was collected by thoroughly scraping the top of the bar after self-quenching of the flame and weighted. For chemical analysis the char was separated for two parts: the black crispy upper char and the dark-brown «transitory» char which lies between non-decomposed polymer and the upper char.

The thermal decomposition of the fire retarded nylon 6 was studied at heating rate of 10°C/min, by DSC or TGA under inert gas flow or air flow at 60 cm^3/min. Du Pont 2100 or Metler TA 3000 thermal analysers were used. To improve separation of weight loss steps in TGA a high resolution TGA 2950 Du Pont system (HRTG) in which the heating rate decreases when the rate of weight loss as monitored by derivative TG tends to increase was applied.

Solid residues of nylon 6 with red phosphorus were collected at different steps of thermal decomposition either in inert atmosphere or in vacuum and characterized by infrared (Perkin Elmer 2000 or Specord 75) or by EPR (ERS 200) or by ^{31}P solid state NMR (Jéol GX 270/89).

3. RESULTS AND DISCUSSION

3.1. Combustion

Pure Ny-6o shows LOI=21 (Figure 1). Its melt flows and drips from the top of specimen. Upon addition of 2.5 - 10 wt. % of PR oxygen index increases to 29, whereas flowing and dripping tend to be suppressed. If the concentration of PR is above 10 wt. % the composition loses its fire retardant effect. This is likely to be because of the inherent flammability of red phosphorus[12].

Nitrous oxide index (NOI) is usually used for distinguishing between condensed phase and gas phase mechanisms of fire retardant action[13].The parallel trend is observed between LOI and NOI with the increasing concentration of PR in Ny-6o which is in favour of condensed phase fire retardant action of PR. This is also confirmed by the correlation between LOI and NOI (insertion in the Figure 1).

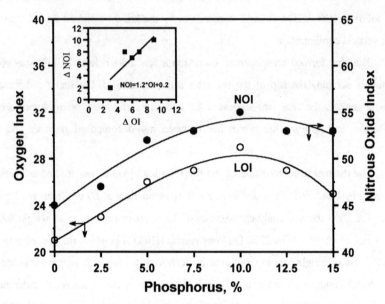

Figure 1. *Dependencies of LOI and NOI on the concentration of red phosphorus in nylon 6.*

In the literature[10,11] it is suggested that red phosphorus is oxidised during combustion and the created phosphoric acids are responsible for the fire retardant effect. Therefore oxidising or reducing co-additives should contribute to the fire retardant performance of PR. These types of co-additives when used alone, do not help with decreasing flammability[14].The results of fire retardant test of Ny-6k with PR and various oxidising or reducing co-additives are shown in the Table 1.

Very few co-additives at high level of loading (e.g. 5 wt. % of MoO_3, WO_3, V_2O_5) help with slight increasing of LOI. In many other cases LOI does not change or even decreases. There is no correlation between LOI and UL94 results, as was found in other systems[15].

Manganese oxides are not effective in nylon 6 + PR independently of the oxidation state. 2% of Mn_2O_3 (formulation 5) even deteriorate fire retardant performance as is shown by UL94 test. Antimony trioxide also shows an antagonistic effect with PR. Antimony pentoxide which is a mild oxidising agent, does not improve fire retardant effect of red phosphorus.

Table 1. *Oxygen index and UL94 test for the mixtures of nylon 6 with red phosphorus +co-additives.*

No	Additive	PR, wt. %	co-add, wt. %	LOI (Vol. %)	t_1/t_2^a	dripping	ranking
					UL94		
1				21.0	burns		NC
2	PR	10		24.3	5/2	D	V2
3	PR	15		24.1	1/20	D	V2
4	PR+MnO	10	5	24.4	2/5	D	V2
5	PR+Mn$_2$O$_3$	10	2	24.8	burns	D	NC
6	PR+Mn$_2$O$_3$	10	5	24.3	1/8	D	V2
7	PR+MnO$_2$	10	2	23.9	8/19	D	V2
8	PR+MoO$_3$	10	2	23.1	2/3	N	V0
9	PR+MoO$_3$	10	5	25.5	0/1	N	V0
10	PR+MoO$_3$	6.7	3.3	22.5	0/0	N	V0
11	PR+Sb$_2$O$_5$	10	2	24.4	15/11	D	V2
12	PR+Sb$_2$O$_3$	10	2	23.7	burns	D	NC
13	PR+V$_2$O$_5$	10	2	23.7	0/1	N	V0
14	PR+V$_2$O$_5$	10	5	25.5	0/4	N	V0
15	PR+V$_2$O$_5$	7.5	2.5	23.2	0/0	N	V0
16	PR+WO$_3$	10	2	23.7	1/3	N	V0
17	PR+WO$_3$	10	5	25.7	0/0	N	V0
18	PR+Al	10	2	24.4	2/1	N	V0
19	PR+Al	10	5	24.4	0/0	N	V0
20	PR+Al	7.5	2.5	22.6	0/20	D	V2
21	PR+Si	10	2	23.2	1/15	D	V2
22	PR+Si	10	5	24.5	3/1	N	V0

[a] *Average combustion time at any single test during the first and the second application of the ignition flame.*

On the other hand molybdenum trioxide, tungsten trioxide and vanadium pentoxide help to improve fire retardant performance of PR. Furthermore, PR with MoO$_3$ or V$_2$O$_5$ behave better with decreasing of the total concentration of the fire retardant additive.

Fine metal powders of Al or Si which are expected to be reducing agents, show a positive effect in terms of the UL94 test. Very tough charred layer consisting of melted or coagulated metal particles with a char is observed after extinguishing of these specimens. No detectable metal oxidation was found by X-ray diffraction. It is likely that melted metal improves the morphology of the char and helps with fire retardancy.

3.2. Thermal analysis

At linear heating 10°C/min in inert atmosphere formulations with nylon 6 + PR show one step of weight loss which is usually at the temperature slightly below that of the thermal decomposition of pure nylon 6. High resolution thermogravimetry distinguishes two steps of weight loss of the formulations Ny-6o + PR (Figure 2). These data are reproducible in few experimental runs. The first step which is about 70 % of weight loss, mostly corresponds to the temperature interval of the thermal decomposition of pure Ny-6o. The weight loss in the second step does not depend on the concentration of FR, however the separation of the steps is better with higher concentration of PR.

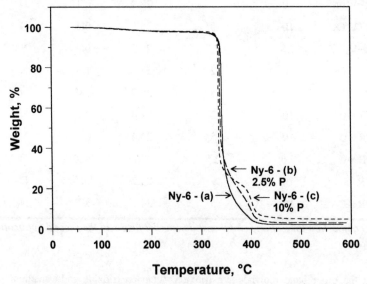

Figure 2. *High resolution thermogravimetry curves of nylon 6o (a) and nylon 6ofire retarded by 2.5 wt. % of PR (b) or 10 wt. % of PR (c). Nitrogen flow 60 cm³/min, heating rate 10 °C/min, resolution factor +5.*

In air Ny-6o gives ca. 20% of solid residue after the main step of weight loss (Figure 3). However, this residue is not stable to oxidation since it is lost above 500°C. In the presence of PR Ny-6o decomposes at lower temperature, because created acidic species probably catalyse degradation of the polymer. PR stabilises solid residue against oxidation. Ny-6o + 2.5 % P produces ca. 15 wt. % of solid residue above 500°C, whereas Ny6o + 10 % PR gives ca. 30 wt. % of solid residue at the same temperature.

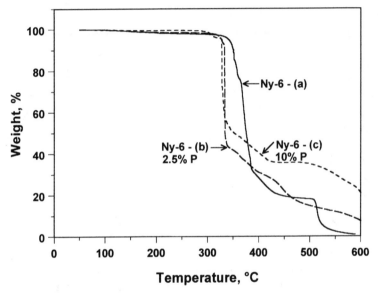

Figure 3. *High resolution thermogravimetry curves of nylon 6o (a) and nylon 6o fire retarded by 2.5 wt. % of PR (b) or 10 wt. % of PR (c). Air flow 60 cm³/min, heating rate 10°C/min, resolution factor +5.*

As Figure 4 shows the yield of solid residue measured in thermogravimetry in air strongly increases at 0 - 12.5 wt. % of phosphorus and then drops down at 15 wt. % of phosphorus. In inert atmosphere the residue slightly increases from 2 to 4% at 0 - 7.5 wt. % of phosphorus and then levels off. The char collected in the combustion tends to increase smoothly from 0 to 7% at addition of 0 - 15 wt. % of phosphorus. This tendency is similar to the thermal decomposition in air, however the levels of chars formed are close to the amount of solid residues in nitrogen. From these results it seems that oxygen or air contributes to the condensed phase processes, but the overall event is closer to the behavior in inert atmosphere.

DSC curves of Ny-6k and its mixtures with PR are shown in Figure 5. Two endotherms are observed for pure polymer in nitrogen: melting at 220°C and thermal degradation at 431°C. In air the pattern of thermal decomposition of Ny-6k is more complex because of overlapping of exothermic oxidation and endothermic degradation. The additive of 10% of PR seriously modifies the DSC curves of Ny-6k. Two exothermic peaks which appear in inert atmosphere (378° and 508°C) are due to interaction between the nylon and red phosphorus. In air, the whole degradation process of Ny-6k + PK is exothermic.

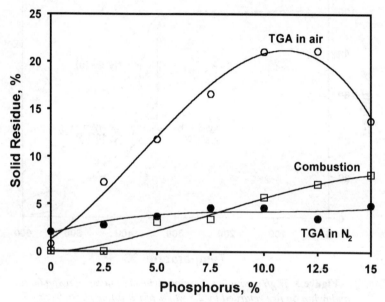

Figure 4. *Dependence of solid residue yield on the content of PR measured in thermogravimetry or in combustion.*

Thus, thermal analysis gives indication that red phosphorus reacts with nylon. The interaction occurs even without oxygen, which is in agreement with the paper of Kuper et al.[16].

3.3. Characterisation of chars and solid residues.

Phosphorus concentration was measured in the bar of the self-extinguished LOI test specimen Ny-6k + 10%PR. The virgin formulation beneath the char contains 10.1 wt. % of

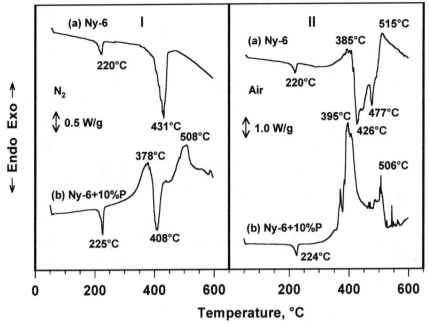

Figure 5. *DSC curves of nylon 6k (a) and nylon 6k fire retarded by 10 wt. % of PR (b) either in nitrogen (I) or in air (II). Gas flow 60 cm³/min, heating rate 10 °C/min.*

phosphorus which is very close to the expected amount. The dark brown layer above the virgin formulation shows 14.1 wt. % of phosphorus, whereas the black char contains ca. 23.6 wt. % of phosphorus. It is evident that phosphorus accumulates in the solid phase, which is in favour of condensed phase mechanism of fire retardant action of PR.

Figure 6 shows infrared spectra of initial formulation of Ny-6k + 10 % PR and of solid residues at different steps of thermal decomposition. In the initial spectrum of the formulation the characteristic pattern of nylon 6 is observed. The main changes of the spectrum at 15% weight loss are observed in the region of NH stretching 3500-3100 cm[-1], which corresponds to the degradation of the polymer[17]. An absorption appeared at 984 cm[-1] is likely to be due to phosphoric esters stretching (P-O-C)[18]. The intensity of this absorption increases with the progress of thermal decomposition which proves accumulation of the phosphorus containing species in the solid residue. The formation of phosphoric esters is also confirmed by solid state ^{31}P NMR where typical for aliphatic esters resonance peak was observed at -10 ppm[19].

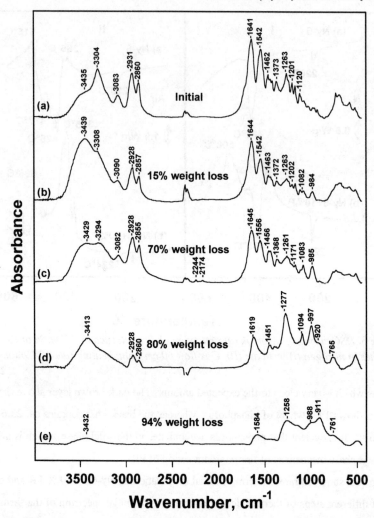

Figure 6. *Infrared spectra of initial Ny-6k + 10 % PR or solid residues collected in thermogravimetry in inert atmosphere at different steps of weight loss. Pellets in KBr.*

Infrared spectrum at 80% of weight loss is similar to the pattern of 5-amidopentyl polyphosphate[20], which is a product of interaction between polyphosphoric acid and nylon 6 upon heating. It seems that PR undergoes oxidation even in inert atmosphere probably somehow using oxygen from the nylon 6 or water present in sample (see TGA curves: ≈ 5 % weight loss) or evolved during the thermal decomposition of the polymer.

Solid residue at 600°C (94% weight loss) shows series of broad absorptions at 1300-700 cm^{-1}. Such pattern is similar to phosphorus nitrides (e.g. phospham) or phosphorus oxynitrides[21] which are very stable at high temperature. A band at 1584 cm^{-1} is due to polyaromatic char. Thus, high temperature residue which is produced from the formulation of nylon 6 with PR is a mixture of phosphorus nitrogen compounds and carbonaceous char.

Recently we reported[22] that red phosphorus generates free radicals upon heating in vacuum. It was suggested that free radical mechanism might be one of the fire retardant action mode of PR. In this work free radical concentration was measured in the solid residues of Ny-6k + 10%PR.

It was found that Ny-6k or its formulation with PR produce similar kind of free radicals which are typical for carbonised organic species[23] (g = 2.0030-2.0032 and h$^{1/2}$ width ca. 7G). In the solid residue obtained at 450°C in vacuum, the concentration of these radicals is three times higher compared to the concentration in the residue from pure nylon 6. However, after keeping for one day at room temperature in air the amount of free radicals in the formulation with PR decreases to the same level as in solid residue from pure nylon 6. It is likely that the formulation with PR generates less stable free radicals which however have very similar g-factor and the resonance curve shape as typical carbonaceous residues. Another explanation might be that PR produces free radical scavengers effective in the condensed phase.

4. CONCLUSIONS

Red phosphorus (PR) is an effective fire retardant additive for nylon 6. It shows maximum efficiency at 10% of loading as measured by oxygen index. The parallel trends of oxygen index and nitrous oxide indices is in favour of condensed phase mechanisms of fire retardant action of PR. Various oxidising or reducing agents contribute to the fire retardant behavior of PR. Vanadium pentoxide, molybdenum trioxide or tungsten trioxide are mildly beneficial co-additives to PR.

As it is seen from thermal analysis, PR reacts with nylon 6 at thermal decomposition in air and in inert atmosphere. Phosphoric esters are formed at interaction of PR with the polymer as it is shown by infrared and by solid state NMR. The chemical mechanism of interaction of

PR with nylon 6 is not clear, however it is shown that PR increases the free radicals concentration in the char formed at thermal decomposition of nylon 6 fire retarded by PR.

References

1. I. C. Williams, *Plastics Today*, 1984 spring, **19**, 14

2. E. N. Peters, A. B. Furtec, D. I. Steinbert and D. T. Kwiatkowski, *J. Fire Retard. Chem.*, 1980, **7**, 69

3. M. T. Huggard, in *"Proc. Conf. Recent Adv. In Flame Retardancy of Polym. Mater."*, Stamford, CT, 1992, 192

4. M. T. Huggard, in *"Proc. Conf. Recent Adv. In Flame Retardancy of Polym. Mater."*, Stamford, CT, 1995, 154

5. A. Granzow, R."G. Ferrillo and A. Wilson, *J. Appl. Polym. Sci.*, 1977, **21**, 1687

6. E. N. Peters, *J. Appl. Polym. Sci.*, 1979, **24**, 1457

7. C. A. Wilkie, J. W. Pettegrew and C. E. Brown, *J. Polym. Sci.: Polym. Lett.*, 1981, **19**, 409

8. T. Suebsaeng and C. A. Wilkie, *J. Polym. Sci.: Polym. Chem.*, 1984, **22**, 945

9. C. E. Brown, C. A. Wilkie, J. Smukalla, R. B. Cody Jr. and J. A. Kinsinger, *J. Polym. Sci.: Polym. Chem.*, 1986, **24**, 1297

10. A. Ballistreri, S. Foti, G. Montaudo, E. Scamporino, A. Arnesano and S. Calgari, *Makromol. Chem.*, 1981, **182**, 1301

11. A. Ballistreri, G. Montaudo, C. Puglisi, E. Scamporino, D. Vitalini and S. Calgari, *J. Polym. Sci.: Polym. Chem.* 1983, **21**, 679

12. J. R. Van Wazer, «*Phosphorus and Its Compounds*», Wiley, New York, 1958, Vol.1, 1958.

13. C. P. Fenimore and G.N. Jones, *Combust. Flame*, 1964, **8**, 133.

14. S. V. Levchik, G. F. Levchik, G. Camino, L. Costa and A. I. Lesnikovich, Fire Mater., 1996, **20**, 183.

15. E. D. Weil, M. M. Hirschler, N. G. Patel, M. M Said and S. Shakir, Fire Mater; 1992, **16**,159.

16. G. Kuper, J. Hormes and K. Sommer, *Makromol. Chem. Phys.*, 1994, **195**, 1741.

17. S. V. Levchik, L. Costa and G. Camino, *Polym. Degrad. Stab.*, 1992, **36**, 229.

18. L.C. Thomas, «*Interpretation of the Infrared Spectra of Organophosphorus Compounds*», Heyden, London, 1974

19. D. G. Gorenstein, in *"Phosphorus-31 NMR. Principles and Application"*, ed. D. G. Gorenstein, Academic Press, Orlando, 1984, Chapter 1.

20. S. V. Levchik, G. Camino, L. Costa and G. F. Levchik, *Fire Mater.*, 1995, **19**, 1.

21. T. N. Miller and A. A. Vitola, «*Inorganic Compounds of Phosphorus*» (in Russian), Zinatne, Riga, 1986, Chapter 8.

22. S. V. Levchik, G. F. Levchik, A. I. Balabanovich, G. Camino and L. Costa, *Polym. Degrad. Stab.*, 1996, **54**, 217.

23. I. C. Lewis and L. S. Singer, in *Chemistry and Physics of Carbon,* eds. P. L. Walker Jr. and P. A. Thrower, Marcel Dekker, New York, 1986, Vol.17, p.1.

Acknowledgements

This research was sponsored in part by the Civil Research and Development Foundation under the Grant BC2-104.

THERMAL DEGRADATION OF AN INTUMESCENT (STYRENE - BUTADIENE COPOLYMER / AMMONIUM POLYPHOSPHATE / PENTAERYTHRITOL / TALC) SYSTEM

E. Gaudin, C. Rossi, Y. Claire, A. Périchaud

Laboratoire de Chimie Macromoléculaire, Université de Provence 3,
Place Victor Hugo, 13331 Marseille Cedex 3, France.

L. El Watik, H. Zineddine

Unité de Chimie de l'Environnement, Département de Chimie,
Faculté des Sciences et Techniques, Université Moulay Ismaïl,
BP 509, Boutalamine Errachidia, Morocco

J. Kaloustian

Laboratoire de Chimie Analytique, Faculté de Pharmacie,
Université de la Méditerranée, 27 Boulevard J. Moulin, 13385 Marseille, France

M. Sergent

Laboratoire de Méthodologie de la Recherche Expérimentale, Université Marseille 3,
Avenue Escadrille Normandie Niemen,13013 Marseille, France

1. INTRODUCTION

The styrene butadiene copolymer (SBR) is one of the polymers most currently used (furniture, packaging, house-hold appliance). The decreasing of its flammability is an important problem which limits its spread use.

Our laboratory has previously studied the fire retardancy (FR) of SBR using an intumescent ammonium polyphosphate (APP) - pentaerythritol (PER) – melamine additive system[1]. The importance of the association APP (non organic acid source) - PER (polyhydroxylated compound) was proved in this formulation[2].

Moreover, the study has shown first the small part played by the melamine in flame retardancy because APP is already a blowing agent by NH_3 evolving, second, the presence of cyanide in the gaseous products of the inflammation[3].

Consequently, we used the talc powder in substitution of melamine which has been shown the main responsible for this formation of cyanide[4]. Talc (TAL) reacts with the excess of polyphosphoric acid formed by esterification and gives inorganic glasses, making APP more effective[5]. The selected system acts via its decomposition under heating, with the formation of a carbonaceous coating (char) with protects the polymer via decreasing the diffusion of the oxygen and of heat to the surface of polymer. As a consequence, the degradation of the material and the formation of very flammable volatile products are decreased. The choice of this system is based on several general criteria in relation to fire retardant effects : stability until the degradation temperature of the polymer, compatibility with the polymer, no loss of the physico-chemical properties of the polymer, induction of the synergistic effects.

The intumescent system shows a two steps thermal degradation under air. The first one which begins at about 210°C, corresponds to the formation of phosphate esters via the reaction of APP and PER with water and ammonia elimination[2]. The second degradation stage begins at about 290°C and leads to the formation of an intumescent material (carbonaceous residue expanded by the volatile products formed at about 300°C[6]).

The formulation of polymer-based flame retardant mixtures depends often on the cleverness of experts. Protocol for experiments with mixtures is a tool to obtain information on the behaviour of studied systems needing a minimum number of experiments. It allows the evaluation of degradation rate of the formulations (quantitative variable, so-called the response) versus the variation of the composition of the intumescent system, the determination of the formulation capable to give the optimal responses (lowest degradation rate and degradation temperature), the evidence of the synergistic and antagonistic effects (interaction) between the constituents of the system.

The present work studies the thermal degradation of the intumescent mixture : SBR - APP - PER - TAL powder, using thermogravimetric analysis (TG). An experimental procedure using protocol for experiments with mixtures, which considers the decomposition of the polymeric formulations, selected as a FR criterion, versus their compositions, is developed to obtain an optimised formulation.

2. EXPERIMENTAL

2.1. Materials

The styrene-butadiene copolymer (lacqrene 70, molecular mass: M_n = 46 000 and M_W = 160 000 determined by chromatography) is used as supplied by Elf-Atochem (France).

Used additives are APP (Hostaflam AP 422, powder as supplied by Clariant), PER (2,2-bis(hydroxymethyl)-1,3-propanediol, powder as supplied by Acros) and TAL (hydrous magnesium silicate, melting point : 800°C ; approximate particle size: 9 μm, as a powder supplied by Acros).

Our experimental strategy considers first the intumescent system used (number and nature of the components). Then the individual and relational constraints imposed to the studied system are considered :

(i)   the composition of the intumescent mixtures. From previous works of this laboratory[7] or managed by G. Camino[8, 9], an additive loading ≤ 30 wt. % in SBR is needed in order to hold the mechanical properties of the polymer[8] (in agreement with economical cost consideration). Moreover, at least 15wt. % additive level is needed to preserve the fireproofing behaviour. So, the retained composition constraints are:

$$\Sigma \text{ (mixture component content)} = 100 \text{ relative \%}$$

$$0 < \text{individual component content} < 100 \text{ relative \%}$$

In fact, the experimental domain of interest may be defined by the following relations:

$$70 \% = X1 \text{ (SBR relative concentration)}$$

$$10 \% < X2 \text{ (APP relative concentration)} < 20 \text{ wt. \%}$$

$$10 \% < X3 \text{ (PER relative concentration)} < 20 \text{ wt. \%}$$

$$0 \% < X4 \text{ (TAL relative concentration)} < 10 \text{ wt. \%}.$$

In order to analyse the thermal behaviour of the FR polymer, we analyse two different responses, hereafter called Y: the degradation temperatures allowing 20 to 50 wt % of weight loss and the degradation rate (wt. % min^{-1}) in the 20 to 50 wt. % weight loss interval.

2.2. TG analysis

A SETARAM TGA 92 apparatus (aluminium sample and reference crucibles) was used. The analyses were carried out in the 30-600°C temperature range (10°C min^{-1} heating rate) under 20 ml min^{-1} air sweeping.

2.3. Protocol for experiments with mixtures

The selected mathematical model has to represent all the experimental field and to take in consideration the individual and relational constraints imposed to the system. Usually, a polynomial model is used. Special protocol for experiments with mixtures, without constraint, are the Simplex - Lattice design (Scheffé lattices[10]).

In this study, the experimental field previously defined is not a simplex, but an irregular polygon with four sides. However, its simplicity allows the use of a simplex and in fact, the selected mathematical models are the true network simplex of Scheffé. They agree well with an empirical reduced polynomial model containing only product terms[11]. In the chosen mathematical model, each response Y is a polynomial function of degree 3 in the 4 variables X_1, X_2, X_3 and X_4 (composition of the mixture) in all the experimental lattice ($\{3,4\}$ polynomial simplex). The experimental matrix is in that particular case, obtained from 12 different experiments (Table 1).

Table 1. *Composition of the four components systems used for experiments with mixtures.*

MIXTURES NUMBER	X1 (PS choc)	X2 (PPA)	X3 (PER)	X4 (TAL)
1	0.70	0.18	0.12	0
2	0.70	0.15	0.10	0.05
3	0.70	0.10	0.20	0
4	0.70	0.10	0.10	0.10
5	0.70	0.10	0.15	0.05
6	0.70	0.14	0.16	0
7	0.70	0.125	0.10	0.075
8	0.70	0.165	0.11	0.025
9	0.70	0.0325	0.13	0.0375
10	0.70	0.118	0.16	0.022
11	0.70	0.114	0.12	0.066
12	0.70	0.156	0.127	0.017

The tests can be run in a random order, but it is necessary to check the validity of the model. In this purpose, the study of the repeatability has been carried five times on the same test. In theory the choice of the test may use random experiments but we have assumed experiment in the centre of gravity of the domain as representative test (t = 2,132 of R. A. Fischer to 10 % error).

3. RESULTS AND DISCUSSION

The first results concern the influence of the composition of the different systems on the degradation temperature. It shows the high dispersion of the values of the weight loss (between 20 to 50 wt. %) according to the experimental conditions. Moreover, the analysis answer present the compositions of the different mixtures at a temperature corresponding to a same weight loss. These results allows us to observe 3 specific behaviours . For example considering a 30% weight loss, we may differentiate:

- mixtures with relatively low degradation temperatures (380 - 390°C range),
- mixtures with slightly upper degradation temperatures (410 - 430°C range),
- intermediary mixtures (390 - 410°C range).

From this results, we may propose that the increasing of talc content and of the ratio (APP)/(PER) act by an antagonism manner on the degradation temperature of the different intumescent mixtures (Figure 1).

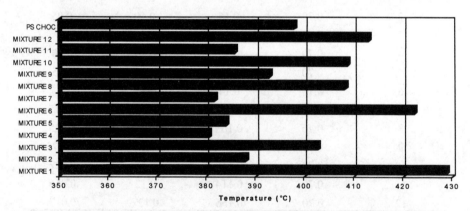

Figure 1. *Degradation temperatures (± 4°C) at 30 % weight loss versus the talc content and the APP/PER ratio.*

This figure shows that the mixtures with the highest degradation temperature are those which present the higher ratio (APP)/(PER) and no talc (mixtures 1 and 6). Furthermore, the

DEGRADATION RATE (%/ min)

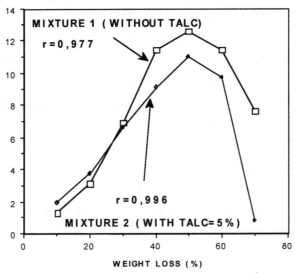

Figure 3. *Degradation rate of the SBR - APP - PER - TAL systems versus the weight loss and the TAL contents (APP/PER = 1.5, r: coefficient of correlation).*

DEGRADATION RATE (%/ min)

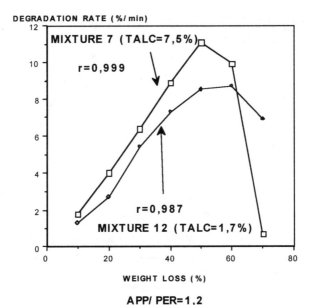

APP/ PER=1,2

Figure 4. . *Degradation rate of the SBR - APP - PER - TAL systems versus the weight loss and the TAL contents (APP/PER. = 1.2).*

mixtures with the lowest degradation temperature contain the highest talc content and their (APP)/(PER) ratio is relatively low (mixtures 4 and 7).

The second series of results concerns the influence of the composition of the different systems on the degradation rate of the SBR - APP - PER - TAL systems. They allows to show the degradation rate of the different systems versus their weight losses.

For every tested mixture a relatively linear increasing of the degradation rate versus the weight loss are observed in the 20 - 50% weight loss range. The comparative study of these linear relations (Figure 2, 3 and 4) shows that the best linear laws are followed when the TAL content is high and when the (APP)/(PER) ratio is low. More, this study confirms that the increases of the TAL content and of the (APP)/(PER) ratio act also with an antagonist manner on the linearity of the increasing of the degradation rate versus the weight loss.

In the highest weight loss range (< 50 wt. %) of every considered system, the degradation rate decreases sharply when the weight loss increases. This behaviour is characteristic of the protection of a polymeric matrix via an intumescent process where the material accept to partially degrade to protect itself[12, 13].

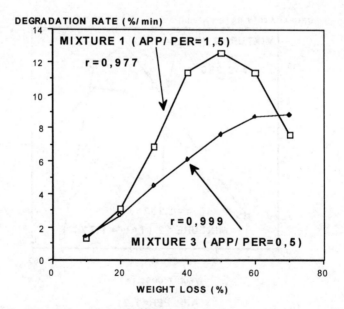

Figure 2. *Degradation rate of the SBR - APP - PER systems versus the weight loss and the APP/PER ratio.*

Comparison of the slopes of the linear relations allows then a direct comparison of the rates of the degradation of the formulations and consequently of the formation of the protective charred material. An increase of the APP/PER ratio from 0.5 to 1.5 (Figure 2) leads to a significant increase of the slope and so, of the self-protective behaviour of the material. This result agrees well with the well known synergy between APP and PER which is classically obtained for APP/PER = 3.

Addition of TAL in the formulation leads to a significant decrease of the slope which value remains similar for 1.7 < TAL content < 7.5 wt. % (Figure 3 and 4). This result confirms that this addition increases the stability of the material when partially degraded (> 10 % weight loss) and, as a consequence, lowers its self-protection. The decrease of the limiting oxygen index (LOI, according ASTM D2863/77) when the talc content of the SBR - APP - PER - TAL systems increases (Figure 5) confirms the partial loss of the protective character of the charring process.

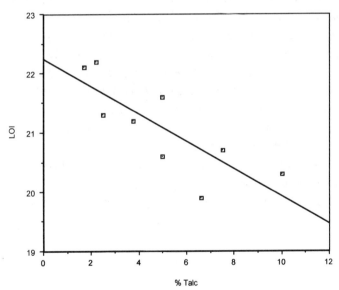

Figure 5. *Limiting oxygen index of the SBR - APP - PER - TAL systems versus talc content.*

Nevertheless, a relation between the continuous relative loss of the FR performances and the values of the corresponding slope (Figure 3 and 4) is not observed. It may be propose that, as previously observed in intumescent systems containing mineral fillers, TAL affects the thermal degradation of the polymeric matrix and/or induce stress and cracks in the intumescent coating[14].

References

1. L. El Watik, Y. Claire, M. Sergent, H. Zineddine, J. Kaloustian, C. Rossi, and A. Périchaud, *J. Therm. Anal.*, under press.
2. G. Camino, L. Costa, L. Trossarelli, *Polym. Deg. Stab.* 1984, **6**, 243.
3. L. El Watik, E. Gaudin, C. Rossi, Y. Claire, J. Kaloustian and A. Périchaud, *"Coll. Intern.sur L'Environnement."*, Kénitra, Morocco, 24 - 25 April 1997.
4. S. V. Levchik, G.F. Levchik, G. Camino and L. Costa, *J. Fire Sci.*1996, **13**,43.
5. S.V. Levchik, G. F. Levchik, G. Camino and L. Costa, *J. Fire Sci.*, 1995, **13**, 43-56.
6. G. Camino, L. Costa, G. Clouet, A. Chiotis, J. Brossas, M. Bert and A. Guyot, *Polym. Deg. and Stab.* 1984,**6**, 105-121, and 1984, **6**, 177-184.
7. E. Gaudin, L. El. Watik, C. Rossi, Y. Claire, J. Kaloustian, and A. Périchaud, in *"Fire Retardancy of Polymers: the Use of Intumescence (Chapter 5)"*, M. Le Bras, G. Camino, S. Bourbigot and R. Delobel ed., Royal Society of Chemistry, Cambridge(1998).
8. G. Camino, L. Costa and G. Martinasso, *Polym. Deg. Stab.* 1989, **23**, 359.
9. G. Camino, L. Costa, and L. Trossarelli, *Polym. Deg. Stab.* 1985, **12**, 213.
10. H. Scheffé, *"Experiments with mixtures"*, in *J. Royal Statistic. Soc.* 1958, 13-20, 344-360.
11. J.A. Cornell, *"Experiments with mixtures"*, J. Wiley, 1990.
12. S. Bourbigot, R. Delobel, M. Le Bras and Y. Schmidt, *J. Chim. Phys.*, 1992, **89**, 1835.
13. M. Le Bras, S. Bourbigot, C. Siat and R. Delobel, in *"Fire Retardancy of Polymers: the Use of Intumescence (Chapter 3)"*, M. Le Bras, G. Camino, S. Bourbigot and R. Delobel ed., Royal Society of Chemistry, Cambridge(1998).
14. M. Le Bras and S. Bourbigot, *Fire & Materials*, 1996, **20**, 39.

ROLE OF MIGRATION PROCESS IN THE EFFICIENCY OF INTUMESCENT FLAME RETARDANT ADDITIVES IN POLYPROPYLENE

Gy. Marosi, I. Ravadits, Gy. Bertalan, P. Anna, M. A. Maatoug

Technical University of Budapest, Department of Organic Chemical Technology,
H-1111 Budapest, Mûegyetem rkp. 3., Hungary
E-mail: marosi.oct@chem.bme.hu

A. Tóth

Hungarian Academy of Sciences, Research Laboratory of Inorganic Chemistry

M. D. Tran

Biophy Research S. A.,
Village d'Entreprises de St. Henri, 6 rue Anne Gacon, F-13016 Marseille, France

1. INTRODUCTION

Increased attention to safety requirements has accelerated the investigation of economic and environmental flame retardant plastics in the recent decade. Considering the well known importance of solid-vapor interfacial processes in flame ignition and spreading, it is surprising that the methods of surface engineering are not utilized more widely. XPS method has been used recently for following the charring process[29], but not for supporting the formulation of flame retardant compositions. Surface engineering, comprising the modern methods of modification and analysis of surfaces, is a promising way for conscious formation of effective flame retardant systems.

The two ways used for achieving flame retardant polymer products are the surface treatment with liquids, solutions or dispersions and modification in bulk:

- the surface modification with solution (or dispersion) of flame retardant compounds is widely used for fibers, but it can not be applied economically for other products and in case the surface layer is removed, it can hardly be recovered later,
- the modification in bulk results in more stable flame retardant effect, but requires higher amount of modifiers,

A method combining the advantages of the two processes is missing. Among the wide range of flame retardant additives the intumescent types are promising to serve as effective and environmental system in polypropylene. The intumescent flame retardants (IFR) are generally applied in higher concentration (~30 wt. %). The additives should meet several requirements simultaneously, i.e. preserving the processability, mechanical properties and chemical resistance in addition to the insurance of self extinguishing character.

The simple intumescent compounds consisting of ammonium-polyphosphate (APP), pentaerythritol (PER), melamine (MEL) can not fulfill these requirements. The substitution of components that cause undesirable side effects has been attempted, such as APP with ammonium-pyrophosphate or melamine-phosphate, PER with dipentaerythritol and MEL with urea, isocyanurates, amine or amide compounds[2-10].

Additives acting as synergist and compensating the negative effects in this system has been applied as well. Examples for such additives were published by Le Bras et al.[11]. Zinc-borate and silicon compounds found to be effective in our earlier works[12-17] and silicon compounds were studied by Russian research group recently[13]. A relationship between the presence of silicon compounds and the mechanical properties of char foam has been proven, and differences has been found between the saturated and unsaturated silicon compounds[13].

We have pointed out recently the importance of the migration of additives for modifying the chemical composition of the surface and consequently the performance of flame retardant polymers[14]. If the migration of additives can be governed by the structure formed during the compounding, it can become an important method for controlling the efficiency of flame retardants.

The formation of appropriate interfacial structure around flame retardant particles may contribute to the realization of this concept as well as improve the stability of additives[12].

In this work we used thermal analytical, surface analytical and FTIR methods for explaining the effect of additives in IFR containing polypropylenes and use the information achieved this way for planning the structure and improving the performance of such systems.

2. EXPERIMENTAL

2.1. Materials

The polyolefin grades used in this study were low density polyethylene (LDPE; Tipolen AE 1716 product of Tisza Chemical Works (Hungary); density: 0.921 g/cm^3; melt index: 7 g/min at 190°C),polypropylene (PP; supplied by Tisza Chemical Works): ethylene-propylene copolymer (as supplied by Tipplen K793; density: 0.9 g/cm^3; melt index: 4 g/min (21.6 N, 230°C)), PP homopolymer (as supplied by Tipplen H348F; density: 0.9 g/cm^3; melt index: 12 g/min (21.6 N, 230°C)), ethylene vinyl acetate copolymer (EVAc: EVA 3325 as supplied by ICI, vinyl acetate content: 33 %, density: 0.98 g/cm^3, melt index: 28.8 g/min (21.6 N, 230 °C)).

Organoboron siloxane elastomer were prepared using OH end group containing siloxane oligomer and boric acid according the method described in literature[28]. Tetra-ethoxy-silane (TES) or vinyl – triethoxy - silane (VTS) was also added to promote the coupling with other components.

Commercial samples of ammonium polyphosphate (APP; Hostaflam 422, powder as supplied by Hoechst), pentaerythritol D/S (PER; powder supplied by Degussa), and melamine (MEL; as supplied by Reanal, Hungary) were used. Their ratio in the intumescent flame retardant (IFR) additive system was APP:PER:MEL=1: 0.2: 0.2 if not stated otherwise.

2.2. Test methods

The mixing of components was carried out using Brabender Plasti-Corder PL2000 (mixing chamber 350) with rotor speed 50 rpm. The compounds were then injection molded at 230°C.

Differential Scanning Calorimetry (DSC) measurements were performed using Setaram DSC 92 (sample weight: 10 mg, heating rate: 10°C/min, under air).

X-ray photoelectron spectroscopy (XPS) measurements were performed using SCIENTA ESCA 200 spectrometer (with monochromatized X-ray radiation from a Al K_α anode (1486,7 eV), energy source calibration: $4f_{7/2}$ line of Au (84.0 eV), control of linearity: Al/Mg method using the $3d_{5/2}$ line of Ag (368.2 eV)). To avoid the sample-charging effect a low energy electron flood gun was used (electron energy: 2.5 eV, emission 30 %). Curve fitting was done using the Scienta WINESCA program package. The spectra were referenced to the hydrocarbon type C1s line (B.E.=255.0 eV).

The conductivity of extraction water during extraction of 3g injection molded samples (surface area: 13 cm^2/g) were determined using a Radelkis (Hungary) conductometer.

3. RESULTS AND DISCUSSION

Migration in polymer systems may occur during processing or at application and it is advantageous or undesirable depending on the circumstances. As a method of surface engineering, migration process can be used for modifying the surface in order to achieve specific, such as antistatic or adherent characteristics. On the other hand, the migration of low molecular additives during processing may lead to processing problems. This is the case in APP+PER containing polymer systems, where the low molecular, low viscosity PER melt may separate from the APP and the polymer matrix and accumulate at the surface. Large PER particles are formed at the die causing surface defects.

APP and PER may, however, react with each other under appropriate circumstances according to the results of model studies published by Camino et al.[18]:

Spectroscopic and thermo-analytical evidences for esterification reaction of APP and PER was given earlier[6,9-10,18-7]. During compounding with polymer, APP and PER may react with each other the same way, but in case of the resulting polymer mixture such spectroscopic methods can not be used for controlling the occurrence of the reaction. In order to be sure that reaction occurs under the real circumstances of compounding a series of PE samples has been studied using DSC method (air atmosphere was applied like in case of compounding). The endotherm melting peak of PER in Figure 1. could be used as an indicator: whether after mixing it remains in the original form or not. In case of the samples prepared without APP (at

230°C) or with APP, but at 170°C compounding temperature (c and d curves), after the melting of LDPE at 125°C, the melting peak of PER appears at 182°C, which is lower than its nominal 255-260°C, probably due to some dipentaerythritol contamination formed during mixing. In case of APP containing sample compounded above 230°C, however, the melting peak of PER could not be detected (e curve). As the melting peak of PER disappeared only in presence of APP one can assume an interaction between the two components, that is probably an ester formation.

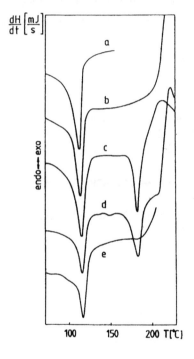

a, pure LDPE
b, LDPE + 36% APP + 11% ME
c, LDPE + 11% PER + 11% ME
d-e, LDPE + 36% APP +
      11% PER+ 11% ME
d, compounded at 170°C
e, compounded at 230°C

Figure 1. *DSC curves of LDPE mixed with IFR additives.*

It means that during compounding up to the applied (about 1:3) PER:APP ratio the two components may react immediately after melting of PER. In case of such an interaction, the surface of APP particle is covered with derivative of PER. Even if the conditions for APP-

PER reaction is ensured, the resistance of the formed flame retardant polymers against water is very poor and in case of humid environment it can lead to undesirable changes at the surface.

We tried to use elastomer additive in order to decrease the water sensitivity. The method of encapsulation by elastomer during the mixing process has been successfully applied earlier in filled polyolefins[19-21] and even in flame retardant PP for improving the compatibility between the phases[14]. The conditions of such an 'in situ' process has been summarized recently[22-23]. Electron microscopic results and indirect evidences confirmed the formation of ethylene-vinyl acetate (EVAc) layer around flame retardant particles in PP matrix[14]. The effect of encapsulation on the water sensitivity of IFR in PP could be followed by measuring the conductivity of water used for the extraction of these samples. The results are presented in the Figure 2.

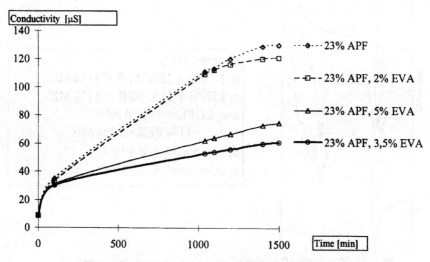

Figure 2. *Conductivity curves of flame retarded PP samples.*

The dissolved flame retardants extracted from the surface of the samples highly increase the conductivity of the water (< 40 μSiemens). The increase in the presence of elastomer is much less dominant, so the sensitivity to environmental humidity can be reduced substantially. The magnitude of effect depends on elastomer content.

On the other hand, the improved compatibility achieved through this modification is coupled with decreased efficiency of the flame retardant system. In the presence of EVAc at least 5% higher amount of IFR additive is required to achieve the V0 rating, as shown in Table 1.

There is no need for increasing the amount of flame retardant, when boron-siloxane elastomer, a typical incompatible additive, is used. Oxygen Index results in our recent papers suggested even some synergetic effect of this elastomer type [14,12]. The results of other authors confirmed also the positive effect of silicone compounds in IFR systems[13,24]. A remarkable difference can be found, however, between the effect of saturated and unsaturated silicone compounds: tetraethoxy-silane (TES) and vinyltriethoxy-silane (VTS). The slight chemical change, the presence of a double bond in case of VTS, increases the amount of IFR additives required for achieving V0 rating significantly (Table 1). No explanation was given for this difference in earlier papers. In order to find probable explanation, the possible differences in crosslinking density and in the interactions with the polymer phase has to be taken into account. In particular, TES will give a much lower crosslink density than VTS and does not incorporate any organic material when VTS does incorporate carbonaceous fuel in the matrix.

Table 1. *The flammability of different PP compositions.*

Polymer matrix	PP homo-polymer	PP +3,5% EVAc		PP +9,7% Boron-siloxane (TES)	PP+9,7% Boron - siloxane (VTS)		
IFR concentration [wt. %]	18	18	23	18	18	30	35
Flammability [UL94, 3mm]	V0	HB	V0	V0	HB	HB	V0

Detailed surface analytical XPS investigation was performed for explaining the effects of silicone compounds in IFR (APP+PER+ME) containing PP. The samples were studied:

- after processing,
- after flame treatment at controlled temperature (310°C),
- and after heat treatment at 300°C for 2 hours (in order to enhance the differences caused by the heat of flame treatment),

the reference samples containing APP alone were treated the same way.

A significant difference can be seen between the XPS survey spectrum of the reference sample (PP+APP model) and that of its silicon-compounded variant containing the IFR additive system (Figure 3). For the first case (Figure 3a) the plain spectrum suggests that no additive migration to the surface occurs during processing, whereas each type of atoms characteristic to the flame retardant additives appears on the surface in the second case (Figure 3b). The atomic concentration data, derived from the corresponding detailed spectra, are given in Table 2. As seen in the first two rows, the increase of oxygen concentration due to oxidation is negligible.

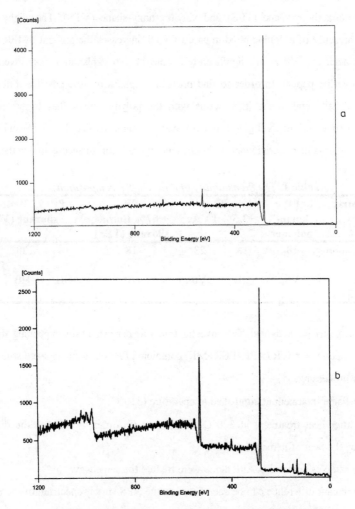

Figure 3. *XPS survey spectra of PP+APP model (a) and Silicone + IFR (APP+PER+ME) containing PP (b).*

Table 2. *Surface composition of flame retarded polypropylene formulations.*

Sample type	Relative concentration (At. %)				
	C	O	N	Si	P
PP+APP	96.9	2.7			0.4
PP+APP heated	96.1	3.1	0.2		0.6
IFR+silicone	84.2	9.7	2.4	2.1	1.6
IFR+silicone heated	67.1	19.6	3.4	5.9	4.0

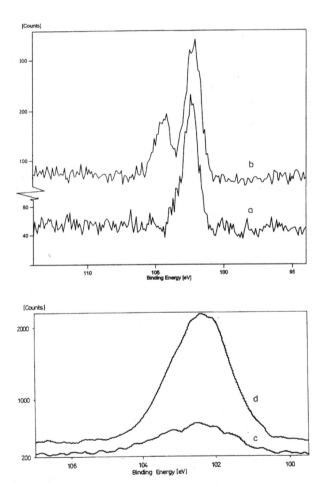

Figure 4. *Si 2p spectra of the IFR + silicone containing PP injection molded samples before (a) and after heat treatment (b), before (c) and after flame treatment (d).*

Changes in the Si-concentration of the surface layer (Figures 4 and Table 2) clearly show the tendency of the organosilicon additives to migrate to the surface: they start to accumulate at the surface already after processing (curves 4a and 4c) and their concentration increases strongly after heat (curve 4b) or flame (curve 4d) treatment. The position of Si 2p peak at about 102.3 eV is characteristic to all the samples studied, showing that the predominant part of the silicon compounds at the surface are still in organic form even after flame treatment. Judging by the literature data[25], they can be identified as poly(dimethyl siloxanes). The spectra prove the possibility of its continuous migration even after the formation of a char layer at the surface. Furthermore, in the Si 2p region of the heat treated sample another component appears at 104.4 eV, assignable[25,26] to Q-type Si atoms, i.e. probably to SiO_2 or to a polycondensate of TES. As shown by plot b in Figure 4, the ratio of this component reaches about 30 % in the surface layer studied.

Interestingly, the organosilicon additive enhances the migration of other flame retardant compounds to the surface. For example, the phosphorus concentration at the surface after processing and after heat treatment is about four and seven times higher, respectively, than in case of the reference APP containing samples (Table 2). The position of the P 2p peak (134-135 eV) did not change significantly after heat treatment, suggesting that the chemical state of P remained practically unaltered. Based on these results, the conclusion can be drawn that an interaction between organosilicon molecules and APP may help to pull the flame retardant additives to the surface, resulting in their enhanced surface concentration and consequently improved efficiency.

The N 1s peak of the untreated samples at 400 eV may be assigned to amine[27] type N, while in case of heat treated samples a peak broadening to the high binding energy side could be observed (with a peak maximum of about 402 eV), suggesting the additional appearance of NH_4^+ groups[27] at the surface.

Further details can be determined from the curve-fitted detailed C 1s spectra of the heat-treated PP + APP and PP + silicone + IFR samples (Figure 5). The comparison of the corresponding spectra clearly shows the appearance of new peak components (C3, C4 and C5) for the latter sample. Remembering that in case of the PP+APP model system only slight oxidation could be estimated, the appearance of these new components can be assigned to the surface segregation of oxygen-containing additives.

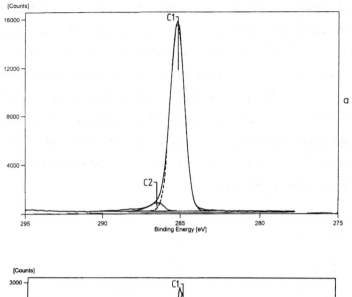

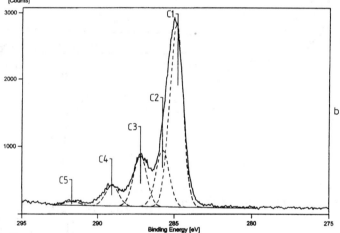

Figure 5. C1s spectra: (a) PP+APP sample; (b) silicone + IFR containing PP sample.

Since peak C5 at 291.7 eV is a shake-up satellite[28] of C 1s, it evidences the presence of C=C double bonds in the IFR-containing, heated PP samples. The C4 peak (289.1 eV) originates from the carboxyl or ester groups. More difficult is the assignment of the C3 (287.28 eV) and C2 (285.84 eV) components, which may be probably ascribed to heteroatomic or miscellaneous bonds (e.g., due to the presence of heteroatoms N, P).

The interpretation of XPS results required the assumption of interaction between silicone molecules and APP. This interaction may occur through the PER-ester layer formed on the surface of APP particles (according to DSC results described above). Model experiments using DSC and FTIR method were carried out in order to determine the type of this interaction. A mixture of PER and TES at ratio of 1:1 was heated up two times in a pressure-tight DSC cell. The curve of the double heating is shown in Figure 6. The melting endotherm peak of PER could be detected only at the first heating, while the flat curve, without any sign of the melting of PER, in case of the second heating proves the interaction (just after melting of PER) between the two compounds.

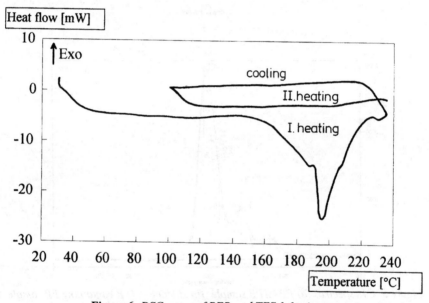

Figure 6. *DSC curve of PER and TES 1:1 mixture.*

In order to prepare sample for FTIR measurements, the PER and TES was boiled at 140°C under reflux cooler for 90 minute. After removal of ethanol white insoluble (polymer) powder was yielded, that was analysed. Comparing the FTIR spectra of the product (Figure 7) to the spectra of PER (Figure 8) the shift of the peak corresponding to the C-OH bond from 3328,4 cm-1 (PER) to 3421,3 cm-1 (product) reflects the decrease of -OH/H associations due to the reaction of OH groups of PER.

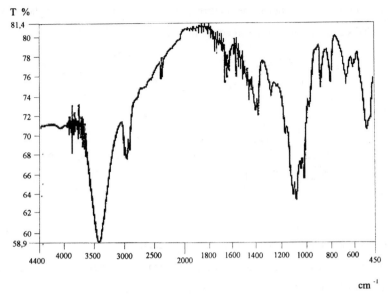

Figure 7. *FTIR spectrum of the product of the PER and TES reaction.*

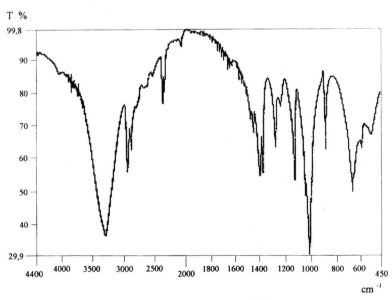

Figure 8. *FTIR spectrum of PER..*

The model measurement suggest the occurring of coupling reaction between the derivatives of PER and TES at the circumstances of compounding according the following equation:

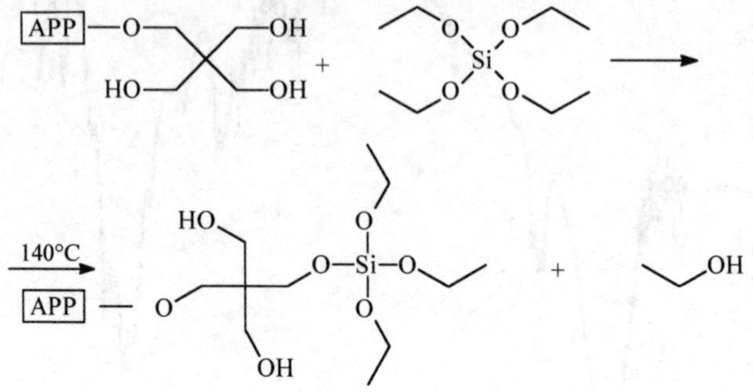

Due to this reaction with TES, chemically bounded interlayer of silicone compounds may form around the APP particles promoting their migration to the surface. In case of applying VTS instead of TES an additional reaction may occur: radical coupling to the PP matrix through the double bonds of VTS:

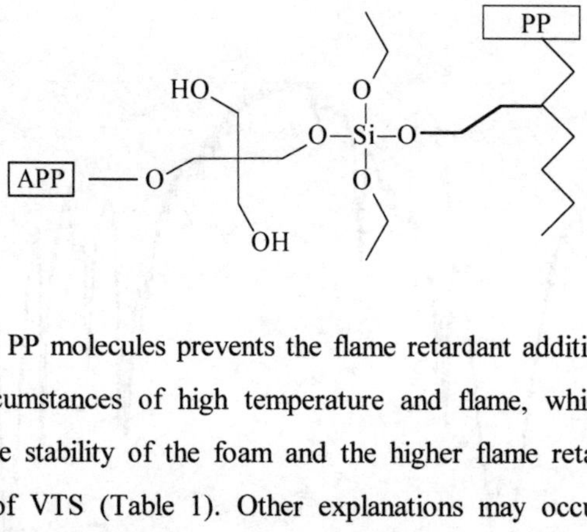

The coupling to the PP molecules prevents the flame retardant additives migration to the surface under the circumstances of high temperature and flame, which may explain the decreased effect on the stability of the foam and the higher flame retardant additive level required in presence of VTS (Table 1). Other explanations may occur as well, so XPS measurement has been started to determine the silicone concentration at the surface in case of VTS containing flame retardant PP samples.

4. SUMMARY

The reaction between APP and PER during the mixing process could be followed by DSC method. A chemically bonded PER layer may form around APP particles due to this reaction.

A layer of elastomer can be formed also, that decreases the water sensitivity of APP. The use of EVAc elastomer improves the compatibility between the phases, but requires higher IFR concentration, while the incompatible polysiloxane elastomer and TES show synergetic effect.

The synergetic effect originates from the higher concentration of IFR additives at the surface, when TES is applied, as it was detected by XPS measurements, especially at higher temperature. A layer of silicone molecules may surround the IFR particles due to the reaction between PER and TES derivatives and this layer promotes the migration of additives to the surface. VTS, being able to bond the additives to the PP matrix, brings about just the opposite effect.

References

1. J. Wang, ", in *"Proceedings of the 6th European Meeting on Fire Retardancy of Polymeric Materials"*, S. Bourbigot, M. Le Bras and R. Delobel eds., AGIR/MITI Pub., Villeneuve d'Ascq, France, 1997, pp. 2.

2. G. Audisio, A. Rossini, F. Severini and R. Gallo, *J. Anal. Appl. Pyrolysis,* 1986, **11(15)**, 263.

3. G. Camino, L. Costa and L. Trossarelli, *Polym. Deg. Stab.*, 1984, **7**, 221.

4. G. Camino and L. Costa, *Polym. Deg. Stab.*, 1986, **8**, 69.

5. G. Camino and L. Costa, *Polym. Deg. Stab.*, 1988, **20**, 271.

6. G. Camino and L. Costa, *Polym. Deg. Stab.*, 1989, **23**, 359.

7. G. Camino, L. Costa and M. Cortemiglia, *Polym. Deg. Stab.* 1991, **33**, 131.

8. F. F. Agunloye, J.E. Stephenson and C.M. Williams, *Flame Retardants Conf.*, London, 1994.

9. R. Delobel, M. Le Bras, N. Ouassou and F. Alistiqsa, *J. Fire Sci.*, 1990, **8**, 85.

10. R. Delobel, M. Le Bras and N. Ouassou, *Polym. Deg. Stab.* 1990, **30**, 41.

11. M. Le Bras and S. Bourbigot; *"Intumescent Fire retardant Polypropylene Formulations"*, in *"Polypropylene: An A-Z Reference"*, J. Karger-Kocsis ed., Chapman and Hall, London, 1998.

12. Gy. Marosi, P. Anna, I. Balogh, Gy. Bertalan, A. Tohl and M. A. Maatoug; *J. Thermal Analysis*, 1997, **48**, 717.

13. I.S. Reshetnikov, M. Yu. Yablokova and N. A. Khalturinskij, Appl. Polym. Sci., 1998, **67(10)**, 1827.

14. Gy. Marosi, Gy. Bertalan, I. Balogh, A. Tohl, P. Anna, G. Budai and Sz. Orbán, *"New flame retardant system for polyolefins."*, in *Flame Retardants'96*, Interscience Communications Ltd., London, 1996, 115.

15. I. Balogh, Gy. Marosi, Gy. Bertalan, P. Anna, A. Tohl, M. A. Maatoug, I. Csontos and K. Szentirmai, Mûanyag és Gumi, 1997, **34**, 209.

16. I. Balogh, Gy. Marosi, Gy. Bertalan, P. Anna, A. Tohl, M. A. Maatoug, I. Csontos and K. Szentirmai, Mûanyag és Gumi, 1997, **34**, 237.

17. Gy. Marosi, I. Balogh, I. Ravadits, Gy. Bertalan, A. Tohl, M. A. Maatoug, K. Szentirmai, I. Bertóti and A. Tóth, Mûanyag és Gumi, 1997, **34**, 265

18. G. Camino, L. Costa, L. Trossarelli, Polym. Deg. Stab. 1985, **12**, 213.

19. Gy. Bertalan, I Rusznák, A. Huszár, G. Székely, L. Trézl, V. Horváth, Z. Kalmár and A. Jancsó, Plaste und Kautschuk, 1978, **25(6)**, 340.

20. Gy. Marosi, Gy. Bertalan, I. Rusznák and P. Anna, *Colloids and Surfaces*, 1986, **23**, 185.

21. Gy. Marosi, Gy. Bertalan, I. Rusznák, P. Anna and, I. Molnár, in *"Polym. Composites"*, Sedlacek B. ed., Waltes de Gruyter and Co., Berlin, 1986.

22. Gy. Marosi, PhD Thesis, 1989.

23. Gy. Marosi, Gy. Bertalan, P. Anna and I. Rusznak, J. Polym. Eng. 1993, **12(1-2)**, 34

24. A. Berlin, N. A. Khalturinskij, I. S. Reshetnikov and M. Yu. Yablokova, in *"Proceedings of the 6th European Meeting on Fire Retardancy of Polymeric Materials"*, S. Bourbigot, M. Le Bras and R. Delobel eds., AGIR/MITI Pub., Villeneuve d'Ascq, France, 1997, pp. 35.

25. C. D. Wagner, J. Vacuum. Sci. Technol., 1982, **21(4)**.

26. F. P. J. Kerkhof, J. A. Moulijn and A. Heeres, J. Electron Spectrosc. Relat. Phenom., 1978, **14**, 453.

27. *"Handbook of X-ray Photoelectron Spectroscopy"*, J. F. Moulder, W. F. Stickle, P. E. Sobol and K. D. Bomben ed., Perkin Elmer Corp., Minnesota, (USA), 1992.

28. J. A. Gardella Jr., S.A. Ferguson and R. L. Chin, *Appl. Spectrosc.*, 1986, **40**, 224.

29. Noll, *"Chemie und Technologie der Silicon"*, Verlag Chemie GmbH, Weirheim, 1968.

Acknowledgement

The support of TEMPUS JEP-09714-95 project and OTKA 014194 is acknowledged with gratitude.

Flame Retarded Intumescent Textiles

FLAME RETARDANT CELLULOSIC TEXTILES

A R. Horrocks and B. K Kandola

Faculty of Technology, Bolton Institute
Bolton, BL3 5AB, UK

1. INTRODUCTION

Cellulosic textiles, while being perhaps one of the most flammable of materials, may be rendered flame retardant by well established means[1].

Cotton is the most commonly used of all textile fibres and with regenerated cellulosics, e.g. viscose, comprise 60% of the world annual fibre consumption. Not surprisingly, therefore, their availability and usefulness coupled with their relative ease of flame retarding makes them the most commonly used flame retardant textile. Flame retarded cotton and viscose probably hold 90% or so of the total flame retardant textiles market.

UK Fire Statistics upto 1993[2], have demonstrated that while about 20% of fires in dwellings are caused by textiles being the first ignited material, over 50% of the fatalities are caused by these fires. Table 1 presents typical data during the last 15 years although since 1993, such detailed data has not been as freely available[3]. Table 1 shows that generally deaths from fires in dwellings have fluctuated around 700 per annum between 1982 and 1988; since then they have fallen to the 500 level.

Fatalities from textile-related fires show a similar pattern and it may be concluded that legislation associated with the mandatory sale of flame retarded upholstered furnishing fabrics into the domestic UK market since 1989 has played a significant factor in these reductions[4]. A significant proportion of flame retarded upholstery and soft furnishing fabrics used are flame retarded cellulosics.

2. FLAME RETARDANT CELLULOSIC TEXTILES

Flame retardant cellulosic textiles generally fall into three groups based on fibre genus:

1. Flame retardant cotton,

2. Flame retardant viscose (or regenerated cellulose),

3. Blends of flame retardant cellulosic fibres with other fibres, usually synthetic or chemical fibres.

Table 1. UK Dwelling Total and Textile - Related Fire Deaths, 1982 – 1994.

Year	Deaths in Dwelling Fires	Textile-related Fatalities in Dwellings				
		Total	Clothing	Bedding	Upholstery	Floor-coverings
1994*	498	235	61	68	86	5
1993	536	269	51	85	105	19
1992	594	322	71	82	134	22
1991	608	293	59	85	127	10
1990	627	346	61	89	157	20
1988	732	456	92	141	195	20
1986	753	485	69	150	219	17
1984	692	396	59	124	167	22
1982	728	426	86	140	152	23

Note: * *denotes values based on sampling procedure*[3]

2.1. Flame retardant cottons

These are usually produced by chemically after-treating fabrics as a textile finishing process which, depending on chemical character and cost, yield flame retardant properties having varying degrees of durability to various laundering processes. These may be simple soluble salts to give non-durable finishes (e.g. ammonium phosphates, polyphosphate and bromide; borate-boric acid mixtures); they may be chemically reactive, usually functional finishes to give durable flame retardancy (e.g. alkylphosphonamide derivatives (Pyrovatex, Ciba; TFR1, Albright & Wilson); tetrakis (hydroxy methyl) phosphonium salt condensates (Proban, Albright & Wilson)) and, more recently, back-coatings which often usually comprise a resin-bonded antimony-bromine flame retardant system. Table 2 summarises the currently popularly-used treatments with selected commercial examples[5].

Most of these treatments have become well-established during the last thirty years[1] and few changes have been made to the basic chemistries since that time; those that have been made often involve minor changes which influence properties such as handle[6] or reduced levels of formaldehyde release during application as seen, in Pyrovatex 7620 (Ciba).

Table 2. Summary of available flame retardant treatments for cotton.[5]

Type	Durability	Structure/formula
Salts :		
(i) Ammonium polyphosphate	Non- or semi durable (dependent on n)	
(ii) Diammonium phosphate	Non-durable	$(NH_4)_2HPO_4$
Organophosphorus :		
(i) Cellulose reactive methylolated phosphonamides	Durable to more than 50 launderings	 e.g. Pyrovatex CP (Ciba) TFR 1 (Albright & Wilson) Afflamit (Thor)
(ii) Polymeric tetrakis (hydroxy methylol) phosphonium salt condensates	Durable to more than 50 launderings	THPC - urea - NH_3 condensate e.g. Proban CC (Albright and Wilson)
(Back) Coatings :		
(i) Chlorinated paraffin waxes	Semi-durable	$C_nH_{(2n-m+2)} \cdot Cl_m$
(ii) Antimony/halogen (aliphatic or aromatic bromine - containing species)	Semi- to fully durable	Sb_2O_3 (or Sb_2O_5) + Decabromodiphenyl oxide or Hexabromocyclododecane + Acrylic resin e.g. Myflam (Mydrin) Flacavon (Schill + Seilacher)

However, during the same period, many other flame retardants based on phosphorus chemistry, in the main, have ceased to have any commercial acceptability for reasons which include toxicological properties during application or during end-use, antagonistic interactions with other acceptable textile properties and cost. The examples cited above may be considered to be those which continue to satisfy technical performance and enable flammability regulatory requirements to be met, while having acceptable costs and meeting health and safety and environmental demands. This last is becoming of particular importance and the fact that the most effective flame retardants contain either phosphorus or antimony-bromine-based systems generates a perception of unacceptable environmental hazard in spite of scientific information which indicates the contrary[7].

2.2. Flame retardant viscose

These FR fibres usually have flame retardant additives incorporated into the spinning dopes during their manufacture, which therefore yield durability and reduced levels of environmental hazard with respect to the removal of the need for a chemical flame retardant finishing process. Additives like Sandoflam 5060[8] are phosphorus-based and so are similar to the majority of FR cotton finishes in terms of their mechanisms of activity (condensed phase), performance and cost-effectiveness. Again, environmental desirability may be questioned and this issue has been minimised by Sateri (formerly Kemira) Fibres, Finland with their silicic acid-containing Visil flame retardant viscose fibre[9]. This fibre not only has removed the need for phosphorus, but also chars to form a carbonaceous and silica - containing mixed residue which offers continued fire barrier properties above the usual 500 °C where carbon chars will quickly oxidise in air.

2.3. Flame retarded cellulosic blends

In principle, flame retardant cellulosic fibres may be blended with any other fibre, whether synthetic or natural. In practice, limitations are dictated by a number of technical limitations including:

1. Compatibility of fibres during spinning or fabric formation; fibres must be available with similar dimensions and be processible simultaneously with other types on the same equipment.

2. Compatibility of fibre and textile properties during chemical finishing; for instance, flame retardant cotton treatments must not adversely influence the characteristics of the other fibres present in the blend during their chemical application.

3. Additivity and, preferably synergy, should exist in the flame retardant blend; it is well known that with some flame retardant blends, antagonism can occur and the properties of the blend may be significantly worse than either of the components alone[1].

2.4. Flame retardant mechanisms

Successful flame retardants for cellulose must interact with and modify the pyrolysis and combustion mechanisms; these latter are not fully understood in spite of considerable research and have been reviewed elsewhere[10,11]. More recently we have reviewed[12] the flame retardant mechanisms operating in flame retarded cellulosic materials including textiles.

Most phosphorus and nitrogen-containing retardants, when present in cellulose, reduce volatile formation and catalyse char formation. They act in this double capacity because on heating they first release polyphosphoric acid, which phosphorylates the $C(6)$ hydroxyl group in the anhydroglucopyranose moiety and simultaneously acts as an acidic catalyst for dehydration of these same repeat units. The first reaction prevents formation of laevogluscosan, the precursor of flammable volatile formation, and this ensures that the competing char-forming reaction is now the favoured pyrolysis route. The acidic catalytic effect of the released polyacid further increases the rate of this favoured route. While considerable research has been undertaken into char formation of flame retarded cellulose, the actual mechanisms of both unretarded and retarded cellulose charring are not well understood[12].

Hirata in his review[10] supports that the general view that for pure cellulose, the major char-forming mechanism is still that based on Kilzer and Broido's competition between dehydration to char and depolymerisation to laevoglucosan[13]. Work in our own laboratories[14] and reviewed previously[5,12] suggests that this mechanism is influenced by presence and type of flame retardants.

Figure 1 shows this modified mechanism and includes the "activated cellulose" intermediate state proposed by Bradbury et al[15]. Evolved gas - DTA studies of a series of flame retarded cotton fabrics identified seven transitions (see Figure 1) associated with formation of activated cellulose (T_2), flame retardant, low temperature-induced formation of char at 200 C (T$_1$), higher temperature (250-300°C) competitive formation of volatiles (and oxidation to CO at T$_3$ and CO$_2$ at T$_5$) and char oxidations above 400°C to CO (T$_6$) and CO$_2$ (T$_7$). Generally, T$_1$ to T$_7$ represent sequentially increasing temperatures of the various transitions.

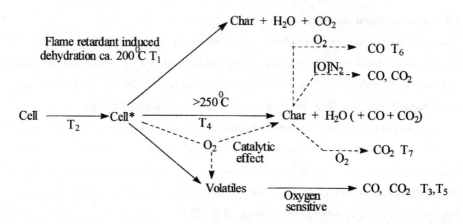

Figure 1. *Schematic Modified Mechanism for Cellulose Pyrolysis*[13-15].

Oxygen- dependence of some stages of the pyrolysis is seen, but of particular interest is the low temperature char formation transition T$_1$ seen only for certain phosphorus-nitrogen-containing flame retarded cottons. Subsequent pyrolysis - GC studies[16] confirmed the char-promoting and volatile fuel-reducing tendencies of the phosphorus and nitrogen-containing flame retardants used. Flame retardants producing greatest char within the 300-500°C range, such as the phosphorus salt condensates, Proban CC (Albright and Wilson), ammonium polyphosphate and the phosphonamide - based Pyrovatex CP, (Ciba) when pyrolysed at temperatures above 400°C, generated high levels of aromatic volatiles including benzene, toluene and phenol. These observations therefore support the proposed aromatisation of carbonaceous chars which occurs following the initial dehydration reactions.

Char studies of flame retardant cellulose have indicated that most phosphorus remains in the char[17]. However, subsequent research by Drews and Barker[18] showed that phosphorus retention in chars related to flame retardant efficiencies and reactivity of phosphorus moieties with cellulose. Oxygen bomb calorimetric studies of all chars produced from different flame retardants at differing phosphoric contents, suggested that they had similar structures possibly derived from phosphorylated cellulose in the first instance. The presence and role of nitrogen in P/N synergistic flame retardants is understood even less than that of phosphorus and evidence exists that char structure is enhanced by formation of P-N bonds[19].

3. INTUMESCENT - APPLICATION TO TEXTILES

Clearly any enhancement of the char barrier in terms of thickness, strength and resistance to oxidation will enhance the flame and heat barrier performance of textiles. Generation or addition of intumescent chars as part of the overall flame retardant property will be of significance here.

The application of intumescent materials to textile materials has been reviewed[5] and is exemplified in the patent literature by the following fibre-intumescent structures;

(i)   Mouldable cellulose or mineral fibre-filled slurries in aluminium or other phosphates which on drying yield a rigid, fire resistant, shaped matrix[20],

(ii)   Sandwich structures in which outer textile fabrics, enclose a layer of intumescent material which is contained following the needlebonding of the outer fabrics together[21]. Typical intumescent systems cited include polyphosphate/amine/polyol systems as above and inorganic expansible materials such as vermiculite, sodium silicate gels and expanded graphites. These patented structures are more textile-like in character than the example in (i) and have preferred area densities in the range 500 - 1500gm^{-2}. Their suggested application areas are flexible barrier and fire blanket materials. Alternatively they may be moulded and heated to about 200°C to yield an intumescent, rigid fire barrier,

(iii)   More conventional, flexible textile fabrics to which an intumescent composition is applied as a coating have been reported[22]. In one example of this patent, the flame and heat resistance of the intumescent coating is complemented by the glass-fibre-cored yarns used in the woven or knitted structure. Presence of sheath fibres of more conventional generic type ensure that the textile aesthetic properties may be optimised,

(iv)   Most recently, we have patented[23] a novel range of intumescent-treated textiles which derive their unusually high heat barrier properties from the formation of a complex char which has a higher-than-expected resistance to oxidation. These require the intumescent to be in intimate contact with the surfaces of flame retarded, char-forming fibres and for char-forming mechanisms to be physically and chemically similar. Exposure to heat promotes simultaneous char formation of both intumescent and fibre to give a so-called "char-bonded" structure.

This integrated fibrous-intumescent char structure has a physical integrity superior to that of either charred fabric or intumescent alone and, because of reduced oxygen accessibility, demonstrates an unusually high resistance to oxidation when exposed to temperatures above 500°C and even as high as 1200°C. Furthermore, these composite structures show significantly reduced rates of heat release when subjected to heat fluxes of 35kW m^{-2}, thus demonstrating additional significant fire barrier characteristics[24].

3.1. Compatible Intumescent - Flame Retardant Combinations

With regard to (iv) above, we have described a number of possible textile structures comprising combinations of compatible intumescents and flame retardant fibres[23-25]. These are shown schematically in Figure 2 and are all based on nonwoven composites in which intumescent - resin systems may be conveniently combined with flame retardant fibrous aggregates to ensure a high degree of randomisation of the former on fibre surfaces. Fibre dimensions are typically those used for short staple textile process machinery, namely 1 – 3.3 dtex in fineness and between 25 and 50 mm length to yield composites having area densities in the range 500 - 900 gm^{-2}.

The introduction of flame retardant cotton woven scrims improves ease of textile processing as well as adding to the final textile physical and flame barrier properties. It is the intention of present research[26] to develop intumescent application methodology which enables more conventional textile fabrics comprising suitable flame retardant fibres to be suitably fibre encapsulated whilst maintaining acceptable textile physical properties. Earlier experiments showed that the intumescent may be applied at up to 50% by weight with respect to the fibre content in order to optimise heat barrier properties via efficient, complex char formation[23,24].

To date, research has demonstrated [23-25, 28-33] that a selection of phosphorus - based intumescents form complex chars when combined with a range of flame retardant cellulosic fibres, namely cotton and viscose variants[34]. Table 3 summarises the combinations found to be effective in this respect. In all composites, intumescents were applied in a standard acrylic textile coating resin at 15% (w/w) with respect to the intumescent[23].

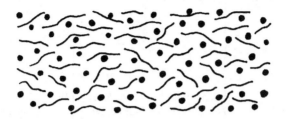

a) Orientated fibre - dispersed intumescent nonwoven

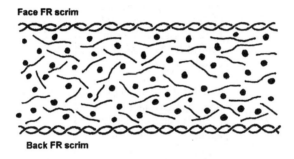

Face FR scrim

Back FR scrim

b) Sandwich composite structure

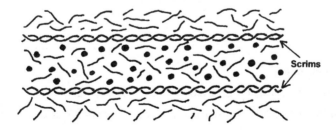

Scrims

c) Sandwich composite with embedded FR fibre scrims

Figure 2. *Schematic Diagrams of Possible Intumescent - Textile Structures[21-25].*

Table 3. *Compatible Flame Retardant Cellulosic Fibres and Intumescents*

Fibre	Flame Retardant	Intumescent (applied to all fibres)	
Visil, hybrid viscose (Sateri (formerly Kemira) Fibres, Finland)	Polysilicic acid, 30% by weight as silica[9]	(i)	Ammonium polyphosphate, pentaerythritol, melamine (3:1:1 mass ratio) (MPC 1000, Albright & Wilson)
Viscose FR (Lenzing, Austria)	Sandoflam 5060 (Sandoz) (2, 2-oxybis) 5,5-dimethyl - 1, 2, 3 - dioxaphosphorinane - 2, 2 - disulphide), 10-15% by weight[8].	(ii)	Melamine phosphate, dipentaerythritol as Amgard NW (formerly MPC 2000), Albright & Wilson.
Cotton	Ammonium polyphosphate (APP) and urea (Amgard LR2, Albright & Wilson), applied 1.7 % P (w/w), heat cured[31].		
Cotton	Tetrakis (hydroxy methyl) phosphonium chloride (THPC) - urea condensate applied via an ammonia cure (Proban CC, Albright & Wilson), 2.5 - 4% P (w/w).		
Cotton	Dimethyl N-methyol phosphono-propionamide (TFRI, Albright & Wilson) applied with trimethylolated melamine resin, 2.3 - 2.7% P (w/w).		

3.2. Char Formation

Initial studies[23,24] showed that exposure of composite fabrics in a furnace in air at temperatures up to 500°C showed intensive char formation and expansion in the sample thickness as the intumescent acted inside the fabric fibrous structure. At temperatures rising to 900°C and in some cases 1200°C, the samples were able to withstand exposure times of at least 10 minutes; under the same conditions, similarly structured polyaramid fabrics were fully oxidised. There observations[25] suggested that the chars formed could withstand oxidising conditions above 500°C. In hybrid viscose (Visil) - containing fabrics, the conversion of the polysilicate present to silica creates a silica - carbonaceous residue which is particularly stable to temperatures above 500°C; in combination with intumescents, these composites yield coherent residues even after exposures at 1200°C[25].

Scanning electron micrographs of chars indicate that the surfaces of the fibrous char component are difficult to distinguish from the intumescent component. This has been reported for hybrid viscose (Visil) - MPC 1000 composites previously[28] and more recent results for Visil and viscose FR - MPC 2000 are shown in Figure 3.

Similar indistinct fibre-intumescent char interfaces have been reported for all flame retardant fibre - intumescent combinations listed in Table 3. Of particular interest here is the observation that the fibrous chars are hollow and that this is especially so for FR viscose. Previous SEM studies of hybrid viscose chars did not indicate this effect[28], however, its presence in FR viscose and FR cotton-intumescent combinations and absence when intumescents are not present merits comment. The presence or absence of polysilicate species and hence silica in the chars seems to reduce any tendency for this hollow char formation. On the other hand, that it occurs only when intumescent is present reinforces the proposed interaction hypothesis.

In addition, these results suggest that not only has an interaction taken place during char formation, but also this char structure may be responsible for the observed high resistance to oxidation. Unpublished, recent SEM-EDAX results show that all charred structures are rich in phosphorus including hybrid viscose chars which are derived from non-phosphorus-containing fibre, Visil.

Subsequent thermal analytical studies[29-31] have demonstrated that DSC and TGA responses of 50 : 50 (w/w) mixtures of fibre-intumescent combinations are different from those calculated from respective, individual component responses. This effect occurs whether the experiments are carried out in air or under nitrogen and again supports the hypothesis that interactive fibre-intumescent char formation is taking place. Figure 4 shows the TGA - derived mass differences of actual versus calculated responses for 50 : 50% (w/w) fibre - intumescent (MPC 1000) mixtures.

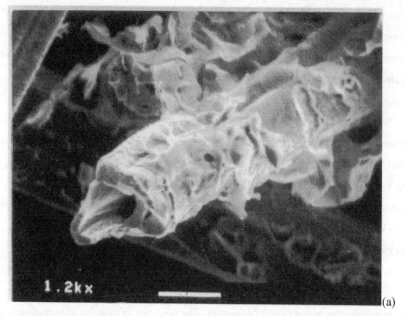

(a)

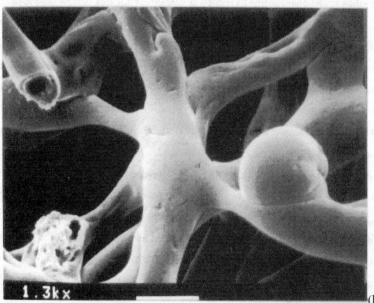

(b)

Figure 3. *Scanning Electron Micrographs of a) (1200 x mag.) Viscose Hybrid, b) (1300 x mag.): Intumescent Chars After Exposure in Air at 500 C for 10 min.*

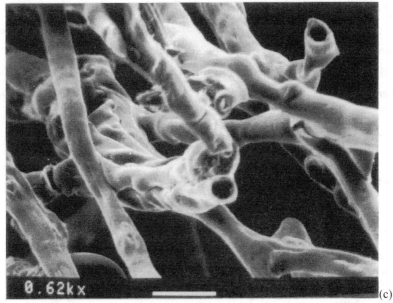

(c)

Figure 3 (continued). *Scanning Electron Micrographs of c) (620 x mag.) FR Viscose - MPC 2000 Intumescent Chars After Exposure in Air at 500 C for 10 min.*

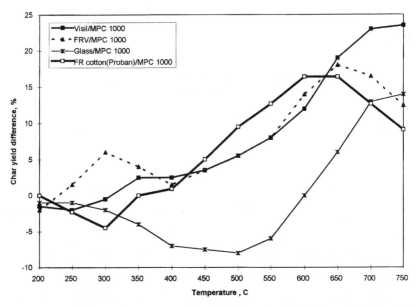

Figure 4. *Mass Differences between actual and calculated residual char yields from TGA in static air for Flame Retarded Viscose and Cotton Intumescent Mixtures.*

The glass - MPC 1000 results represent a non-interactive mixture, although the negative values between 200 and 600°C and its positive values above 600°C suggest that some form of lesser interaction possibly has occurred. On the other hand above 300°C, all flame retarded cellulosic - intumescent mixtures[30,31] indicate enhanced char formation. The particularly high percentage values above 500°C suggest that oxidative stability is greater than expected with respect to those of respective components. In addition, shifts in DSC and DTG maxima have also been observed as illustrated in Table 4[30,31] for the latter.

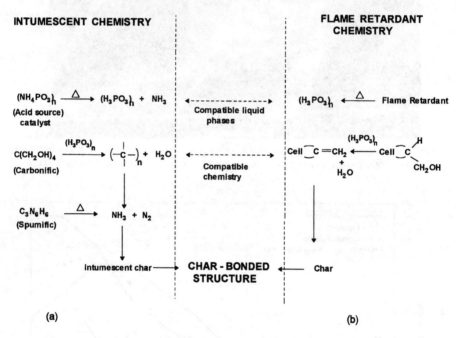

Figure 5. *Simplified Mechanism of Complex Char Formation[31].*

The transitions in Table 4 are grouped into six ranges: 241 - 299, 306 - 348, 402 - 409, 426 - 470, 525 - 568 and 663 – 699°C. Those within the first three groups relate to reactions involving decomposition of flame retardant/intumescent species, namely dehydration of cellulose and char-forming reactions[30,34,35]. Above 400°C and especially in the region of 450-470°C, exothermic oxidation is observed in unretarded fibres and intumescent when heated alone. Flame retarded cotton and Viscose FR have char oxidative exothermic transitions in the 500 - 600°C range indicating that probably the presence of covalently - bonded phosphorus increases the char oxidative resistance. However, intumescent and flame retardant - intumescent combinations show exotherms above 650°C. These observations coupled with the enhanced char residues noted in Figure 4 above 500 °C confirm that the enhanced chars are also particularly oxidation-resistant.

Table 4. *Temperatures of DTG Maxima in Static Air*[30,31].

Sample	DTG maxima, °C				
Cotton		318	470		
Cotton (Proban) (2.5%P)		317		559	
FR Cotton (APP-Amgard LR2)(1.7%P)	288			568	
Viscose	299	314	454		
Visil	285	306	426		
Viscose FR	241			525	
Melamine phosphate/dipentaerythritol (MPC 2000)	260	343	456		699
FR Cotton (Proban)/MPC 2000	255	328	409		663
FR Cotton (APP)/MPC 2000	261	305	408		681
Visil/MPC 2000	242	348	408		
Viscose FR/MPC 2000	266		402		

Concurrent FTIR reflectance spectra of chars[30,34] provide the expected evidence that char formation is accompanied by loss of cellulosic - OH and C-H bonds through dehydration coupled with presence of C = C bond formation. The IR absorption behaviour of char of flame retardant fibres above is similar to that in the presence of intumescent, suggesting that the chemistry of char formation is not greatly changed.

The above results, therefore, confirm the general hypothesis that the fibre and intumescent components may char in an interactive manner if their physical and chemical thermal behaviours are similar. SEM studies[28] have also indicated that flame retardant cellulosics may char via a semi-liquid stage as the Lewis acids polysilicic acid (in Visil) or polyphosphoric acid (in viscose FR and FR cotton) are released. Similarly, thermal analytical studies[30,31] show that the chosen intumescents pass through liquid phases over the same temperature range (200-300°C) as components melt and/or release phosphoric acid. This causes the char-forming reaction stages to physically interact and hence become chemically homogeneous giving rise to complex chars in higher-than expected quantities. Their resistance to oxidation above 500°C is more difficult to explain, although it could be a consequence of high levels of C-O-P cross-linking and reduced oxygen accessibility.

Figure 5[31] attempts to show schematically this series of interactive stages although the actual flame retardant fibre[12] and intumescent[35] pyrolysis chemistries are far more complex..

3.3. Char Oxidative Resistance

More recently, mass loss calorimetric experiments have been undertaken which have attempted to quantify the enhanced char formation and resistance to oxidation of more realistic textile fabric samples comprising flame retardant fibre - intumescent combinations. Initial studies using flux rates upto 100 kW m^{-2} on 100 x 100 mm samples, using a Fire Testing Technology mass loss calorimeter, indicated that these combinations give rise to enhanced char oxidative resistance[29]. Figure 6 shows the typical increased oxidative resistance in terms of reduced mass loss rate when the intumescent is present in the flame retarded textile, in this case Proban-treated cotton. This same effect is shown also in Figure 7 for flame retardant viscose and cotton fabrics in the presence and absence of intumescent. Here the residual char remaining after 5 minutes under 50kW m^{-2} in air shows the magnitude of the char-enhancing and oxidation-resistance effects that the intumescent confers. Further unpublished research and analysis of mass loss versus time curves show that three regions of mass loss exist, namely initial volatilisation, a char-formation stage (both occurring during the first 4 minutes exposure period) and a char oxidative stage continuing over a much longer period.

Analysis of these apparently first order oxidative stages for each fibre - intumescent combination at four different temperatures corresponding to the heat fluxes of 35, 50, 75, and 100 kW m^{-2}, yields Arrhenius - derived activation energies as shown in Table 5. It is probable that each of the mass loss regions relate to the groups of DTG peaks in Table 4, namely 241-348°C for volatilisation, 402-470°C for full char formation and first stage of oxidation and > 500°C for full oxidation of the wholly carbonised char.

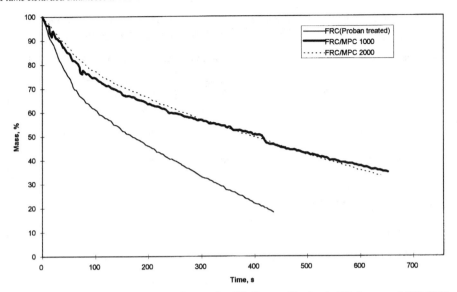

Figure 6. *Mass Loss Calorimetric Curves for FR Cotton (Proban), FR Cotton - MPC 1000 and FR Cotton - MPC 2000 at 50 kW m^{-3} in air[32].*

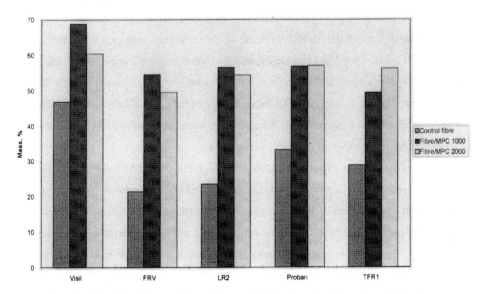

Figure 7. *Residual Chars of Intumescent-Fabric Combinations after 5 Minutes (Exposure at 50 kW m^{-2}).*

Table 5. *Arrhenius Activation Energies for First Order Char Oxidative Stage from Selected FR Cellulosic - Intumescent Combinations.*

Sample	Arrhenius Activation Energies, kJ mol^{-1}		
	0-1 min	1-3.5 min	3.5-10 min
Hybrid viscose (Visil)	33	18	84
Visil-MPC 1000	24	50	87
Visil-MPC 2000	85	69	99
FR Cotton (Amgard LR2)	39	42	104
LR2 - MPC 1000	74	95	104
LR2 - MPC 2000	93	71	102
FR Cotton (Proban)	41	18	14
Proban - MPC 1000	69	53	83
Proban - MPC 2000	78	66	102

Clearly the general sudden increase in the activation energy of the volatilisation (0 - 1 min) (except for Visil combinations) and char formation stage (1 - 3.5 min) in the presence of intumescent is a measure of the interaction between flame retardant fibre and intumescent. The slight increase in activation energy of the char oxidation stage which occurs after 3.5 min exposure (seen also in other fibre-intumescent combination char results not as yet published) is indicative of this increased resistance to oxidation.

This observation corroborates those derived earlier via SEM[28] and thermal analysis[30] and discussed above. Currently the mass loss calorimetric behaviour of the other fibre-intumescent combinations is being investigated. The cause of the increased char oxidative resistance is believed to be a consequence of a combination of reduced oxygen diffusivity of the char physical structure and the presence of phosphorus in the cross-linked char chemistry. Further research is being undertaken to explore this cause and also to more effectively quantify the thermal barrier characteristics of the chars.

References

1. A. R. Horrocks, *Revs. Prog. Coloration*, 1986, **16**, 62.
2. "UK Fire Statistics, 1993", Home Office, London, 1995.
3. G. Goddard and J. Poole, *"Summary Fire Statistics, United Kingdom, 1994"*, Home Office, London, 1996, p.61.
4. A .R. Horrocks, *"Flammability '93 - The Inside Story"*, Textile Institute, Manchester, 1993.
5. A .R. Horrocks, *Polym. Deg. Stab.*, 1996, **54**, 143.
6. M. Zakikhani and X. Lei, *"Niches in the World of Textiles - 7th World Conference"*, Volume 2, Textile Institute, Manchester, 1996, 427.
7. A. R. Horrocks, M. E. Hall and D. L. Roberts, *Fire Materials*, 1997, **21**, 229.
8. M. E. Hall and A. R. Horrocks, *Trends Polym. Sci.*, 1993, **1**, 550.
9. S. Heidari and A. Paren, *Fire Materials*, 1993, **17**, 65.
10. T. Hirata, Natl. Bur. Stand., NBSIR, 85-3218, 1985.
11. G. Varhegi, E. Jakab and M. J. Antal, *Energy Fuels*, 1994, **8**, 1345.
12. B. Kandola, A. R. Horrocks, D. Price and G. Coleman, *Rev. Macromol. Chem. Phys.*, 1996, **C36**, 721.
13. F. J. Kilzer and A. Broido, *Pyrodynamics*, 1965, **2**, 151.
14. D. Price, A. R. Horrocks and M. Akalin, *Brit. Polym. J.*, 1988, 20, 61.
15. A. G. W. Bradbury, Y. Sakai and F. Shofizadeh, *J. Appl. Polym. Sci.*, 1979, **23**, 327.
16. A. A. Faroq, D. Price, G. J. Milnes and A. R. Horrocks, *Polym. Degrad. Stab.*, 1991, **33**, 155.
17. K. Yeh and R. H. Barker, *Text. Res. J.*, 1971, **41**, 932.
18. M.J. Drews and R. H. Barker, *J. Fire Flammability*, 1974, **5**, 116.
19. E. D. Weil., in *"Flame Retardancy of Polymeric Materials"*, Volume 3, eds. W. C. Kuryla and A. J. Papa, Marcel Dekker, New York, 1975, p. 185.
20. J. H. Morris, W. E. Smith and P. G. Parkins, *UK Patent* 17646/74, 23 April 1974.
21. W. von Borrin and U. von Gizycki, Bayer A.G., *Europ. Patent*, 91121065.6, 9 December 1991.
22. T. W. Tolbert, J. S. Dugan, P. Jaco and J. E. Hendrix., Springs Industries Inc., *US Patent*, 333174, 4 April 1989.
23. A. R. Horrocks, S. C. Anand and B. J. Hill, *UK Patent*, GB 2279084 B, 20 June 1995.
24. A. R. Horrocks, S. C. Anand and D. Sanderson, in *"Interflam '93"*, Interscience Communications, London, 1993, p. 689.

25. A. R. Horrocks, S. C. Anand and D. Sanderson, *Textile à Usages Techniques*, 1994, **11**, 56.

26. A. R. Horrocks, P. Davies, R. Hallos, and S. Clegg, *"6th European Meeting on Fire Retardancy of Polymeric Materials"*, Lille, 1997.

27. A. R. Horrocks, S.C. Anand and D. Sanderson., in *"Flame Retardants '94"*, Interscience Communications, London, 1994, p.117.

28. A. R. Horrocks, S. C. Anand and D. Sanderson, *Polymer*, 1996, **37**, 3197.

29. A. R. Horrocks and B. K. Kandola., in *"Flame Retardants '96"*, Interscience Communications, London, 1996, p.145.

30. B. K. Kandola and A. R. Horrocks, *Polym. Deg. Stab.*, 1996, **54**, 289.

31. B. K. Kandola, S. Horrocks and A. R. Horrocks, *Thermochimica Acta*, 1997, **294**, 113.

32. A. R. Horrocks and B. K. Kandola, *Textile Res. J.*, to be published.

33. A. R. Horrocks and B. K. Kandola, in *"Recent Advances in Flame Retardancy of Polymeric Materials, Volume 8"*, ed. M. Lewin, BCC, Stamford, 1997

34. B. K. Kandola and A. R. Horrocks, *to be published.*

35. G. Camino and L. Costa, *Rev. Inorg. Chem.*, 1986, **8**, 69.

Acknowledgements:

The authors wish to acknowledge the support of the Cotton Industry War Memorial Trust, UK, the Engineering Physical Science Research Council, UK and Albright and Wilson, Ltd, Oldbury, U.K.

NEW INTUMESCENT SYSTEMS: AN ANSWER TO THE FLAME RETARDANT CHALLENGES IN THE TEXTILE INDUSTRY

C. Cazé, E. Devaux

Laboratoire de Génie et Matériaux Textiles (GEMTEX), E.N.S.A.I.T.
2, Place des Martyrs de la Résistance, F-59070 ROUBAIX Cedex 1, France

G. Testard and T. Reix

Thor S.A.R.L., Z.I.P.
325, rue des Balmes, F-38150 SALAISE SUR SANNE, France

1. INTRODUCTION

The increasing need for flame retardant protection of textile fabrics constrains the technologists' ability to fulfil all necessary requirements. The European standards and directives tend to limit for instance the presence of halogens and heavy metals in the flame retardant formulations. The existing market is indeed dominated by phosphorus, chlorine, bromine and antimony containing additives, and also to some extents hydrated alumina. Moreover, beside these standardising restraints, the ease of application of the coatings and the transparency of the films realised are important parameters for the end-user. In this context, the concept of intumescent systems allows us to satisfy these requirements. Their action is principally based on the development of an expanded carbonated layer onto the surface of the textile support, which prevents oxygen diffusion and the heat transfer to the core of the material[1]. On the other hand, the carbonised layer reduces the transfer of the combustible gases which can propagate the flame. The intumescent systems are generally made of an inorganic acid playing the role of dehydration agent of a polyhydric compound with a high carbon content in order to form the char layer[2,3], and of a swelling or expansion agent which ensures the expanded character to the carbonised structure[2,4]. The intumescent additives classically used for the thermoplastic polymers are based on ammonium polyphosphates[5] and pentaerythritol[6].

The Thor Company develops and markets flame retardant coating formulations for the textile industry. These systems are principally constituted of an aqueous dispersion of poly(vinyl acetate) (commercial name: Rhenappret RA) which plays the role of binder for the flame retardant additives. The latter are products based on phosphorus and nitrogen, with the commercial names respectively Aflamman PCS and Aflamman IST. The mixtures are applied onto the textile surface by padding, then dried at 150°C during two minutes. As part of this study, the flame retardant coatings were applied to polyester fabrics or glass fibre based structures. The latter substrate has as a major feature the fact that the observation of the intumescence may be carried out independently of the thermal behaviour of the fabrics. Three kinds of formulations have been investigated: the first is only composed of the binder without any flame retardant fillers; the two others contain the different additives (Table 1). The volume fraction of the coating applied was approximately 15 %.

Table 1. *Compositions of the different coating formulations (g/l).*

Sample	1	2	3
Rhenappret RA	700	500	350
Aflamman IST	-	125	-
Aflamman PCS	-	-	150

After drying, the coatings have been characterised by differential scanning calorimetry (DSC) in order to appraise the additive's influence on the physico-chemical behaviour of the products. The evolution of the glass transition temperature has been examined in this way. The flame retardant properties have been studied by thermogravimetric analysis (TGA), by the electric burner test (standard NF P 92-503) and by cone calorimetry (standard ASTM E 1354-90, ISO 5660). Finally, the physico-chemical investigations on the surfaces of the materials after burning have been carried out by infrared spectrometry and by atomic force microscopy (AFM). During discussion of the results, we will consider the possible industrial applications of these intumescent systems.

2. EXPERIMENTAL PART

2.1. Determination of the glass transition temperature of the different formulations

After drying at 150°C of the aqueous dispersion of poly(vinyl acetate) Rhenappret RA, the glass transition of the product was observed using DSC (Perkin Elmer DSC-7, and Polymer Laboratories PL-DSC), by heating the sample under nitrogen with a heating rate of 20°C/min.

A value close to 38°C has been recorded, showing that the polymer is in a glassy state at room temperature.

The addition of the different flame retardant additives does not change the values of the glass transition in a significant way. The coating filled with the Aflamman IST shows a glass transition temperature at 38-39°C, and this is the case even if the additives content is increased. On the contrary, a decrease of 5°C is noticed when the Aflamman PCS is incorporated in the system. However, this value remains constant, whatever the concentration used (100 to 250 g/l of Aflamman for a dispersion with 350 g/l of Rhenappret). Despite this slight difference in the T_g values, the aspect and the handle of the coatings remains similar to the unfilled one.

2.2. Fire performance by the electric burner method

This test (standard NF P 92-503) is suitable for the evaluation of the fire performance of flexible materials and wall coverings. Standard methods consist in submitting the samples, in defined conditions, to the action of a heat and hot gases radiation scanning the samples surface. The samples have the dimensional characteristics 600 mm × 180 mm. A flame is used in order to promote the pyrolysis gases ignition. The ignition delay, the flame persistence, the nature and the effects of the combustion are recorded. In France, the results of these tests allow to classify the behaviour of the samples in the presence of fire from M4 for a highly flammable material, to M1 for a non-flammable material. (see Table 2).

Tests were carried out with the different coatings with or without Aflamman applied to glass fibre fabrics. This non-flammable substrate allows characterisation of the only behaviour of the coating in the presence of fire. The samples are placed on the holder in an oblique position at 30°, then heated by an electric burner (power = 500 W). They are submitted in a repetitive way to the action of a burning vertical flame during 5 seconds, followed by an interruption of 30 seconds. This operation is carried out again over a 5 minutes period.

The results obtained show that only the fabric coated with the Rhenappret RA obtains the classification M3, whereas the formulations filled with the Aflamman IST and with the Aflamman PCS according to the filler contents presented in table 1, obtain the classification M1. These first results confirm the flame retardant action of the fillers for the polymeric coatings.

Table 2. *Standards for the electric burner test.*

Evaluation	Classification			
Flame remanence	≤ 5 s	≤ 5 s	≥ 5 s	> 5 s
Length destroyed	< 35 cm	35 ≤ L ≤ 60	< 35 cm	35 ≤ L ≤ 60
Width destroyed	≤ 9 cm	≤ 9 cm	≤ 9 cm	≤ 9 cm
Absence of droplets	M1	M2	M2	M3
Non-flaming droplets	M1	M2	M2	M3
Flaming droplets	M2	M3	M3	M4

2.3. Thermogravimetric analysis (TGA)

The TGA (Shimadzu TGA-51) does not give any information directly correlative to the behaviour of a material in the presence of fire, but it allows evaluation of the temperature domain over which the sample remains stable. The mass loss of a sample is measured during a heating at 10°C/min under nitrogen atmosphere. In order to understand the way that the treated fabrics and their coatings transform during the thermal treatment, we have studied the behaviour of an untreated polyester-based fabric alone and then coated with the three previous formulations.

Figure 1 presents the curves obtained for the different materials. It can be noticed that the untreated polyester does not undergo any mass loss until 370°C, then it rapidly decomposes and loses more than 80% of its mass by 470°C.

When the textile structure is covered by the poly(vinyl acetate)-based coating, a shoulder can be observed as early as 200°C. This shoulder corresponds to the transformation of the polymer to volatile acetic acid. After this shoulder, the curve superimposes exactly with the polyester response which remains the only material present. The addition of flame retardant additives does not significantly modify the behaviour of the system until 500°C. As previously, the transformation of the coating binder is followed by the degradation of the polyester substrate. However, it appears that the mass losses of the samples filled with the different Aflamman compounds begin at a temperature slightly lower than for the fabrics treated with the Rhenappret. This phenomenon is probably linked to the formation of inert gases which are responsible of the development of the intumescent layer.

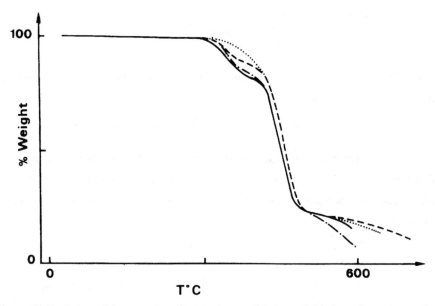

Figure 1. *Evolution of the mass loss for a polyester fabric, and for the polyester coated with the different formulations (— for the Rhenappret RA; — - - for the Rhenappret RA filled with the Aflamman IST; - - - for the Rhenappret RA filled with the Aflamman PCS; ······ for the untreated polyester).*

2.4. Cone calorimetry studies

The cone calorimeter allows evaluation, in a dynamic way, of the rate of heat release produced by a sample during its combustion. This parameters is considered today as the most representative one in order to quantify the size of a fire[7]. Developed during the nineteen eighties, the cone calorimeter is recognised on an international scale, and is the object of the ASTM standard E 1354-90 and of the ISO standard 5660. It allows the measurement of the rate of heat release, the mass loss, the effective heat of combustion, the smoke opacity, and the carbon monoxide and carbon dioxide contents. From the knowledge of the oxygen consumed during a trial, it is possible to calculate the heat quantity q involved during the combustion[8], according to the equation:

$$q = \Delta H_c \times m_{comb} = E \times m_{O_2} \qquad (1)$$

where ΔH_c is the molar heat of combustion, m_{comb} is the combustible mass, E is the Huggett constant (13.1 kJ.kg⁻¹) and m_{O_2} the mass of oxygen consumed.

The heat emission q' can now be calculated as:

$$q' = E \frac{d\left(m_{O_2}\right)}{dt} \qquad (2)$$

The rate of heat release (R.H.R.) corresponds to q' divided by the area A of the sample surface submitted to an external irradiation:

$$R.H.R. = \frac{q'}{A} \qquad (3)$$

The irradiation flux, adjustable between 0 and 100 $kW.m^{-2}$, is emitted by a truncated cone in order to not disturb the flame. The cone calorimeter allows work to be undertaken in open conditions, and thus to be closer to the real conditions of a fire. The ignition is promoted by an electric spark source.

Different trials have been carried out with polyester fabrics covered by the binder Rhenappret, and by this binder filled with the two Aflamman supplied. Several layers of fabrics are superposed in order to work with a sufficient quantity of material. The dimensional characteristics of the different samples are given in Table 3. The sample surface area submitted to the irradiation is 100 cm^2. The irradiation chosen is 25 $kW.m^{-2}$, and it corresponds to the power emitted when a fire begins (smouldering fire).

Table 3. *Dimensional characteristics of the different samples used.*

	Rhenappret RA	Aflamman IST	Aflamman PCS
Area density (g/m^2)	260	255	240
Number of layers	8	5	8
Sample thickness (mm)	4	2.5	4
Sample mass (g)	20.7	12.8	19.1

Figure 2 shows the typical evolution of the rate of heat release as a function of time for an intumescent system. At the beginning of the irradiation, inert gases are created, and the rate of heat release remains zero. When the ignition of the combustible material appears, a significant increase in the rate of heat release is noticed, during which the intumescence phenomenon occurs. During this step, the carbonised protection layer develops. The increase of the rate of heat release carries on with the combustion of the protected material. Further, the combustion of the residue occurs until the sample extinguishes when the rate of heat release becomes again zero.

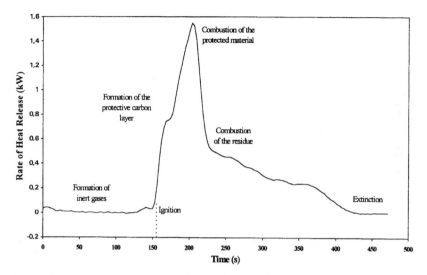

Figure 2. *Rate of heat release as a function of time for a typical intumescent system.*

The experimental results obtained on our different systems are presented on figure 3. The ignition of the polyester fabric only covered by the Rhenappret coating is observed after 70 seconds, and the combustion continues for 413 seconds. The formulations containing the flame retardant fillers Aflamman delay the ignition which may be 105 seconds for the Aflamman PCS, and after 155 seconds in the case of the Aflamman IST. Moreover, the intumescence phenomenon, directly observable during the experiment, is confirmed by the courses of the different curves. At last, the extinction occurs at the same time than for the polyester treated with the Rhenappret, when the coating is filled with the Aflamman IST. On the contrary, the time of extinction is considerably decreased (293 seconds) with the Aflamman PCS. As a consequence, it can be noticed that the system based on the Aflamman PCS burns clearly earlier than the second product, but that it also extinguishes distinctly faster.

The integration of the previous curves allows to determine the total heat release (THR). When this value is divided by the sample mass, it appears that the presence of the Aflamman fillers decreases the total heat release of 20%. At the end of the combustion, the

THR is 12 kJ.g^{-1} for the polyester treated with the Rhenappret, and is only 10 kJ.g^{-1} for each coating based on Aflamman.

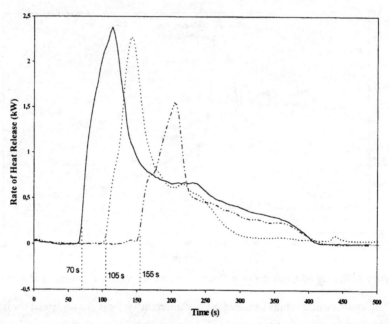

Figure 3. *Rate of heat release versus time for a polyester fabric coated with the different formulations (—— for the Rhenappret RA; — - - for the Rhenappret RA filled with the Aflamman IST; - - - for the Rhenappret RA filled with the Aflamman PCS).*

Figure 4 shows the evolution of the mass loss for the three samples considered as a function of time. It can be noticed that for the sample coated with the Rhenappret, the mass undergoes an important decrease from its ignition. For the two fabrics treated with the Aflammans, a slight mass loss occurs before the ignition. This phenomenon is due to the formation of the inert gases. This first decomposition provides the gases necessary for the appearance of the expanded layer characteristic of the intumescent systems. Incidentally, we confirm that the growth of this layer occurs at a temperature clearly lower than the other decompositions. Moreover, after the extinction of the material (293 s for the Aflamman PCS), the fabric keeps on losing mass. This aspect is of major interest in a flame retardant application, because this degradation occurs without emission of flames. The character of

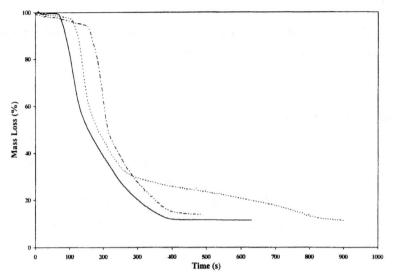

Figure 4. *Mass loss of the polyester fabric coated with the different formulations (——— for the Rhenappret RA; —- - - for the Rhenappret RA filled with the Aflamman IST; - - - for the Rhenappret RA filled with the Aflamman PCS).*

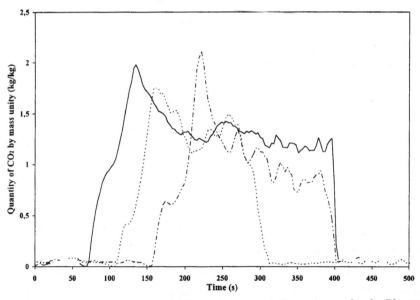

Figure 5. *Evolution of the emission of CO_2 as a function of time (——— for the Rhenappret RA; —- - - for the Rhenappret RA filled with the Aflamman IST; - - - for the Rhenappret RA filled with the Aflamman PCS).*

extinction is also important if we consider the emission of fire gases (CO_2 particularly, Figure 5). We indeed establish that the emission of this gas is also stopped when the extinction occurs, and this confirms the interest of systems having a marked character of auto-extinction.

2.5. Study by atomic force microscopy (AFM)

The atomic force microscopy (Nanoscope III from Digital Instruments) allows to observe a sample clamped on a piezoelectric conductor, scanned by a metallic point. The surface relief is recorded by reflection of a laser beam onto an optic position sensor. Besides the visual clarity and resolution attained by this technique, its principal interest is that it can be used with non-conducting materials. Samples have not to be metal-coated before their observation, thus their superficial topography remains unmodified. We have chosen to carry out our observations on a glass fibre-based fabric, in order to give evidence for the consequences of the intumescence phenomenon, independently of an eventual degradation of the substrate. Figure 6 shows the glass fibre-based fabric coated with the binder Rhenappret alone. It can be noticed that after combustion, the glass fibres are clearly apparent. On the contrary, the coating containing the Aflamman IST shows that the fabric is entirely covered by an intumescent layer which does not allow us to distinguish the substrate anymore (see Figure 7). So the latter is protected by the creation of this layer.

3. INDUSTRIAL APPLICATIONS OF THE INTUMESCENT SYSTEMS

The intumescent systems supplied by the Thor company, based on the formulations described previously, have a large range of application. Moreover, their use is not dependent on problems associated with normal product processing. In the case of the automotive industry for instance, the classical intumescent systems based on bromide-nitrogen or nitrogen-phosphorus are generally not convenient because a noticeable modification of the handle occurs, and because there exists some risks for the material corrosion. In addition, during the coating's processing, there may be a problem of dirt accumulation, and eventually of extraction of the colorants due to too high a processing temperature. These disadvantages do not permit us to satisfy particularly the American standard MVSS 302 related to the automotive products. One of the interests of the formulations studied in this paper is to pass

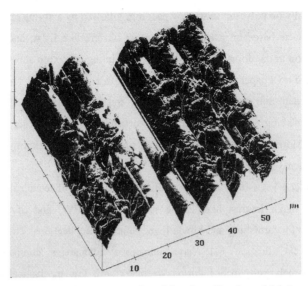

Figure 6. *Surface topography of the glass fibre based fabric coated with the binder Rhenappret RA.*

Figure 7. *Surface topography of the glass fibre based fabric coated with the binder Rhenappret filled with the Aflamman IST.*

this previous standard using lower additive contents. Since coatings for the automotive industry are usually on polyester fabrics by padding, followed by a drying at 150°C.

The products based on Rhenappret or other binders like EVA, and Aflamman have also an application in the domain of textile transfer printing on cotton substrate. The printing pattern is realised on a paper support, then transferred to a cotton fabric. The latter has in this case flame retardant and soil repellent properties, and would have principal interest in the clothing and furnishing domains. The transfer, carried out by calendering at 150°C, is very easy to realise. Moreover, these systems are free of halogens and antimony oxides, and they enable achievement of the French classification M1.

The binder EVA is also used for coating formulations of technical fabrics based on glass fibres. A major interest here is the substitution of PVC and of the halogens and antimony oxides for furnishing and blinds flame retardant applications. These coatings while achieving the classification M1, have numerous advantages compared to classical formulations, namely permanence during washings with low mechanical action, identical handle, and amelioration of the striking. The application of these products is also carried out by padding followed by drying.

The flame retardant systems we have investigated have, additionally, an interesting application in the hot melt glues. The product obtained contains neither halogens nor antimony, and can be used for automotive, aeronautics or electronic industries. These formulations do not modify the specifications of the glue, and combine excellent flame retardant properties and ecological compatibility.

4. CONCLUSION

The intumescent coating formulations supplied by the Thor company for textile applications have been studied, and their flame retardant properties have been evaluated. These formulations are based on a binder called Rhenappret RA which is an aqueous dispersion of polyvinyl acetate, filled with different flame retardant fillers (Aflamman IST and Aflamman PCS). These systems have an intumescent behaviour in the presence of a fire, and present the major advantage to be free of halogens or antimony oxides. The different experiments carried out with these products show that the ignition time is considerably increased, whereas the extinction time is lowered. The intumescence phenomenon is clearly confirmed by the thermogravimetric analysis, the observation of the samples surfaces by atomic force microscopy, and the behaviour under cone calorimetric conditions. Moreover, the application of these formulations on textile fabrics (polyester or glass fibres) does not radically change the fabric's handle.

References

1. Camino, G., Costa, L., and Trossarelli, L., *Polym. Deg. & Stab.*, **7**, 25 (1984).
2. Vandersall, H. L., *J. Fire & Flammability*, **2**, 97 (1971).
3. Le Bras, M., Bourbigot, S., Delporte, C., Siat, C. and Le Tallec, Y., *Fire & Materials*, **20**, 191 (1996).
4. Jones, G., *U.S. Patent* 2,628,946 (assigned to Albi Manufacturing Company) (1953).
5. Kishore, K. and Mohandas, K., Combustion & Flame, **43**, 145 (1981).
6. Camino, G., "*1ᵉʳ Colloque Francophone sur l'Ignifugation des Polymères*", Saint-Denis (France), p. 36 (1985).
7. Babrauskas, V., Fire & Materials, **8(2)**, 81 (1984).
8. Huggett, C., *J. Fire & Flammability*, **12** (1980).

THE EFFECT OF FABRIC STRUCTURE ON THE FLAMMABILITY OF HYBRID VISCOSE BLENDS

S.J. Garvey, S.C. Anand, T. Rowe, A.R. Horrocks

Faculty of Technology, Bolton Institute
Deane Road, Bolton BL3 5AB, UK

D. Walker

S Frankenhuis & Son (Man-made Fibres) Ltd
Broadgate, Broadway Business Park, Oldham OL9 9XA, Lancashire, UK

1. INTRODUCTION

Throughout the past three decades, there has been an increasing demand for flame retardant fabrics made from variously blended fibres[1]. End-use properties can be maximised through the blending of different fibres - one fibre's defects may be compensated by the positive features of another totally different fibre. However, many problems have arisen in trying to achieve effective fire retardancy within these blended fabrics.

Fire remains a hazard even in today's modern society, with accidents and fatalities increasing each year[2]. The latest available government statistics show that in 1994, the total number of dwelling fires in the UK was 63,600. Of this number, 16,480 resulted in non-fatal casualties, but unfortunately, 676 resulted in fatal casualties[3]. People use flammable materials for clothing and they are also used in everyday life at home, in the office and in public buildings. The problems associated with fire have increased due to the present day phenomenon of higher concentrations of large numbers of people in confined spaces. Society is now more aware of the hazards of fires, and in particular, the danger of smoke and toxic gases. Recent fire statistical evidence shows that over half of the fatalities in dwelling fires are due to smoke and toxic gas inhalation.

Textiles are not only rendered flame retardant by the introduction of chemical species, but these treatments may also enhance the toxicity of fire gases involved. All organic materials emit carbon monoxide and carbon dioxide during combustion. Carbon monoxide is a highly toxic gas and its ratio to carbon dioxide is influenced by the amount of air present during combustion. Smoke and fire gas inhalation causes a significant proportion of deaths associated with fire. Breathing airborne combustion products reduces a person's ability to function and thus hinders escape. At present, while the use of FR treatments prevents injury from ignition, in most cases the retardant effect is insufficient to hinder the production of toxic gases.

Visil is a hybrid of organic-inorganic components to give a cellulosic fibre containing polysilicic acid. Viscose on its own will burn, but polysilicic acid, which is present in Visil when heated causes the cellulose to char, and the silica formed to provide a flame and heat barrier. The way in which flame-retardants influence the character of toxic gas formation during combustion has recently been studied for finished cotton fabrics[4]. When examining the mechanism of flame-retardancy, a fibre or an additive to a fibre that chars rather than burns tends to give a lower concentration of flammable volatiles, therefore reducing flammability. Like normal regenerated cellulose, Visil fibres have a low density and a high degree of flexibility. Furthermore, like many conventional flame retardant fibres, they are processible using existing textile technology, but are produced via a modified viscose process using the following raw materials: cellulose and silica as polysilicic acid[5]. There are two modifications of these hybrid viscose fibres available: Visil containing 30 - 33% pure silicic acid, and Visil AP containing 33% aluminosilicate modified silica. The latter has improved flame retardancy and laundry performance. An aesthetic advantage of both these fibres is that their handle, appearance and durability are not sacrificed for enhanced flame retardancy. The resultant textiles in conventional woven, knitted or nonwoven forms already have good comfort properties. However, the fibre strength is rather low, the aesthetic character of a typical viscose is not always desirable and the high cellulose content can give rise to surface flash and reduced flame retardancy in certain lightweight fabric structures. Improvement in these features may be conferred by the use of blends with other flame-retardant fibres such as wool and modacrylic, both of which may offer an improved aesthetic character. However, both these fibres are known to generate significant levels of smoke and toxic fire gas components such as hydrogen cyanide and, additionally in the case of modacrylic, hydrogen chloride.

The aims and objectives of this research are to utilize the environmental acceptability of Visil, to construct a series of yarns which have the properties of adequate flame retardancy combined with reduced toxic gas emission plus desirable aesthetic properties, to produce potentially commercial apparel fabrics from blends of Visil fibre with either wool or modacrylic, to investigate the relationship between blend variables on flammability characteristics of a standard knitted structure comprised of blended yarns, and to study the effects of fabric structural variables on the flammability of variously blended yarns containing

Visil/wool and Visil/modacrylic fibres. This paper reports the work carried out so far in this research program.

1. EXPERIMENTAL

The work so far has involved using a systematic approach to achieve optimum blend ratios within a series of Visil/modacrylic yarns and Visil/wool yarns with regard to their processibility and flame retardancy. The fibres were intimately blended at the drawframe stage as opposed to using two-fold yarns within the fabric structure. Two sets of yarn were produced using both the ring spinning and rotor spinning routes, with the following blend proportions:

100%	Modacrylic	79/21%	Wool/Visil
		60/40%	Wool/Visil
83.3/16.7%	Modacrylic/Visil	43/57%	Wool/Visil
66.7/33.3%	Modacrylic/Visil	27/73%	Wool/Visil
50.0/50.0%	Modacrylic/Visil	13/87%	Wool/Visil
33.3/66.7%	Modacrylic/Visil		
16.7/83.3%	Modacrylic/Visil	Modacrylic	51mm, 3.3dtex
		Visil	40mm, 1.7dtex
100%	Visil	Wool	38mm, 21μm diameter

Each set of yarns was tested for strength and regularity, and the results can be seen in Table 1 and Figures 1, 2 and 3.

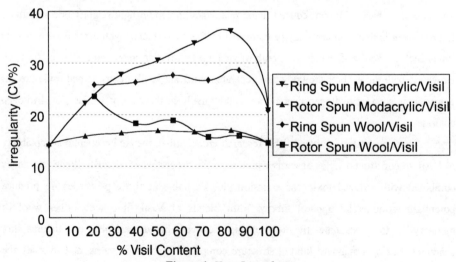

Figure 1. *Yarn Irregularity*

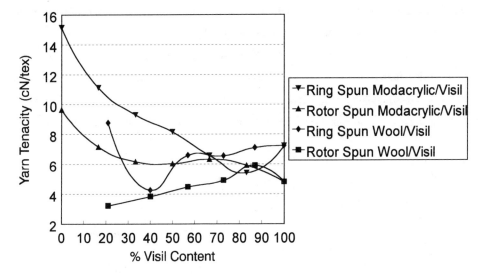

Figure 2. *Yarn Tenacity.*

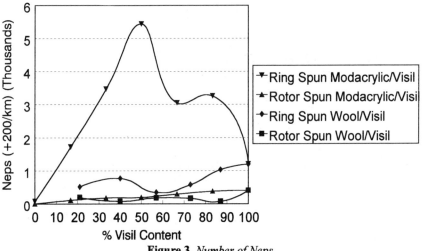

Figure 3. *Number of Neps.*

The above yarns were knitted into plain single-jersey structures on a Camber Versanit machine (26" diameter, E18, 36 feeds), at 20r.p.m. 100% Visil, 100% wool and 100% modacrylic yarns of similar linear densities were also knitted on an E7 flat machine to produce plated plain single-jersey, and plated 1x1 rib structures by presenting two yarns in different combinations in a plating feeder. The aim was to investigate the relative performances of intimately blended yarn-containing structures and plated yarn structures.

The fabrics were washed (modacrylic/Visil at 55°C, wool/Visil at 40°C) and conditioned before being tested for flame retardancy using the vertical strip test BS 5438: 1989 (test 2 conditions: face and bottom edge ignition) and the limiting Oxygen Index (LOI) ASTM D2863-77 method, the results of which can be seen in Tables 2,3,4 and 5, and Figures 4 and 5.

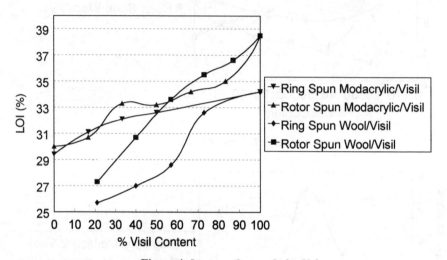

Figure 4. *Limiting Oxygen Index Values.*

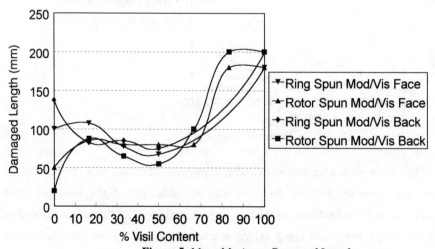

Figure 5. *Mean Maximum Damaged Length.*

A detailed investigation was also carried out in order to study the effect of fabric structure on the flammability characteristics constructed from 44% Visil and 56% modacrylic rotor spun yarn.

Preliminary studies had shown that the fabrics made from the above yarn type with the above fibre content and blend proportion produced the best results as regards the flame retardancy.

Table 1. *Summary of yarn properties.*

YARN TYPE	RING SPUN YARN			ROTOR SPUN YARN		
	Linear Density (tex)	Tenacity (cNtex^{-1})	Yarn Evenness (CV%)	Linear Density (tex)	Tenacity (cNtex^{-1})	Yarn Evenness (CV%)
100% Modacrylic	31	15.2	13.8	35	9.6	14.3
83.3/16.7% Modacrylic/Visil	32	11.1	22.1	35	7.2	15.9
66.7/33.3% Modacrylic/Visil	24	9.3	27.7	34	6.2	16.5
50.0/50.0% Modacrylic/Visil	21	8.2	30.5	35	6.0	16.9
33.3/66.7% Modacrylic/Visil	44	6.6	33.9	34	6.3	16.7
16.7/83.3% Modacrylic/Visil	44	5.4	36.2	37	5.9	16.9
100% Visil	40	7.2	20.9	40	4.8	14.5
79/21% Wool/Visil	38	8.8	25.8	38	3.2	23.5
60/40% Wool/Visil	38	4.3	26.3	38	3.8	18.3
43/57% Wool/Visil	39	6.3	27.6	39	4.5	18.9
27/73% Wool/Visil	39	6.6	26.7	38	4.9	15.6
13/87% Wool/Visil	39	7.1	28.6	44	5.9	15.9

Table 2. *Char lengths after testing to BS 5438: 1989 conditions.*

| | CHAR LENGTH (mm) | | | |
| | RING SPUN SAMPLES | | ROTOR SPUN SAMPLES | |
YARN BLEND	Face Ignition	Bottom Edge Ignition	Face Ignition	Bottom Edge Ignition
100% Modacrylic	100	137.5	50	20
83.3/16.7% Modacrylic/Visil	107.5	82.5	85	87.5
66.7/33.3% Modacrylic/Visil	77.5	85	80	65
50.0/50.0% Modacrylic/Visil	67.5	75	80	55
33.3/66.7% Modacrylic/Visil	-------	-------	80	100
16.7/83.3% Modacrylic/Visil	-------	-------	180	200
100% Visil	180	200	180	200

Table 3. *Limiting oxygen Index Values.*

| Yarn Type / Blend | Ring Spun Samples | | Rotor Spun Samples | |
	Fabric Area Density (gm^{-2})	LOI (Vol. %)	Fabric Area Density (gm^{-2})	LOI (Vol. %)
100% Modacrylic	205	29.4	194	30.0
83.3/16.7% Modacrylic/Visil	176	31.1	177	30.7
66.7.33.3% Modacrylic/Visil	139	32.1	174	33.3
50.0/50.0% Modacrylic/Visil	162	32.6	176	33.2
33.3/66.7% Modacrylic/Visil	-------	-------	174	34.2
16.7/83.3% Modacrylic/Visil	-------	-------	201	35.0
100% Visil	163	34.2	195	38.5
100% Wool				
79/21% Wool/Visil	186	25.7	177	27.3
60/40% Wool/Visil	221	27.0	178	30.7
43/57% Wool/Visil	199	28.6	190	33.6
27/73% Wool/Visil	193	32.6	178	35.5
13/87% Wool/Visil	-------	-------	217	36.6

(------- *Unknittable Yarns*)

The following fabrics were produced using an E7 Dubied flat machine and an E18 Camber Versanit single-jersey circular machine:

E7 DUBIED FLAT MACHINE

C1	- plain single-jersey	C4	- thermal
C2	- 1 x 1 rib	C5	- quilted
C3	- milano rib		

E18 CAMBER VERSANIT SINGLE-JERSEY CIRCULAR MACHINE

D1	- plain single-jersey	D3	- double lacoste
D2	- knit and miss	D4	- thermal

The following parameters and properties (Full fabric analysis: courses per cm; wales per cm; stitch density; stitch length; tightness factor; actual and theoretical area densities; Fabric thickness; Air permeability; and Bulk density).were measured and recorded in order to study whether or not different fabric structures have an effect on the flammability behaviour of fabrics containing yarns of the same blend composition and produced by the same processing methods. i.e. rotor spinning system.

Table 4. Flammability test results (Control fabrics).

Yarn Type	Yarn Linear Density	Fabric Structure	Preparation	BS 5438:1989 Char Length(mm)		LOI (Vol.%)
				Face Ignition	Bottom Ignition	
100% Visil	R.tex 150/2	Plain Single Jersey	Washed 55°C Zetex 30 min	126.4	200	37.5
		1x1 Rib		45	200	39.0
100% Wool	R.tex 150/2	Plain Single Jersey	Washed 40°C Zetex 30 min	180	35	26.4
		1x1 Rib		117.5	107.5	28.1
100% Modacrylic	R.tex 150/2	Plain Single Jersey	Washed 55°C Zetex 60 min	50	72.5	33.8
		1x1 Rib		37.5	45	34.4

3. RESULTS AND DISCUSSION

3.1. Yarn Irregularity

In terms of irregularity, the rotor spun yarns produced are more regular than ring spun in all proportions, the difference being dependent upon blend levels (Figure 1).

3.2 Yarn Elongation

The results for modacrylic/Visil blends show a marked decrease in elongation as the blend proportion of Visil is increased; this being irrespective of the yarn formation technique used. With Visil/wool blends, the opposite is the case. Elongation improves as the percentage content of Visil is increased. The level of twist in these yarns was kept to a minimum during spinning in order to reduce the yarns' tendency to spirality, as these yarns are primarily intended for the knitting process. This factor may also affect the elongation properties of a yarn.

Table 5. *Flammability test results (Plated fabrics).*

Yarn type	Yarn Linear Density	Fabric Structure	Preparation	BS 5438:1989 Char Length (mm)		LOI (Vol. %)
				Face Ignition	Bottom Ignition	
100% Visil 100% Wool	75 tex 75 tex	Plain Single Jersey	Washed 40°C Zetex 30 min	180 (tech. face)	200 (tech. face)	26.5
		Plain Single Jersey		180 (tech. back)	200 (tech. back)	
100% Visil 100% Wool	75 tex 75 tex	1x1 Rib	Washed 40°C Zetex 30 min	50.3	200	32.3
		Reverse Plated 1x1 Rib		55*	170*	33.2
100% Visil 100% Modacrylic	75 tex 75 tex	Plain Single Jersey	Washed 55°C Zetex 60 min	63 (tech. face)	35 (tech. face)	35.8
		Plain Single Jersey		63 (tech. back)	25 (tech. back)	
100% Visil 100% Modacrylic	75 tex 75 tex	1x1 Rib	Washed 55°C Zetex 60 min	50	13.5	32.0
		Reverse Plated 1x1 Rib		44*	35*	38.9

(*Visil yarn plated outside)

3.3. Yarn Tenacity

As the percentage of Visil is increased within the modacrylic/Visil ring spun yarn, the yarn tenacity reduces up to a percentage content of about 80%, when a further increase in Visil content indicates a slight improvement in tenacity (Figure 2). With the modacrylic/Visil rotor spun yarn, the trend is slightly different in that as the percentage content of Visil increases, the yarn tenacity reduces with no obvious improvement as seen in the ring spun blends.

A comparison of ring and rotor yarns produced from the same raw material shows that rotor spun yarns in general are weaker than ring spun yarns, the difference being dependent on the blend. As the percentage content of Visil is increased, the differences become less obvious. A similar situation can be seen with the wool/Visil blends, showing ring spun yarns to be stronger than rotor spun.

3.4. Yarn Imperfections

Figure 3 illustrates the level of neps in the experimental yarns produced. The numbers of neps in ring-spun yarns are higher than in the same blends produced by rotor spinning. Ring-spun modacrylic/Visil blended yarns showed much higher neps than ring-spun wool/Visil blend yarns. Rotor spinning is much superior than ring spinning as regards yarns imperfections in both fibre blends.

3.5. Flammability of Visil/Modacrylic Blends

Table 2 shows that when Visil is the sole yarn component knitted into fabric samples, it does not display good flammability properties under BS 5438 test 2 conditions in that char lengths are in excess of 180mm even for the less intense face ignition test (Figure 6). The same can be observed for the 100% modacrylic ring-spun samples where char lengths are 100mm or greater. However, the 100% modacrylic rotor-spun fabric samples show significantly reduced char lengths. It can also be noted that when blended together by either of the yarn formation techniques used in this study, again the char lengths measured are significantly reduced. In Figure 7 it can be seen that the ring-spun results suggest a synergistic effect for both face and bottom edge ignition of blended yarn fabrics since all blend char lengths are much less than the expected values calculated from the weighted pure component values . The rotor-spun yarn-containing fabric results indicate slightly lower char lengths than expected and so suggest a more additive and hence less synergistic effect.

The results in Table 3 and plotted in Figure 4 show that for the same blend proportions as above, the LOI results for fabrics containing rotor-spun yarns were higher than those for ring-spun yarns having similar blend contents; this could be a consequence of increased area densities (see Table 3) of rotor-spun yarn-containing fabrics especially at higher Visil contents. For the ring-spun yarn-containing fabrics at each blend level, observed LOI values are greater than the additive, expected values which again suggests a synergistic flame retardancy effect which has been reported previously for other blended fibre constructions[6]. However, Figure 4 also demonstrates an antagonistic effect for the same rotor-spun blends at high Visil content in spite of their generally higher LOI results compared to ring-spun analogues.

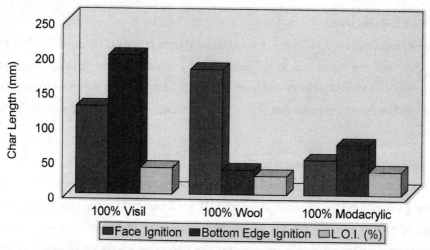

Figure 6. *Flammability of Plain Single-Jersey Fabrics.*

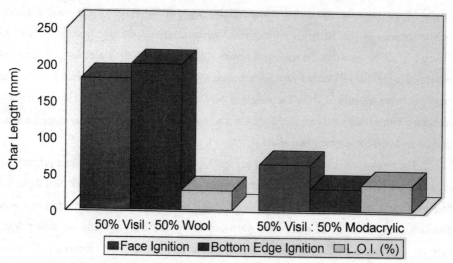

Figure 7. *Flammability of Plated Yarn Single-Jersey Fabrics.*

3.6 **Flammability of Visil/Wool Blends**

When wool fibre was blended with Visil, surprisingly the resultant fabrics all failed the BS 5438 strip tests despite good LOI test results ranging from 25.7% to over 36% (see Table 3, figures 7 and 8). Normally, such high LOI results would suggest that fabrics would pass a vertical strip test. Lofty wool structures are known to burn intensely because of high oxygen accessibility to fibre surfaces. Such a condition may exist in these blends where the wool fibres are highly randomised, thus allowing them to burn more easily. Although the LOI of the respective blends would indicate high flame retardancy, it should be noted that the LOI test has a downward burning geometry whereas the vertical strip test has an upward burning condition that would favour the enhanced access of oxygen from the convective effects of the flame.

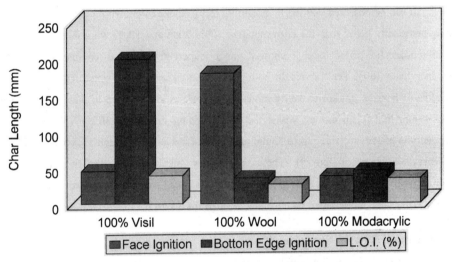

Figure 8. *Flammability of 1 x 1 Rib Fabrics.*

Another explanation could be the problem of the small flame that is used in the BS 5438 test method, not to enable the flame retardant system which is present in Visil to activate at relatively low temperatures. Wool decomposes at around 250°C to form volatiles that may ignite above this temperature: Visil, however, decomposes significantly only above 300°C and so the flame retardant properties (which do not transfer to the wool) are activated after fuel formation by the wool.

3.7 Flammability of Plated Structures

Plain single-jersey and 1 x 1 rib fabrics were produced by using two different yarns plated in the same structure in a plating relationship. Table 4 shows the flammability and LOI (%) results of 100% Visil, 100% wool and 100% modacrylic yarns in plain single-jersey and 1x1 rib structures. Table 5 shows the corresponding results of fabrics produced by plating two yarns of different fibre types in the same fabric. The resultant yarn linear densities in all fabrics, as well as the fabric area densities of plain single-jersey and 1x1 rib fabrics were exactly the same in the two sets of experiments. A number of interesting and significant trends can be observed in Tables 4 and 5. In most cases 1x1 rib fabrics exhibited lower char lengths and higher LOI (%) values than corresponding plain single-jersey fabrics, which means that 1x1 rib fabrics are much more flame retardant than plain single-jersey fabrics constructed from same raw material.

It can be observed that the flammability test results of Visil/wool plated structures are only marginally better than the corresponding 100% Visil and 100% wool fabrics, where as Visil/modacrylic plated fabrics showed much superior results as compared to 100% modacrylic or 100% Visil fabrics for both plain single-jersey and 1x1 rib structures. The char lengths of bottom ignition of 100% modacrylic fabrics were 72.5mm in plain single-jersey and 45mm in 1x1 rib fabrics, whilst the corresponding figures for 100% Visil and 100% modacrylic plated structures were 35mm and 25mm (face and back) and 13.5 and 35 (plated and reverse plated; Figure 9). These results have demonstrated that the flammability properties of knitted fabrics can be substantially improved by combining two different yarn types in the same fabric rather than using a blended fibre yarn.

3.8 The Effect of Fabric Structure

A wide range of single- and double-jersey structures were knitted on two different machine types: (1)E7 v-bed flat machine; and (2) E18 single-jersey circular machine, in order to study the effect of fabric structure on flammability characteristics of fabrics containing 44% Visil and 56% modacrylic rotor spun yarn.

It can be observed from Figures 10 and 11 that percentage LOI values have a linear relationship with both fabric area density and fabric thickness irrespective of the fabric structure. The linear regression equations and linear correlation coefficients are also indicated in Figures 10 and 11. It was also found that percentage LOI values are also related to fabric air permeability in such a manner that LOI decreases initially as air permeability increases up to a certain point after which the LOI values tend to increase (Figure 12).

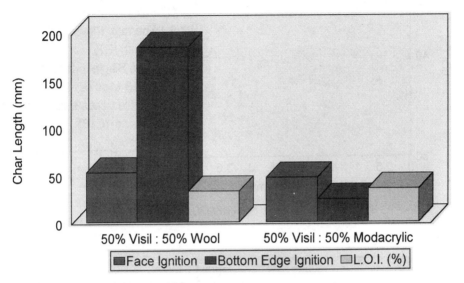

Figure 9. *Flammability of Plated 1 x 1 Rib Fabrics.*

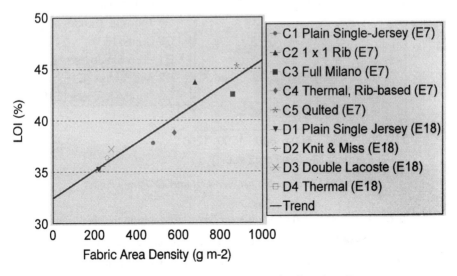

Figure 10. *Relationship between LOI and Fabric Area Density.*

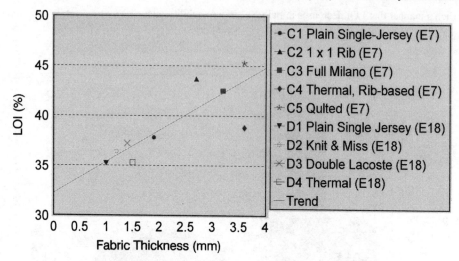

Figure 11. *Relationship between LOI and Fabric Thickness.*

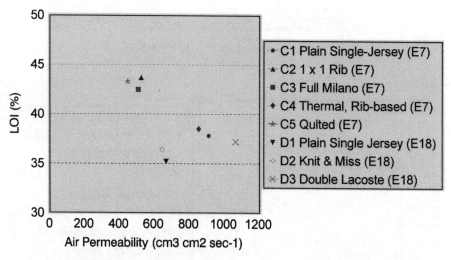

Figure 12. *Relationship between LOI and Air Permeability.*

Figure 13 illustrates the effect of different fabric structures on the char lengths obtained after testing to BS 5438:1989 test 2 conditions, face and bottom edge ignition. It can be clearly seen that the fabric structure has a significant influence upon the flammability of this particular yarn type and blend. The different structures were ranked in order of the extent of damage which occurred during this particular test method as follows: the structure exhibiting the shortest char length was ranked as number one; whilst the one with the longest char length was ranked as number nine.

RANKING ORDER	FABRIC STRUCTURE	
1	milano rib	(E7 flat)
2	quilted	(E7 flat)
3	1 x 1 rib	(E7 flat)
4	plain single-jersey	(E7 flat)
5	knit and miss	(E18 circular)
6	thermal	(E7 flat)
7	double lacoste	(E18 circular)
8	thermal	(E18 circular)
9	plain single-jersey	(E18 circular)

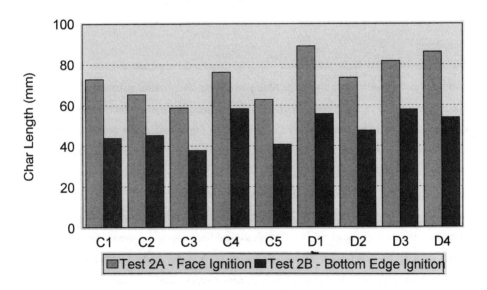

Figure 13. *Char Lengths for Different Fabric Structures.*
KEY: *C1:E7 plain single-jersey; C2: E7 1 x 1 rib; C3: E7 milano rib; C4: E7 thermal; C5: E7 quilted; D1: E18 plain single-jersey; D2: E18 knit and miss; D3: E18 double lacoste; D4: E18 thermal.*

It will be noticed from the above ranking order that four out of the five structures produced on the 7-gauge v-bed flat machine exhibited the best flame retardant properties due to the following reasons. They were produced with much thicker yarn, and were heavier and thicker than those produced on the 18-gauge single-jersey circular machine. It is also interesting to observe that double-jersey structures produced on the flat machine such as milano rib, quilted and 1 x 1 rib, showed the minimum char lengths in comparison to thermal and plain single-jersey structures produced under similar conditions. This is because they are double-layered structures and hence trap a higher volume of dead air within the structure.

A close examination of single-jersey structures produced on the circular machine reveals that the knit and miss sequence produces less flammable fabric than either the knit and tuck sequence, as used in double lacoste and thermal structures, or all knit plain single-jersey structure. This can be explained due to the fact that miss stitches make the fabric narrower in width and longer in length tending to make the fabric denser with relatively smaller holes. Whilst on the other hand, tuck stitches make the fabric wider in width and shorter in length which creates bigger holes due to the nature of tuck stitches. This work has clearly demonstrated that the fabric structure has a significant effect upon the flame retardancy of weft knitted fabrics.

4. CONCLUSIONS

The results observed from the Visil/modacrylic blends demonstrate that the level of flame retardancy is dependent not only on the fibre content, but the physical structure of the yarn also. The properties of a yarn are partly determined by the spinning method that is used. Rotor-spun yarns are reported to have an improved distribution of fibres[7] due to the doubling action of the rotor. Different twist levels cause the fibres to be randomised on the surface of the yarn, hence changing the bulk or covering properties of a yarn. The bulk density of rotor-spun yarns has been studied closely and many workers claim that rotor-spun yarns are 5-10% bulkier than ring- spun, and so produce better fabric cover. The result can be an increase in cover but at the expense of other properties, for example, hairiness. Minh studied fibre migration[8] during rotor spinning as a factor affecting the mechanism of yarn formation. This study covered different types of spun yarn and dealt with the influence of yarn twist, fibre fineness, and fibre length on yarn structure and some practical conclusions with regard to the properties of the yarns, mainly hairiness, were derived. The amount of hairiness can be regulated by modifying the yarn twist or by selecting the fibre fineness and fibre length (fibre migration parameters)[9], and might be expected to influence flammability behaviour. However, there is little evidence to suggest that hairiness of the rotor-spun yarn-containing fabrics increased their flammability since they gave higher LOI values than ring- spun analogues which, as already mentioned, could be due partly to their increased area densities.

Ring- spun yarns, on the other hand, have a lower bulk density and fibre-fibre contact is enhanced. This could give rise to greater fibre-fibre interactions in terms of flame retardancy; for example, transferred vapour phase activity of modacrylic towards Visil may be enhanced as has been noted previously in the case of cotton/modacrylic blends[10]. Thus blended ring-spun fabrics exhibit synergism relative to rotor-spun yarn behaviour at each blend ratio in both char and LOI value studies. However, while char length studies still indicate a possibly slight positive reaction in rotor-spun samples, this is less than the obvious synergy for ring-spun analogues in Table 2. Clearly, more research needs to be undertaken to more fully explain these results.

The Visil/wool blends obviously behave in a much more complex fashion because of the disparity between LOI and vertical strip test results. The non-transferability of flame retardant effect between components may be a factor and so each is behaving more as individual fibres.

The most significant aspect of this work so far has been the improved performance of plated single- and double-jersey structures as regards their flammability and limited Oxygen Index test results in comparison to those of the fabrics made from the individual component yarns.

It has also been demonstrated that percentage LOI values have a linear relationship with both fabric area density and fabric thickness irrespective of the fabric structure. It is also related to fabric air permeability, but the relationship is such that LOI values decrease initially as air permeability increases up to a certain point, after which they tend to increase somewhat.

It was also found that double-jersey structures such as milano rib, quilted and 1 x 1 rib produced on an E7 flat machine were more flame retardant than plain single-jersey or thermal (which contains knit and tuck stitches). Of the different single-jersey structures produced on E18 single-jersey circular machine the fabric containing knit and miss sequence produced the most desirable effect as regards flame retardancy. On the other hand, single-jersey structures containing knit and tuck stitches or all knit stitches produce structures exhibiting lower flame retardant properties.

References

1. Garvey, S.J., "*Visil - the hybrid viscose fibre*", *Textiles Magazine*, **3**, (1996), 21.
2. Allen, J., "*The effect of hard water on the wash durability of flame retardant cotton fabrics*", Volume 1, Ph.D. Thesis, Bolton Institute, (1991).
3. *Home Office Statistical Bulletin, Summary Fire Statistics*, United Kingdom, (1994).
4. Horrocks, A. R., Akalins, M. and Price, D., *J. Fire Sci.*, **8**, 135, (Mar/Apr 1990).
5. Heidari, S., "*Hybrid fibres by viscose technology*", "*Physico-Mathematicae et Chemico-Medicae*", University of Helsinki, (1993).
6. Horrocks, A. R., Tunc,M. and Price,D., *Textile Progress*, **18**, 156, (1989).
7. Bancroft, F. and Lawrence, C. A., "*Progress in O-E spinning*", Shirley Institute Publication S16, World Literature Survey, (1968-1974).
8. Minh, H. V., *Industrie Text.*, **53**, (Apr 1989).
9. Barella, A., *Textile Progress*, No. 3, **24**, (1993).
10. Reeves, W. A., *J. Fire Retard. Chem.*, **8**, 209, (1981).

FLAME RETARDANT COMPOSITES, A REVIEW:
THE POTENTIAL FOR USE OF INTUMESCENTS

B. K. Kandola and A. R. Horrocks
Faculty of Technology, Bolton Institute
Deane Road, Bolton BL3 5AB, UK

1. INTRODUCTION

A composite may be defined as a material having two or more chemically distinct constituents or phases, separated by distinct interface. The properties of composites are noticeably different from the properties of the constituents. Fibre-reinforced composites nowadays have successfully replaced many conventional metallic and other polymeric materials because of their favourable mechanical, physical and chemical properties and high strength to weight ratios.

The major advantages include low density, high specific strength and stiffness, good corrosion resistance and improved fatigue properties. The ability to tailor properties of composites makes them the only materials capable of transforming new design concepts into reality. Their major applications are in load - bearing structures in aircraft, automobiles, ships, pipelines, storage tanks, sports equipment, etc. as bulkhead, framework and panel components. However, with the organic nature of both the matrix (mainly epoxy and polyester types) binder and in some cases, fibres (e.g., polyester, aramid), one would be replacing non-flammable materials (e.g., aluminium, steel) with materials which may be thermoplastic, will support combustion and evolve large quantities of smoke while burning. Even if inorganic fibres like E-glass are the reinforcing structures, the composite fire resistance will be determined by that of the organic matrix and the low melting point of these fibres. Hence, a major current concern is the evaluation and improvement of the heat and flame resistance of these structures and, more specifically, the need to investigate the potential for use of intumescents.

To understand the performance of composite structures in fire and intensive heat situations, it is necessary to discuss the nature and chemistry of their constituents.

2. COMPOSITES: CONSTITUENTS

As mentioned earlier, a composite is a mixture of two or more chemically distinct phases of which the continuous phase, often present in the greater quantity, is termed the matrix, which can be ceramic, metallic or polymeric in character. The second constituent is referred to as reinforcing phase, as it enhances or reinforces the mechanical properties of the matrix and this can be either fibrous or particulate.[1, 2] The matrix resin provides uniform load distribution to the fibre and protects or safeguards the composite surface against abrasion or environmental corrosion, either of which can initiate fracture. Adhesion between two dissimilar phases is necessary to allow uniform load distribution between the dissimilar phases. Various types of constituents used for these matrices are shown in Table 1. Our main interest lies with fibre-reinforced polymeric composites, so, in this review only polymer matrix and fibre reinforcements are discussed.

Table 1. *Classification of Composite Materials Based on Constituents*

2.1. Matrix

The most common matrix materials for composites are polymeric which can be thermosets, thermoplastic or rubber. Within any class there are many different polymers (see Table 1). Thermosets are resins which readily cross-link during curing. Curing involves the application of heat and the addition of a curing agent (hardener) and sometimes the addition of a catalyst. Commonly used thermoset resins are epoxies, polyesters, phenolics and polyimides (Table 1) and their chemical structures are shown in Table 2. Thermoset resins are more commonly used in load - bearing composite structures, so the main emphasis in this paper will be on these type of polymers, although brief references to others will be given as well. Thermoplastics readily flow under stress at elevated temperatures, can be fabricated into required components and become solid and retain their shape when cooled to room temperature. However, the reversibility of this process generates composites having a thermoplastic property and hence, poor physical resistance to heat. Most common thermoplastics are acrylic, nylon (polyamide), polystyrene, polyethylene and polyetherether ketone (PEEK). Structures of some of them are shown in Table 2.

2.2. Reinforcements (Fibres)

Although the use of textile fabrics as reinforcing phase started about 100 years ago,[3] only in last 25 years, with the development of so-called high performance fibres, has this technology gained momentum. While natural fibres such as cotton, silk, wool, jute, hemp, sisal, etc and asbestos are still used for composites, synthetic organic and inorganic fibres have been a more popular choice because of their superior textile properties and uniformity of physical characteristics. Only those fibres which possess high strength, very high modulus or stiffness and preferably have moderately low densities are suitable as reinforcing agents in composites.[2] Examples are aramids, poly (para-aramids) e.g., Kevlar (DuPont), Twaron (Akzo); polyaramid copolymers, e.g. Technora (Teijin), ultra high molecular weight polyethylene (UHMWPE), S-2 and E-glass, carbon, alumina and boron fibres[1] and structures of some of them are shown in Table 3.

The interphase between the fibre and matrix significantly affects the properties of composites, e.g., shear and compression strength, fatigue, environmental durability, tensile strength, modulus and impact resistance. For good adhesion, in some cases fibres are surface - treated with reactive media, which is specific to the fibre and matrix involved.[4]

Table 2. *Structures of Most Common Resins Used for Composites.*[1,2]

Generic type	Structure
Unsaturated polyester	Unsaturated linear polyester crosslinked with styrene
Epoxy	The structure of resin depends on the curing agent.
Phenolic	
Polyimide	
Acrylic	Polyacrylate
Polyamide	Polyamide 6 Polyamide 6.6

Table 2. *(continued)*

Generic type	Structure
Poly(styrene)	
Poly(ethylene)	
Polycarbonate	
Polyetheretherketone	
Poly(butadiene) (synthetic rubber)	
Poly(isoprene) (natural rubber)	

Fibre-reinforced composites may be single or multilayered. In single layered composites long fibres with high aspect ratio (ratio of the length to the cross-sectional dimension of the fibre) form continuous fibre reinforced - composites, while short fibres with low aspect ratio form discontinuous fibre composites. The orientation of the discontinuous fibres can be random or preferred. Multilayered composites sub-divide into two types, laminate or hybrids. Laminates are sheet constructions, made by stacking layers (also called plies or laminae and usually are unidirectional) in a specified sequence. One laminate can have 4 to 40 layers with fibre orientation changes from layer to layer in a regular manner. Hybrids are usually multilayered composites with mixed fibres in a ply or layer - by - layer.[1]

Table 3. *Structures of Some Reinforcing Fibres Used for Composites.*[1,2]

Generic type	Structure
Aramid	Poly p-phenylene terephthalamide e.g., Kevlar (DuPont) and Twaron Akzo)
Polyethylene	$\left[CH_2-CH_2 \right]_n$ UHMWPE e.g., Spectra (Allied Chemicals) and Dyneema (Akzo)
Carbon	Graphite form Properties of carbon fibres depend upon the precursors used for their production i.e., (i) cellulose or rayon fibres; (ii) PAN (polyacrylonitrile) filaments; (iii) petroleum and coal-tar pitches
Alumina	Al_2O_3 (Crystalline form e.g., Nextel (DuPont))
E - Glass	$SiO_2 - CaO - Al_2O_3$ with some B_2O_3 (Calcium alumino borosilicate)
S - Glass	$SiO_2 - Al_2O_3 - MgO$ (Magnesium alumino silicate)

3. FLAMMABILITY OF COMPOSITES

Since composites are widely used in aircraft and marine vessels where fire hazards are a major issue, there are guidelines from various government and regulatory bodies regarding selection of materials, flammability tests and tests to monitor the evolution of smoke and toxic gases during burning. Flammability parameters most frequently considered for rating and ranking composite materials[5] are:

1. Flammability :	Ability to withstand high radiant energy
	High ignition temperature
	Low flame spread rate
	Low rate of heat release/fuel evolution/fuel contribution
	Ease of extinguishment
	No violent reaction in proximity to heat or ignition source
2. Smoke :	High evolution temperature
	Low optical obscuration
	Low soot conversion factor
3. Toxicity:	High evolution temperature for toxicants
	Low rate of toxic gas released
	Low toxicity of gases

There are various standard tests to measure flammability. Some common tests are limiting oxygen index (LOI - ASTM D-2863), the UL-94 test, the radiant panel test, the Ohio State University (OSU) test, the flame spread test (ASTM E-162, BS ISO 4589-2), smoke density (ASTM E-662), toxic gas analysis (Bombadier SMP 800) and, more recently, the cone calorimeter (ASTM E-1354 (ISO 5660)) . These tests provide information on surface flammability, heat release rates and flame spread. Smoke evolution is often measured using NBS smoke density chambers (ASTM E-662) and from the collected gases, toxicity can be calculated. Cone calorimetry can also be used to measure smoke density and evolution of different gases. These are well known tests, so are not discussed further here.

Brown et al.[6] have conducted and reviewed cone calorimetric tests for the flammability of various composite materials. Kourtides et al[5, 7] have reported LOI values for various thermoplastic resins, thermoset resins and their composites, as shown in Figure 1 a and b, respectively. Based on LOI, smoke and heat release tests, the ranking in decreasing order of fire resistance of resins was: phenolic, polyimide, bismaleimide and epoxy[5] in presence and absence of glass fibre reinforcement.

Gilwee et al.[8] have ranked flammability of some thermoset resins according to oxygen index and char yields from TGA results in a nitrogen atmosphere. The results, listed in Table 4, show a correlation between char yields and LOI values, indicating that LOI can be estimated from char yields of the samples as discussed comprehensively by Van Krevelen[9] for polymers in general. The results in Table 4 show the poor performance of epoxy resins.

The inherent fire-retardant behaviour of phenolic resins can be explained due to their chemical structure (see Table 2). A recent review by Brown et al.[10] discusses the fire performance and mechanical strengths of phenolic resins and their composites. Owing to the large proportion of aromatic structures in the crosslinked cured state, phenolic resins carbonise in a fire and hence, extinguish once the source of fire is removed. They may thus be said to encapsulate themselves in char and therefore do not produce much smoke[11]. Epoxy and unsaturated polyesters on the other hand carbonise less than phenolics and continue to burn in a fire situation, and structures based on these aromatic structures produce more smoke.

Table 4. *Limiting Oxygen Index and Char yields from TGA curves in Nitrogen of Thermoset Resins* [5,8]

Resin	Char yield at 800^0C	LOI (%)			
		23^0C	100^0C	200^0C	300^0C
Benzyl	63	43	36	32	31
Melamine	58	27	26	25	21
Phenolic	54	25	23	19	13
Polyimide	53	27	26	23	19
Epoxy	10	23	22	18	12

Ballard et al.[12] have reported that radiant panel pyrolysis tests on epoxy/carbon fibre composites under non-flaming and flaming conditions produce large quantities of smoke, CO and HCN among toxic products. Although phenolics have inherent flame retardant properties, their mechanical properties are inferior to other thermoset polymers - polyester, vinyl ester and epoxies[10]. Hence, they are less favourable for use in load-bearing structures. However, the mechanical properties can be enhanced with improving curing techniques[10].

4.   TECHNIQUES   FOR   INTRODUCING   OR   ENHANCING   FLAME RETARDANCY

At present, there are three main methods to render composite structures flame retardant. The first one is to use mineral and ceramic wool particularly for composites used for deck and bulkhead structures for naval applications[13], but the main disadvantage is that they add significant weight

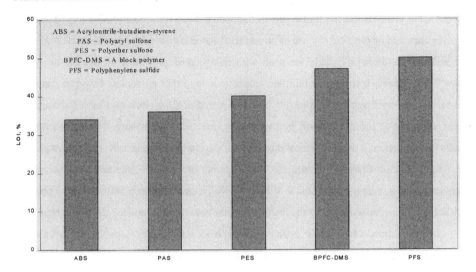

Figure 1(a). *Limiting oxygen index for thermoplastic resins[5,7] : ABS (acrylonitrile - butadiene - styrene), PAS (Polyaryl sulfone), PES.(Polyether sulfone), BPFC-DMS (9,9 Bis -(4-hydroxyphenyl) fluorene/polycarbonate - poly(dimethyl siloxane) block polymer), PFS (Polyphenylene sulfide).*

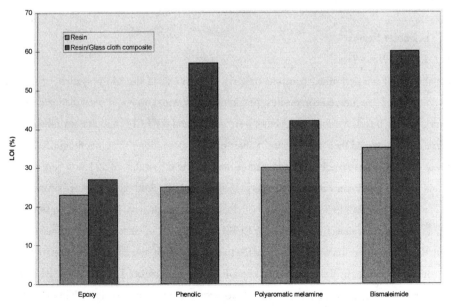

Figure 1(b). *Limiting oxygen index for thermoset resins in the presence and absence of reinforcement[5,7].*

and can act as an absorbent for spilled fuel or flammable liquid during a fire. The second method involves chemical or physical (by use of an additive) modification of the resin matrix. Additives like zinc borate, antimony oxide are used with halogenated polyester, vinyl ester or epoxy resins[10, 14]. Alumina hydrate and bromine compounds are other examples. However many of these resins and additives are ecologically undesirable and in a fire situation increase the amount of smoke and toxic fumes given off by the burning material. Furthermore, this method usually results in a reduction in the mechanical properties of the composite structure. The third way is to use flame retardant (usually intumescent based) paints or coatings. Intumescent systems are chemical systems, which by the action of heat evolve gases and form a foamed char. This char then acts as an insulative barrier to the underlying structural material against flame and heat. One very effective intumescent coating is fluorocarbon latex paint[15]. Alternatively, the reinforcing fibre phase can be rendered flame retardant by appropriate treatment or by the use of high heat and flame resistant fibres, such as aramids, although the flame retardancy levels desired should really match those of the matrix if high levels of fire performance are to be realised. Based on above methods there have been many attempts over the years to make flame retardant composites and test them accordingly. Results of some of the referenced material from the literature are discussed here.

4.1. Thermoset Resins

4.1.1. Effect of Resin Type.

Brown et al.[16] have performed cone calorimetric tests on extended-chain polyethylene (ECPE) and aramid fibre - reinforced composites containing epoxy, vinyl ester and phenolic matrix resins at various heat fluxes. Various parameters were determined for ECPE and aramid fabrics only, matrix resins only and their composites. Values of time to ignition (TTI), maximum RHR (rate of heat release), time to reach RHR and sample mass loss at a heat flux of 50 kWm^{-2} are reported in Table 5. The reinforcement properties have a significant effect on the fire-hazard properties of the composite material. ECPE/epoxy and vinyl ester composites have lower ignition times than resin alone, but the values match those for ECPE fabric alone. In each case the surface layer of resin, despite its lower flammability, is ineffective in providing any significant improvement in the ignition properties of the composites. Phenolic resin has a higher TTI and lower RHR value than others showing its inherent fire retardant property. However, this low ignitability property is not maintained in ECPE/phenolic composites demonstrating non-transferability to the fibres present. Aramid fabric is far less flammable than ECPE, having a higher time to ignition and lower RHR values. Aramid reinforcement with these three resins had little effect on time to ignitions (compared to resin alone) except for the phenolic, but reduced RHR, CO emission and mass loss values. In general, resin and reinforcement contributions to the composite rate of heat release behaviour as a function of time are generally discernible and depend on respective flame retardant mechanism operating and levels of their transferability .

Table 5. *Fire Performance of Fibre Reinforcements, Matrix Resins and Composite Materials at 50 kWm^{-2} Cone Irradiance*[16].

Material	TTI (s)	Max. RHR (kWm^{-2})	Max. RHR time (s)	Mass Loss (wt. %)
Fibres				
ECPE	31	691	195	84.0
Aramid	185	63	255	47.0
Resins				
Epoxy	87	998	180	86.4
Vinyl ester	74	628	210	91.1
Phenolic	281	123	400	26.1
Composites				
ECPE/Epoxy	24	608	260	85.9
ECPE/Vinyl ester	29	812	260	83.1
ECPE/Phenolic	20	186	580	83.2
Aramid/Epoxy	89	207	250	57.1
Aramid/Vinyl ester	73	184	275	52.6
Aramid/Phenolic	418	89	700	37.2

Kourtides has published a number of papers on thermal stability and flammability of graphite composites[7, 17, 18]. Resin matrices evaluated were epoxy, phenolic xylok, bismaleimide, phenolic novolac, polyethersulfone and polyphenylsulphone. Graphite-reinforced laminates were prepared and their flammability was evaluated using LOI, ease of ignition, flame spread and smoke emission and thermal stability by TGA in nitrogen. Some of results are reproduced in Table 6.

It was concluded that the ignition tendency of a polymer, as determined by the LOI test, is a linear function of the resin char yield. Similar relationships were observed for other vapour phase, rate-dependent flammability properties such as smoke production, flame spread and toxicity. Epoxy composites demonstrated lowest fire resistance properties and bismaleimide components showed excellent properties. Phenolic-xylok have lower mechanical properties than the epoxy composites, but they retain them at elevated temperatures. The phenolic-novolac, polyether sulphone and polyphenyl sulphane composites have high LOI and low smoke evolution values.

Table 6. *Flammability Properties of Graphite Composites*[17].

Resin	LOI (Vol. %) at 23^{0}C	Resin char yield (wt. %) from TGA	Optical Transmission (%)
Epoxy	41	21	1.8
Phenolic - Xyloc	46	46	1.5
Bismaleimide A	47	50	23
Phenolic - Novolac	50	46	92
Polyether sulfone	54	40	74
Polyphenyl sulfone	52	47	92

In an another study[17, 18] four resins - vinylpolystyryl pyridine/bismaleimide, bismaleimide, phenolic, polystyryl pyridine reinforced with graphite tapes and graphite fibres were evaluated for flammability and compared with epoxy composites. All the graphite composites exhibited lower smoke evolution, lower mass losses, lower CO evolution and higher oxygen index values compared to baseline epoxy composite panels. A blend of vinylstyryl pyridine and bismaleimide resinated composites showed the best results.

Nyden et al.[19] have discussed the flammability properties of honeycomb composites consisting of Nomex (DuPont) honeycomb fibre backing and phenol-formaldehyde resin. Analyses of gases evolved (mainly CH_4 and CH_3OH) during their thermal degradation indicated that the formulation of phenol-formaldehyde resin, i.e. how they are prepared and cured, affects the flammability of these composites. A reduction in the flammability was demonstrated in the case of resins synthesised from reaction mixtures containing an excess of phenol.

Nir et al.[20] have studied the mechanical properties of brominated flame retarded and non-brominated epoxy (tris-(hydroxyphenyl)-methane triglycidyl ester)/graphite composites. The resin was modified with synthetic rubber (carboxy-terminated butadiene acrylonitrile, CTBN) to increase the impact resistance. The incorporation of bromine did not change the mechanical properties within ± 10% of those of the non-brominated resin. The addition of bromine helped in decreasing water absorption and increasing environmental stability, thereby indicating that this is an easy method to flame retard and increase impact strength of graphite - reinforced composites.

Table 7. *Thermal Response Parameter (TRP), OI, Flame Propagation Index (FPI) and Flame Spread measured by Tewarson and Macaione[21-23] using the FRMC 50kW - Scale Apparatus.*

Composite Material	TRP $(kW\text{-}s^{1/2}/m^2)$	LOI (%) at 25^0C	FPI	Flame Spread
Glass/Polyester	275			
Glass/Vinylester	281			
Glass/Epoxy	388			
Graphite/Phenolic	400			
MLT#1 S2/Polyester (Baseline material)	382	23	13	Yes
MLT#2 S2/Polyester	406	28		
MLT#3 S2/Polyester	338	52	10	Yes
MLT#4 Kevlar/Phenolic PVB	403	28	8	No
MLT#5 S2/Phenolic	601	53	3	No
MLT#6 S2/Epoxy	420	38	9	No
MLT#7 S2/Epoxy	410	50	11	Yes
MLT#8 S2/Epoxy	400	43	10	Yes
Glass/Vinylester/ceramic coating	676			
Graphite/Phenolic/ceramic coating	807			
Graphite/Epoxy/intumescent coating	962			
Glass/Vinylester/intumescent coating	1471			
Graphite/Phenolic/intumescent coating	1563			
Graphite/Epoxy/ceramic coating	2273			

MTL#1-8 composite materials tested and named by Tewarson and Macaione[21-23]; S2 : composition of glass fibre; PVB : polyvinyl butyral

Tewarson and Macaione[21-23] have evaluated flammability of different S2-glass -polyester, -epoxy, -phenolic and Kevlar/phenolic- PVB (polyvinyl butyral) composites by oxygen index, TGA, NBS smoke chamber, GC-MS and FRMC (Factory Mutual Research Corporation) 50kW - scale apparatus (discussed in references 21-23) methods. From the FRMC test, the critical heat flux (CHF) and thermal response parameter (TRP) values were calculated. Apart from these samples, they also tested some standard glass/resin composite samples and some selected results are reproduced in Table 7. Higher CHF and TRP values for a material suggest that it is harder to ignite and resist flame spread or have slower flame spreads beyond the ignition zone. CHF values for these samples ranged from 10 to 20 kWm^{-2} and define heat flux threshold values necessary to cause ignition and sustained burning. Table 7 shows that these same samples have higher TRP values than ordinary combustible materials such as wood and most fire retarded plastics, and thus show higher resistance to ignition and flame spread. The FPI (flame propagation index) values for these samples ranged from 3 to 13, indicating that for samples with values <10, self-sustained flame spread beyond the ignition zone would be difficult, whereas for samples with FPI values >10, flame spread beyond the ignition zone would be expected, although at a slower rate. They were, however, found to generate higher amounts of products associated with incomplete combustion such as CO and smoke, compared to ordinary combustibles. It is also clear from the results in Table 7 that ceramic and intumescent coatings on composite materials improve fire resistance.

As is generally seen from all the results, phenolics have inherent flame retardant properties, but have low mechanical strength. Culberston et al[24, 25] have developed a new class of thermosetting resins based on the chain - extending and crosslinking reactions of phenolic resins with bisoxazolines. These, when reinforced with carbon fibres, produced composites with high mechanical strength, low flammability, low smoke and heat release properties.

4.1.1.   Effect of Flame Retardant Pigments.

Scudamore [26] has studied the fire performance of glass-reinforced polyester, epoxy and phenolic laminates by cone calorimetry. The effect of flame retardants and of gel coats on both non FR and FR grades was also observed. FR polyester consisted of brominated resin whereas FR epoxy and phenolic resins contained ATH (alumina trihydrate). The same polyester and epoxy gel coats were used on both non-FR and FR products. ATH was used in the FR phenolic laminate. Selected cone results at 50 kWm^{-2} are listed in Table 8.

It was concluded that the fire properties depend on type of resin and flame retardant, the type of glass reinforcement, and for thin laminates, the thickness. Flame retardants for all resins seem to be effective in delaying ignition and decreasing heat release rates. The presence of a gel-coat on FR laminates has a negative effect in delaying ignition and it increased the RHR. This is because the gel-coat itself does not contain a flame retardant additive. Phenolic laminates have lower flammability than FR polyester or epoxy resin, but addition of ATH further enhances this property.

Morchat and Hiltz[27, 28] have studied the effect of the FR additives antimony trioxide, alumina trihydrate and zinc borate on the flammability of FR resins - polyester, vinyl ester and epoxy resins by TGA, smoke production, toxic gas evolution, flame spread and OI studies. Except for epoxy resin, the others contained halogenated materials from which they derived their fire-retardancy characteristics through the vapour phase activity of chlorine and/or bromine. In most cases, with a few exceptions, the additives lowered the flame spread index (2-70%), increased LOI (3-57%) and lowered specific optical density (20-85%) depending on the fire retardant and the resin system evaluated. However, for the majority of resins, the addition of antimony trioxide resulted in an increase in smoke production. The best performance was observed upon addition of zinc borate to the epoxy resin.

Table 8. *Fire Performance of Glass-Reinforced, Plastic Laminates using a Cone Calorimeter at 50 kWm^{-2} : Effect of Flame Retardants and Gel-Coats*[26].

Material	TTI (s)	Max. RHR peak (kWm^{-2})	Av. RHR (kWm^{-2})
Polyester			
NFR	25	374	115
FR	62	159	49
NFR - gel coat	31	371	108
FR - gel coat	53	176	66
Epoxy			
NFR	26	492	144
FR	88	363	77
NFR - gel coat	30	478	146
FR - gel coat	45	409	98
Phenolic			
NFR	147	127	30
FR	206	81	29

4.1.2. Recent Developments - High Performance Composites

Hshieh and Beeson[29] have tested flame retarded epoxy (brominated epoxy resin) and phenolic composites containing fibre glass, aramid (Kevlar) and graphite fibre reinforcements using the NASA upward flame test and the controlled atmosphere, cone-calorimeter test. The upward flame propagation test showed that phenolic/graphite had the highest and epoxy/graphite composites had the lowest flame resistance as shown in Figure 2. The controlled - atmosphere, cone calorimeter test showed that phenolic composites had lower values of time of ignition, peak heat release rate, propensity to flashover and smoke production rate.

Egglestone and Turley[30] have used the cone calorimeter to evaluate flammability of glass reinforced panels with an isophthalic polyester resin, the same resin with 50% w/w aluminium trihydrate (ATH) as flame retardant, two different vinylester resins and resole-phenolic resin matrices. The resole-phenolic composite showed best performance, i.e., longer ignition time, low

RHR, low effective heat of combustion and low smoke yield. Even the addition of 50% w/w of aluminium trihydrate to polyester resin could not improve its flammability rating equal to that of the latter.

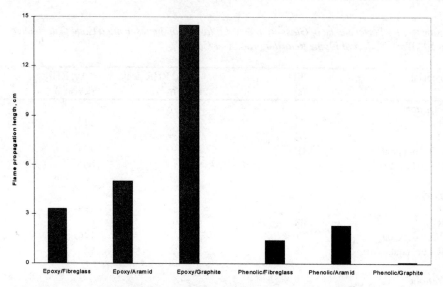

Figure 2. *Flame Propagation Lengths of Composites*[29].

Kim et al.[31] have studied the behaviour of poly(p-phenylene-benzobisoxazole) (PBO) resins. PBO polymers are rigid-rod, lytotropic liquid crystal polymers which can be fabricated into fibre, film and composites[31]. PBO exhibits ignition resistance, low heat release rate and very low smoke and toxic combustion product generation. Their results indicated that carbon fibre/PBO composites have a unique combination of thermal, mechanical and physical properties. These composites meet most of the US Navy's critical fire, smoke and toxicity requirements for applications inside submarines.

4.2 Thermoplastic Resins

Although thermoset polymers are more popular for load-bearing composite structures because of their superior thermal properties, for some specific applications thermoplastic resins are also used. To flame retard aliphatic and aromatic thermoplastic polymers - polypropylene, nylons, polyester, styrene - acrylonitrile, etc., halogenated organic compounds in combination with antimony synergistic agents are used. Theberge et al.[32] have studied glass fibre reinforced composites of these samples and also fluoropolymers such as ETFE (ethylene-tetrafluoroethylene) and PFA (perfluoroalkoxy modified tetrafluoroethylene).

Vanderschuren et al[33] and Shue[34] have shown the use of polyphenylene sulfide in high performance, advanced composites. Lem et al.[35, 36] have used UHMWPE (Spectra, Allied chemicals) fibres for construction of UHMWPE (Spectra) composites, where each filament is embedded in a thermoplastic resin matrix filled with flame retardant. Kutty and Nando[37] have studied the flammability of short Kevlar (DuPont) aramid fibre-thermoplastic polyurethane with respect to fibre loading and flame retardant additives. The latter comprised halogen-containing polymers, antimony oxide/chlorine donor combinations, zinc borate and aluminium hydroxide. It was observed that by using short poly para-aramid fibres, flammability of composite increases marginally but rate of flame spread and smoke generation reduce significantly because of high levels of char formation. The former is possibly a consequence of the non-thermoplastic Kevlar fibres behaving as a support for the melting polyurethane prior to its charring.

4.3 The Use of Thermal Barriers and the Role of Intumescents

The use of thermal barriers to protect both matrix and fibre components in a composite has been discussed above in terms of ceramic or mineral wool barriers[13] or paints or coatings. The latter often contains intumescent materials[15] and their effectiveness as demonstrated by Tewarson and Macaione[21-23] is shown in Table 7. Recently, more complex thermal barriers have involved greater integration of the barrier with the minor composite core components.

Sorathia et al.[38] have evaluated the relative flammability characteristics of conventional and advanced glass or graphite-reinforced organic matrices based on thermoset and thermoplastic composite materials. The thermoset materials included fire retardant vinyl ester, epoxies, bismaleimides, phenolics and polyimides, and thermoplastic included polyphenylene sulfide (PPS), polyether sulfone (PES), polyaryl sulfone (PAS) and polyether ether ketone (PEEK).The parameters evaluated were flame spread index (ASTM E-162), specific optical density of smoke (ASTM E-662) and combustion gas generation, residual flexural strength (ASTM D-790), heat release and ignitability as measured by cone calorimeter (ASTM E-1354). Except for those containing vinyl ester, all composite systems passed the MIL-STD-2202 requirements (i.e. for qualifying as a composite material system for use inside a US Navy submarine) of flame spread index (maximum 20) and specific optical density (maximum 100). Phenolics, polyimide and polyetheretherketone resin-based composites also met the requirements of heat release (maximum 50, 65, 100 and 150 kWm^{-2} at 25, 50 75 and 100 kWm^{-2} cone irradiances, respectively). However, all systems failed the ignitability requirements (minimum 300, 150, 90 and 60 seconds at 25, 50 75 and 100 kWm^{-2} cone irradiances, respectively). This study indicated that the phenolic-based composite materials offer benefits in fire performance over many existing polymer - based composites. In another study[39] they evaluated phenolic composites reinforced with glass, graphite, UHMWPE (Spectra) and aramid (Kevlar) fibres for mechanical and flammability characterisation. Spectra fibre-reinforced phenolic composite was excessively voided due to poor compatibility with the selected phenolic resin. The carbon and glass panels yielded fairly good and consistent results whereas, Kevlar and Spectra - reinforced samples yielded inconsistent and poor behaviours. The latter composite gave high peak heat release rate and low time to ignition results.

These same Authors[40, 41] then extended their work to explore the use of integral, hybrid thermal barriers to protect the core of the composite structure. These barriers function as insulators and reflect the radiant heat back towards the heat source. This delays the heat-up rate and reduces the overall temperature on the reverse side of the substrate. Thermal barrier treatments evaluated include ceramic fabrics, ceramic coatings, intumescent coatings, hybrid of ceramic and intumescent coatings, silicone foams and a phenolic skin. The composite systems evaluated in combination with thermal barrier treatments include glass/vinyl ester, graphite/epoxy, graphite/bismaleimide and graphite/phenolic. All systems were tested for flammability characteristics mentioned before. Selected cone results at 75kWm^{-2} are reproduced in Table 9. Without any barrier treatment, all composites failed to meet the ignitability and peak heat release requirements, whereas all treated ones passed.

5. ROLE AND POTENTIAL OF INTUMESCENTS

Results in Tables 7[21-23] and 9[40] show that an intumescent coating and a hybrid of an intumescent and a ceramic coating were the most effective fire barrier treatments for all composite systems evaluated. As mentioned previously, intumescent coatings on exposure to flame initially produces a viscous carbonaceous or inorganic mass that expands into a foam by the gas generated. This thermally stable insulating char protects the substrate from the thermal effects of the flame. In a later study Sorathia et al.[42] studied the residual strength of these composite materials (mentioned above and in Table 9) during and after fire exposure and also residual flexural strength retained (%RSP) after exposure to 25 kWm^{-2} for a duration of 20 minutes.

Again the best performance was shown by composite panels treated with intumescent coating and ablative protective material.

While the use of intumescent coatings are quite effective in providing flame and heat barriers, Kovlar[13] has used an intumescent component as an additive in the phenolic matrix. A novel composite structure was developed consisting of phenolic resin and intumescent in 1:1 ratio, reinforced with glass fabric. Upon exposure to fire the intumescent composite panel immediately begins to inflate, foam, swell and char on the side facing the fire, forming a tough, insulating, fabric-reinforced carbonaceous char that blocks the spread of fire and insulates adjacent areas from the intense heat. The prepared samples were tested by a so-called 'Direct Flame Impingement Test', in which each sample was mounted on a fixture and subjected to the direct propane flame upon the centre of its front face. A thermocouple was mounted on to the back face and the temperature rise was recorded for 5 minutes or until the sample had burned through.

The intumescent containing samples showed marked improvement in the insulating properties than control phenolic or aluminium panels. After 3 min of testing, the back surface temperatures of these three samples were 148, 363 and 263^{0}C, respectively. The low back surface temperature for the intumescent-phenolic system is a consequence of foaming and charring of the front face on exposure to fire, hence, insulating the remaining plies from the heat of the flame for the entire exposure duration. Kovlar is extending this technology to other resins such as vinylester, polyester, epoxy and BMI based composites.

Clearly this review shows that there has not been much use made of intumescents for providing flame retardant effect to composites in general. But whether as coatings or additives, they have significantly enhanced fire performance in spite of their little usage. The greater challenge most probably lies in using intumescents as an integral component of the composite rather than their addition whether as a coating or additive.

Given the long history associated with intumescents and the surprising lack of scientific understanding of their physical and chemical mechanism until recently,[43, 44] they merit considerable reinvestigation. Recent research by Camino, Levchik et al.[43, 44] and Le Bras et al.[45, 46] has shown their effectiveness in enhancing the fire performance of highly thermoplastic and flammable polymers like polypropylene and aliphatic polyamides by their ability to generate char structures.

Table 9. *Heat Release Rate and Ignitability of Composites with Thermal Barrier Treatment at 75kWm^{-2} Cone Irradiance*[40].

Composite/Fire Barrier	Weight loss (wt. %)	T$_{ig}$ (s)	PHR (kWm$^{-2)}$)	AHR (kWm$^{-2)}$)
Glass/Vinyl ester				
Control	41	22	498	220
Ceramic fabric	-	55	213	146
Ceramic coating	33	88	344	193
Intumescent coating (water based)	41	450	204	150
Ceramic/Intumescent hybrid coating	39	445	139	105
Silicone foam	38	34	435	268
Intumescent coating (solvent based)	42	80	288	200
Ceramic coating (RT)	5	170	289	234
Phenolic skin	34	88	146	140
Graphite/Epoxy				
Control	25	53	197	90
Ceramic fabric	-	45	156	71
Ceramic coating	15	105	179	80
Intumescent coating (water based)	29	628	76	60
Ceramic/Intumescent hybrid coating	23	264	97	69
Silicone foam	22	32	270	134
Intumescent coating (solvent based)	23	90	171	98
Ceramic coating (RT)	2	140	119	74
Graphite/Bismaleimide				
Control	30	66	172	130
Ceramic coating	14	147	141	98
Intumescent coating (water based)	32	472	84	70
Ceramic/Intumescent hybrid coating	32	510	52	50
Graphite/Phenolic				
Control	28	79	159	80
Ceramic fabric	14	79	155	50
Ceramic coating	16	185	184	49
Intumescent coating (water based)	10	NI	34	10
Ceramic/Intumescent hybrid coating	8	NI	51	3

Char structures in polypropylene are possibly associated with the intumescent char only, whereas in the more reactive aliphatic polyamides, work by Levchik et al.[44] and Bourbigot et al.[46] suggests that intumescence presence may promote polymer char formation. Such an interaction has been definitely demonstrated by Horrocks et al[47, 48] in intumescent - flame retarded cellulosic fibre - containing, flexible composites. This work has demonstrated that interactive char-formation can occur between selected intumescents and flame retarded viscose[49, 50] and cotton[51, 52] fibre composites as evidenced by SEM, thermal analysis and mass loss calorimetry.

The potential for more effective use of intumescents in reinforced composite systems may be realised only if the following improvements in our understanding occurs :

(i)   research is undertaken into factors which enable char-forming mechanisms in both intumescent and resin components to be compatible and synergistic;

(ii)   means of incorporating intumescents into resin matrices without reducing composite mechanical properties are discovered ;

(iii)   char formation, char structure and thermal behaviour relationships are more fully understood and, perhaps, mathematically modelled.

References

1. F. L. Matthews and R. D. Rawlings (Eds.), '*Composite Materials : Engineering and Science*', Chapman and Hall, London, 1994.
2. L. A. Pilato and M. J. Michno (Eds.), '*Advanced Composite Materials*', Springer Verlag, London, 1994.
3. W. S. Hearle, "*Proc. 7th European Conference on Composite Materials*", London, 1996, **2**, 377.
4. M. R. Piggot, *Proc. Int. SAMPE Symps.* 1991, **36 (2)**, 1773.
5. J. E. Brown, J. J. Loftus and R. A. Dipert, '*Fire Characteristics of Composite Materials - A Review of the Literature*', Report 1986, NBSIR 85-3226.
6. J. E. Brown, E. Braun, W. H. Twilley, '*Cone Calorimetric Evaluation of the Flammability of Composite Materials*', Report 1988, NBSIR - 88 - 3733.
7. D. A. Kourtides et al., *Polym. Eng. Sci.*, 1979, **19(1)**, 24 ; 1979, **19(3)**, 226.
8. W. J. Gilwee, J. A. Parker and D. A. Kourtides, *J. Fire Flamm..*, 1980, **11(1)**, 22.
9. D. W. van Krevelen, *Polymer*, 1975, **16**, 615.
10. J. R. Brown and N. A. St John, *TRIP*, 1996, **4 (12)**, 416.
11. -J. Gabrisch and G. Lindenberger, *SAMPE J.*, 1993, **29 (6)**, 23.
12. R. Ballard et al., "*Radiant Panel Tests on Epoxy/Carbon Fibre Composites*", NASA Technical Memorandum 81185, 1980. (Taken from ref. 5)
13. P. F. Kovlar and D. E. Bullock, in "*Recent Advances in Flame Retardancy of Polymeric Materials*", Volume IV, ed. M. Lewin, BCC, Stamford, 1993, 87.
14. J. L. Stevart, O. H. Griffin, Z. Gurdal and G. A. Warner, *Naval Engn. J.*, 1990, **102 (5)**, 45.
15. D. R. Ventriglio, *Naval Engn. J.*, October 1982, 65.

16. J. R. Brown, P. D. Fawell and Z. Mathys, *Fire Mater.*, 1994, **18**, 167.

17. D. A. Kourtides, *Polym. Compos.* , **5 (2)**, 1984, 143 ; **9(3)**, 1988, 172; *J. Thermoplast. Compos. Mater.*, 1988, **1(1)**, 12.

18. D. A. Kourtides, *J. Fire Sci.*, 1986, **4 (6)**, 397.

19. M. R. Nyden, J. E. Brown and S. M. Lomakin, *'Flammability Properties of Honeycomb Composites and Phenol-Formaldehyde Resins,* Chapter 16 in *'Fire and Polymers'*, ACS Symp. Res. 1995, Volume 599, 245 ; *Polym. Mater. Sci.*, 1994, **71**, 301.

20. Z. Nir, W. J. Gilwee, D. A. Kourtides and J. A. Parker, *SAMPE Q.*, 1983, **14 (3)**, 34.

21. D. P. Macaoine and A. Tewarson, *'Flammability Characteristics of Fibre-Reinforced Composite Materials'*, Chapter 32 in *'Fire and Polymers Hazards Identification and Prevention'* G. L. Nelson ed., *ACS Symp. Ser.* 425 ACS p 542.

22. D. P. Macaoine, *'Flammability Characteristics of Fibre-Reinforced Epoxy Composites for Combat Vehicle Applications'*, Report, 1992, MTL TR 92-58

23. A. Tewarson and D. P. Macaoine, *J. Fire Sci.*, 1993, **11**, 421 ; in *Recent Advances in Flame Retardancy of Polymeric Materials,* Volume III, ed. M. Lewin ; Proc. of the 1993 Conf, BCC, Stamford, 1992, 307.

24. B. M. Culberston, M. L. Diviney, O. Tiba and D. D. Carlos, Proc. 23rd *Int. SAMPE Symp. Exhib.*, 1988, **33**, 1531.

25. M. L. Diviney, O. Tiba, B. M. Culberston, R. J. Schafer, D. D. Carlos, D. C. Boyer, J. H. Newman and K. R. Friley, *Polym. Prepr.* (Am. Chem. Soc., Div. Polym. Chem.) 1991, **32 (2)**, 282.

26. M. J. Scudamore, *Fire Mater.*, 1994, **18**, 313.

27. R. M. Morchat and J. A. Hiltz, *'Fire Composites for Marine Applications'*, Proc 24th Int. SAMPE Tech. Conf., 1992, **24**, T153.

28. R. M. Morchat, *'The Effects of Alumina Trihydrate on the Flammability Characteristics of Polyester, Vinylester and Epoxy Glass Reinforced Plastics'* Techn. Rep. Cit. Govt. Rep. Announce Index (U.S.) 1992, **92 (13)**, AB NO 235, 299.

29. F. Y. Hshieh and H. D. Beeson, *'Proc. Int. Conf. Fire Saf.*, 1996, **21**, 189.

30.   G. L. Egglestone and D. M. Turley, *Fire Mater.*, 1994, **18**, 255.

31. P. K. Kim, P. Pierini and R. Wessling, *J. Fire Sci.*, 1993, **11**, 296.

32. J. E. Theberge, J. M. Crosby and K. L. Talley, *Plast. Eng.*, 1988, **44 (8)**, 47.

33. J. Vanderschuren, B. R. Bonazza, R. L. Hangenson and D. A. Soules, *Mater. Sci. Monogr.*, 1991, **72**, 439.

34. R. S. Shue, *Proc. Int. SAMPE Symp. Exhib.*, 1988, **33**, 626.

35. K. W. Lem, Y. D. Kwon and D.C. Prevorsek, *Int. J. Polym. Mater.*, 1993, **23 (1-2)**, 87 ; in *Recent Advances in Flame Retardancy of Polymeric Materials,* Volume IV, ed. M. Lewin ;, BCC, Stamford, 1993, 99.

36. K. W. Lem, Y. D. Kwon, H.B. Chin, H. L. Li and D. C. Prevorsek, *Polym. Eng. Sci.,* 1994, **34 (9),** 765.

37. S. K. N. Kutty and G. B. Nando, *J. Fire Sci.,* 1993, **11,** 66.

38. U. Sorathia, T. Dapp and J. Kerr, *Proc. 36th Int. SAMPE Symp.,* 1991, 1868.

39. U. Sorathia, H. Telegadas and M. Bergen, *Proc. Int. Conf. Fire Saf.,* 1995, **20,** 82 ; *Proc. Int. SAMPE Symps. Exhib.,* 1994, **39,** 2991.

40. U. Sorathia, C. M. Rollhauser and W. A. Hughes, *Fire Mater.,* **16,** 1992, 119 ; *Proc. Int. Conf. Fire Saf.,* 1991, **16,** 300.

41. U. Sorathia, *SAMPE J.,* 1996, **32 (3),** 8.

42. U. Sorathia, C. Beck and T. Dapp, *J. Fire Sci.,* 1993, **11,** 255.

43. G. Camino, L. Costa and G. Martinasso, *Polym. Deg. Stab.,* 1989, **23,** 359.

44. S. V. Levchik, L. Costa and G. Camino, *Polym. Deg. Stab.,* 1992, **36,** 31 ; 1992, **36,** 229 ; 1994, **43,** 43.

45. M. Le Bras, S. Bourbigot, C. Delporte, C. Siat and Y. Le Tallec, *Fire Mater.,* 1996, **20,** 191.

46. S. Bourbigot, M. Le Bras and C. Siat, in *"Recent Advances in Flame Retardancy of Polymeric Materials"*, BCC, Stamford, 1997.

47. A. R. Horrocks, S. C. Anand and D. Sanderson, *"Flame Retardants '94"*, London, Interscience Communication Ltd., UK, 1994, 117 ; 1996, 145.

48. A. R. Horrocks, S. C. Anand and D. Sanderson D, *Polymer,* 1996, **37(15),** 3197.

49. B. K. Kandola and A. R. Horrocks, *Polym. Deg.. Stab.,* 1996, **54,** 289.

50. B. K. Kandola and A. R. Horrocks, *"Proc of the 17th International Congress of the IFATCC"*, Vienna, June 5-7, 1996 p 265.

51. B. K. Kandola and A. R. Horrocks, *Thermochim. Acta,* 1997, **294,** 113.

52. A. R. Horrocks and B. K. Kandola, in *"Recent Advances in Flame Retardancy of Polymeric Materials"*, BCC, Stamford, 1997.

Intumescence: An Environmentally Friendly Process?

ECOLOGICAL ASPECTS OF POLYMER FLAME RETARDANCY

G. E. Zaikov and S.M. Lomakin

Institute of the Biochemical Physics of the Russian Academy of Sciences
Kosygin Street 4, 117334 Moskow, Russia

1. INTRODUCTION

Plastics have played an important role in sharping our history. After the invention of the celluloid billiard ball in 1860's, plastics demonstrated their endless possibilities. Since then, plastics have become an integral part of our everyday life. The interest in flame retarding plastics goes back to the mid nineteenth century with the discovery of highly flammable cellulose nitrate and celluloid. In more recent times the conventional large volume of plastics such as phenolics, rigid PVC and melamine resins possess adequate flame retardancy. By the 1970's the major flame retardant polymers were the thermosets, namely, unsaturated polyesters and epoxy resins which utilised reactive halogen compounds and alumina hydrate as an additive. There was also a large market for phosphate esters in plasticized PVC, cellulose acetate film, unsaturated polyesters and modified polyphenylene oxide. Alumina trihydrate (ATH) was the largest volume flame retardant into unsaturated plastics.

Consumption of halogen-containing flame retardant additives in 70th's was much less than the other additives. The halogenated flame retardant additives were Dechlorane Plus, a chlorinated acyclic (for polyolefins), tris-(dibromopropyl) phosphate, brominated aromatics, pentabromochloro-cyclohexane and hexabromocyclododecane (for polystyrene). The next five ears was to see a number of new brominated additives on the market. There were produced a number of chlorinated flame retardant products under the Dechlorane trade name. The products included of two moles of hexachlorocyclopentadiene and contained 78 % chlorine, Dechlorane Plus, a Diels-Alder reaction product of cyclooctadiene and hexachlorocyclopentadiene with 65 % chlorine, a Diels-Alder product with furan and a product containing both bromine and chlorine with 77 % halogen developed for the polystyrene and ABS materials.

In 1985-86, a German study detected brominated dioxins and furans from pyrolysis of a brominated diphenyl oxide in the laboratory at 510-630°C[1]. The relevance of these pyrolysis studies to the real hazard presented by these flame retardants under actual use conditions has been questioned. Germany and Holland have considered a ban or curtailed use of brominated diphenyl oxide flame retardants because of the potential formation of highly toxic and potentially carcinogenic brorninated furans and dioxins during combustion. The issue spread to other parts of Europe where regulations were proposed to restrict their use. Nevertheless, demand for brominated flame retardants including decabromo- and pentabromodiphenyl oxides continues to be strong and growing.

1.1. Halogenated diphenyl ethers, dioxins

Chlorinated dibenzo-*p*-dioxins and related compounds (commonly known simply as dioxins) are contaminants present in a variety of environmental media. This class of compounds has caused great concern in the general public as well as intense interest in the scientific community. Laboratory studies suggest the probability that exposure to dioxin-like compounds may be associated with other serious health effects including cancer. Recent laboratory studies have provided new insights into the mechanisms involved in the impact of dioxins on various cells and tissues and, ultimately, on toxicity[2]. Dioxins have been demonstrated to be potent modulators of cellular growth and differentiation, particularly in epithelial tissues. These data, together with the collective body of information from animal and human studies, when coupled with assumptions and inferences regarding extrapolation from experimental animals to humans and from high doses to low doses, allow a characterisation of dioxin hazards.

Polychlorinated dibenzodioxins (PCDDs), polychlorinated dibenzofurans (PCDFs), and polychlorinated biphenyls (PCBs) are chemically classified as halogenated aromatic hydrocarbons. The chlorinated and brominated dibenzodioxins and dibenzofurans are tricyclic aromatic compounds with similar physical and chemical properties, and both classes are similar structurally. Certain of the PCBs (the so-called coplanar or mono-ortho- coplanar congeners) are also structurally and conformationally similar. The most widely studied of these compounds is 2,3,7,8-tetrachlorodibenzo-*p*-dioxin (TCDD). This compound, often called simply dioxin, represents the reference compound for this class of compounds. The structure of TCDD and several related compounds is shown in Figure 1.

2,3,7,8-Tetrachlorodibenzo-p-dioxin

1,2,3,7,8-Pentachlorodibenzo-p-dioxin

2,3,7,8-Tetrachlorodibenzofuran

2,3,4,7,8-Pentachlorodibenzofuran

3,3',4,4',5,5'-Hexachlorobiphenyl

3,3',4,4',5,'-Pentachlorobiphenyl

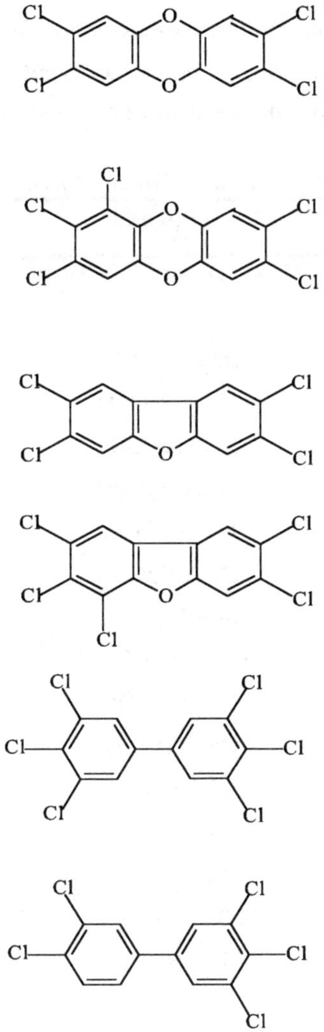

Figure 1. The structures of dioxin and similar compounds.

These compounds are assigned individual toxicity equivalence factor (TEF) values as defined by international convention (U.S. EPA, 1989). Results of in vitro and in vivo laboratory studies contribute to the assignment of a relative toxicity value. TEFs are estimates of the toxicity of dioxin-like compounds relative to the toxicity of TCDD, which is assigned a TEF of 1.0. All chlorinated dibenzodioxins (CDDs) and chlorinated dibenzofurans (CDFs) with chlorine substituted in the 2,3,7, and 8 positions are assigned TEF values. Additionally, the analogous brominated dioxins and furans (BDDs and BDFs) and certain polychlorinated biphenyls have recently been identified[2] as having dioxin-like toxicity and thus are also included in the definition of dioxin-like compounds.

Table 1. *Toxicity Equivalency Factors (TEF) for CDDs and CDFs*

Compound	Toxicity Equivalency Factors
Mono-, Di-, and Tri-CDDs	0
2,3,7,8-TCDD	1
Other TCDDs	0
2,3,7,8-PeCDD	0.5
Other PeCDDs	0
2,3,7,8-HxCDD	0.1
Other HxCDDs	0
2,3,7,8-HpCDD	0.01
Other HpCDDs	0
Mono-, Di-, and Tri-CDFs	0
2,3,7,8-TCDF	0.1
Other TCDFs	0
1,2,3,7,8-PeCDF	0.05
2,3,4,7,8-PeCDF	0.5
Other PeCDFs	0
2,3,7,8-HxCDF	0.1
Other HxCDFs	0
2,3,7,8-HpCDF	0.01
Other HpCDFs	0
OCDF	0.001

Generally accepted TEF values for chlorinated dibenzodioxins and dibenzofurans are shown in Table 1[3].In general, these compounds have very low water solubility, high octanol-water partition coefficients, and low vapor pressure and tend to bioaccumulate. Although these compounds are released from a variety of sources, the congener profiles of CDDs and CDFs found in sediments have been linked to combustion sources[6].

The Hazards Substance Ordinance in Germany specifies the maximum level of chlorinated dibenzodioxins and furans that can be present in materials marketed in Germany. This has been extended to the brominated compounds. The two largest volume flame retardants decabromodiphenyl oxide and tetrabromo bis-phenol A are said to meet these requirements[4].

The International Program for Chemical Safety (IPCS) of the World Health Organisation has made several recommendations. Polybrominated diphenyls production (France) and use should be limited because of the concern over high persistency, bioaccumulation and potential adverse effects at low levels. There is limited toxicity data on deca- and octabromodiphenyls. Commercial use should cease unless safety is demonstrated. For the polybrominated diphenyl oxides, a Task Group felt that polybrominated dibenzofurans, and to a lesser extent the dioxins, may be formed[5]. For decabromodiphenyl oxide, appropriate industrial hygiene measures need to be taken and environmental exposure minimised by effluent and emission control. Controlled incineration procedures should be instituted. For octabromodiphenyl oxide, the hexa and lower isomers should be minimised. There is considerable concern over persistence in the environment and the accumulation in organisms, especially, for pentabromodiphenyl oxide.

There are no regulations proposed or in effect anywhere around the world banning the use of brominated flame retardants. The proposed EU Directive on the brominated diphenyl oxides is withdrawn. Deca- and tetrabromo bis-phenol A, as well as other brominated flame retardants, meet the requirements of the German Ordinance regulating dioxin and furan content of products sold in Germany. The European search for a replacement for decabromodiphenyl oxide in HIPS has led to consideration of other bromoaromatics such as Saytex 8010 from Albermarle and a heat-stable chlorinated paraffin from Atochem. The former product is more costly, and the latter, if sufficiently heat stable, lowers the heat distortion under load (HDUL) significantly. Neither approach has been fully accepted.

In September 1994, the US Environmental Protection Agency (EPA) released a final draft of exposure and risk assessment of dioxins and like compounds[7]. This reassessment finds the risks greater than previously thought. Based on this reassessment, a picture emerges that tetrachlorodiphenyl dioxins and related compounds are potent toxicants in animals with the potential to produce a spectrum of effects. Some of these effects may be occurring in humans at very low levels and some may be resulting in adverse impacts on human health. The EPA also concluded that dioxin should remain classified as a probable human carcinogen.

Polymer producers have been seeking non-halogen flame retardants and the search has been successful in several polymer systems. Non-halogen flame retardant polycarbonate/ABS blends are now commercial. They contain triphenyl phosphate or resorcinol diphosphate (RDP) as the flame retardant. Modified polyphenylene oxide (GE's Noryl) has used phosphate esters as the flame retardant for the past 15-20 years and the industry recently switched from the alkylated triphenyl phosphate to RDP. Red phosphorus is used with glass-reinforced nylon 6/6 in Europe and melamine cyanurate is used in unfilled nylon. Magnesium hydroxide is being used commercially in polyethylene wire and cable. The non-halogen solutions present other problems such as poor properties (plasticizers lower heat distortion temperature), difficult processing (high loadings of ATH and magnesium hydroxide), corrosion (red phosphorus) and handling problems (red phosphorus). In this paper we have tried to look at some new trends in the search of the new ecologically-friendly flame retardants.

2. MODES OF ACTION OF FLAME RETARDANTS

The main flame retardant systems for polymers currently in use are based on halogenated, phosphorous, nitrogen, and inorganic compounds (Scheme 1). All of these flame retardant systems basically inhibit or even suppress the combustion process by chemical or physical action in the gas or condensed phase.

The main processes are:

1. physical process acting by cooling (endothermic process) or by gas dilution (aluminium and magnesium hydroxides, melamine),

2. chemical process taking place in the Gas Phase are based on radical chain mechanisms supported by high-energy H and OH radicals. These radicals can be removed by chemical species like halogen halides, metal halides and P-containing fragments (flame poisoning) (Halogenated flame retardants and antimony trioxide),

3. carbonisation of polymer material in the condensed phase which protect the native polymer against heat and oxygen attack (phosphorous compounds and another intumescent systems).

Scheme 1.

POLYMER FLAME RETARDANTS

HALOGENATED FR

PHOSPHOROUS FR, PHOSPHATE ESTERS

ANTIMONY OXIDE

Mg HYDROXIDE, ALUMINA TRIHYDRATE, BORON FR,
AMMONIUM POLYPHOSPHATE, MELAMINES, MOLYBDENUM FR

Ecologically friendly FR

POLYMER-ORGANIC CHAR FORMERS:
PVA, STARCH, GLUCOSE DERIVATIVES, POLYFUNCTIONAL ALCOHOLS

LOW-MELTING GLASSES AND GLASS-CERAMICS:
$K_2CO_3 - SiO_2$, $K_2SO_4 - Na_2SO_4 - ZnSO_4$

POLYMER MORPHOLOGY MODIFICATION,
POLYPROPYLENE-POLYETHYLENE COMPOSITIONS

The empirical approach to fire retardancy can no longer meet alone the present requirements either in terms of effectiveness or hazard. The above mentioned, commercially the most effective flame retardants so far are halogen, phosphorous, antimony and heavy metal based compounds which however have the most environmental impact either while they perform their action or when burned in incineration of waste containing fire retardant polymeric materials.

The search of new ecologically-friendly flame retardants requires cooperation between industrial and academic research. Much remains to be solved in complete replacement of ecologically harmful flame retardants. This is a problem for many years. It is not our goal today to try to complete that approach, but its our aim to point out only some new trends in this field.

2.1. Polymer - organic char former systems.

They are the first type of ecologically-friendly flame retardant systems. Our recent study[8, 9] has been directed at finding ways to increase the tendency of plastics to char when they are burned. There is a strong correlation between char yield and fire resistance[9]. This follows because char is formed at the expense of combustible gases and because the presence of a char inhibits further flame spread by acting as a thermal barrier around the unburned material.

It has been studied polymeric additives (polyvinyl alcohol systems) which promote the formation of char in polyvinyl alcohol-nylon 6,6 system[8]. These polymeric additives usually produce a highly conjugated system - aromatic structures which char during thermal degradation and/or transform into cross-linking agents at high temperatures.

$$(-CH-CH_2\ )_n-\ CH-CH_2\ - \quad \longrightarrow \quad (-CH=CH_2\ )_n-CH-CH_2- \quad + \quad H_2O$$
$$|||$$
$$OHOHOH$$

$$-CH-CH_2-(-CH=CH_2\ )_n-CH-CH_2- \quad \longrightarrow \quad -CH-CH_2-(-CH=CH_2\ )_n-CH \quad + \quad H_3C-CH-$$
$$|||\||$$
$$OHOHOHOOH$$

Scission of several carbon-carbon bonds leads to the formation of carbonyl ends. For example, aldehyde ends arise from the reaction.

The identification of a low concentration of benzene among the volatile products of PVA has been taken to indicate the onset of a crosslinking reaction proceeding by a Diels-Alder addition mechanism[8]. Clearly benzenoid structures are ultimately formed in the solid residue, and the IR spectrum of the residue also indicated the development of aromatic structures.

Acid-catalysed dehydration promotes the formation of conjugated sequences of double bonds (a) and Diels-Alder addition of conjugated and isolated double bonds in different chains may result in intermolecular crosslinking producing structures which form graphite or carbonization (b)

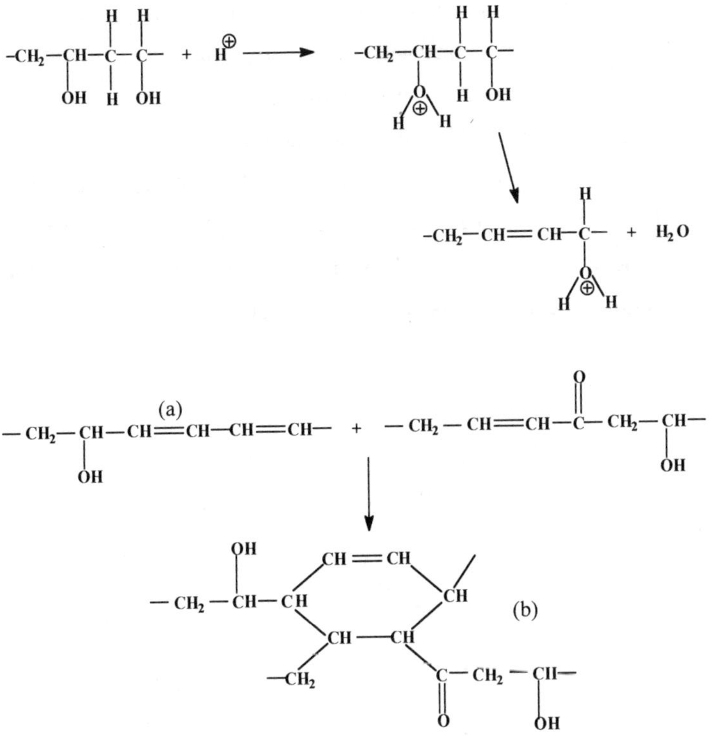

$$-CH_2-CH-\underset{\underset{OH}{|}}{\overset{\overset{H}{|}}{C}}-\underset{\underset{OH}{|}}{\overset{\overset{H}{|}}{C}}- \; + \; H^{\oplus} \longrightarrow \; -CH_2-CH-\underset{\underset{H}{|}}{\overset{\overset{H}{|}}{C}}-\underset{\underset{OH}{|}}{\overset{\overset{H}{|}}{C}}-$$

$$\longrightarrow \; -CH_2-CH=CH-\overset{\overset{H}{|}}{C}- \; + \; H_2O$$

(a)

$$-CH_2-\underset{\underset{OH}{|}}{CH}-CH=CH-CH=CH- \; + \; -CH_2-CH=CH-\overset{\overset{O}{\|}}{C}-CH_2-\underset{\underset{OH}{|}}{CH}-$$

(b)

In contrast to PVA, nylon 6,6 which was subjected to temperatures above $300\,^{\circ}C$ in an inert atmosphere it completely decomposed. The wide range of degradation products, which included several simple hydrocarbons, cyclopentanone, water, CO, CO_2 and NH_3 suggested that the degradation mechanism must have been highly complex. Further research has led to a generally accepted degradation mechanism for aliphatic polyamides[8].

The idea of introducing PVA into nylon 6,6 composition is based on the possibility of high-temperature acid- catalysed dehydration. This reaction can be provided by the acid products of nylon 6,6 degradation hydrolysis which would promote the formation of intermolecular crosslinking and char. Such a system we have called "synergetic carbonization" because the char yield and flame suppression parameters of the polymer blend of poly(vinyl alcohol) and nylon 6,6 are significantly better than pure poly(vinyl alcohol) and nylon 6,6 polymers.

The next step in our plan to improve the flame resistant properties of poly(vinyl alcohol) - nylon 6,6 system was the substitution of pure poly(vinyl alcohol) by poly(vinyl alcohol) oxidised by potassium permanganate (PVA-ox). This approach was based on the fire behavior of the (PVA-ox) itself. It was shown experimentally (Cone Calorimeter) the dramatic decrease of the rate of the heat release and significant increase in ignition time for the oxidised PVA in comparison with the original PVA (Table 1). It was reported that the oxidation of PVA by in alkaline solutions occurs through formation of two intermediate complexes (1) and/or (2) (7). The reactions (a) and (b) lead to the formation of poly(vinyl ketone) (3) as a final product of oxidation of the substrate.

Preliminary Cone tests for PVA and PVA oxidised by $KMnO_4$ (Table 2) clearly indicated the substantial improvement of fire resistance characteristics for PVA oxidised by $KMnO_4$ in comparison with PVA. PVA oxidised by $KMnO_4$ gives about half the peak of heat release rate (Peak R.H.R. kW/m^2), when compared with pure PVA. Even at 50 kw/m^2, the yield of char residue for PVA oxidised by $KMnO_4$ was 9.1%. One reason for this phenomenon may be explained by the ability for PVA oxidised by $KMnO_4$ - (polyvinyl ketone structures) to act as a neutral (structure 1) and/or monobasic (structure 2) bidentate ligand[10].

$$\text{+CH}_2\text{-C-CH}_2\text{-C+}_n \quad \rightleftharpoons \quad \text{+CH}_2\text{-C-CH=C+}_n$$

1 2

The experimental results of others (IR and electronic spectra)[11] provided strong evidence of coordination of the ligand (some metal ions Cu^{2+}, Ni^{2+}, Co^{2+}, Cd^{2+}, Hg^{2+}) through the monobasic bidentate mode (structure 2). Based on the above the following structure can be proposed for the polymeric complexes:

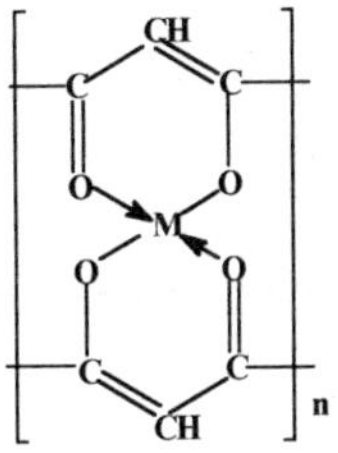

M - metal

The result of elemental analysis of PVA oxidised by $KMnO_4$ indicated the presence of 1.5% of Mn remaining in this polymeric structure. Thus, we suggested that this catalytic amount of chelated Mn-structure incorporated in the polymer may provide the rapid high-temperature process of carbonization and formation of char.

The fire tests at 50 kW/m^2 for Nylon 6,6 and PVA (80:20 %) compositions (typical rate of heat release shown in Table 2) confirmed the assumption of the synergistic effect of carbonization[8]. Each of the individual polymers is less fire resistant than their composition. The sample with PVA oxidised by $KMnO_4$ displayed even a better flame retardant properties due to the catalytic effect of Mn-chelate fragments on the formation of char.

Thus, the polymeric char former such as PVA , starch, glucose derivatives an polyfunctional alcohols may present a new trend in the global search for the type of ecologically-safe flame retardant systems.

Table 2. *Cone Calorimeter Data of nylon 6,6 – PVA.*

Material, Heat flux, kW/m^2	Initial wt., g	Char yield, wt. %	Ignition time, sec.	Peak R.H.R., kW/m^2	Total Heat Release, MJ/m^2
PVA, 20	7.6	8.8	39	255.5	159.6
PVA, 35	28.3	3.9	52	540.3	111.3
PVA, 50	29.2	2.4	41	777.9	115.7
PVA-ox KMnO$_4$, 20	27.9	30.8	1127	127.6	36.9
PVA-ox KMnO$_4$, 35	30.5	12.7	774	194.0	103.4
PVA-ox KMnO$_4$, 50	29.6	9.1	18	305.3	119.8
nylon 6,6, 50	29.1	1.4	97	1124.6	216.5
nylon 6,6 + PVA(8:2),50	26.4	8.7	94	476.7	138.4
nylon 6,6 + PVA-ox(8:2) KMnO$_4$, 50	29.1	8.9	89	399.5	197.5

2.2. Low - melting glasses

The second example has been developed on the basis of an original concept of a low-melting phases (glasses) as a polymer additives[12-13].In principle, low-melting glasses can improve the thermal stability and flammability characteristics of polymers by:

1. providing a thermal barrier for both undecomposed polymer and the char, if any, which forms as a combustion product,

2. providing an oxidation barrier to retard oxidation of the thermally degrading polymer and its combustion char residue,

3. providing a "glue" to hold the combustion char together and gives it structural integrity,

4. providing a coating to cover over or fill in voids in the char, thus providing a more continuous external surface with a lower surface area,

5. creating the potentially useful components of intumescent polymer additive systems.

W. J. Kroenke developed a low-melting glass flame retardant system for PVC based on $ZnSO_4$ - K_2SO_4 - Na_2SO_4 [12]. This system proves to be an excellent char former and a smoke depressant. There were also tested the FR - low-melting - glasses systems with transition metals: Al, Ca, Ce, Ni, Mn, Co, V.

J. Gilman et al. has shown that silica gel in combination with potassium carbonate is an effective fire retardant for a wide variety of common polymers (at mass fraction of only 10 % total additive) such as polypropylene, nylon, polymethylmethacrylate, poly(vinyl alcohol), cellulose, and to a lesser extent polystyrene and styrene-acrylonitrile[13]. The Cone Calorimeter data indicated that the peak heat release rate is reduced about two times without significantly increasing the smoke or carbon monoxide levels during the combustion. The efficiency of the additives in each of these inherently non-char-forming thermoplastic polymers (PP, PS, Nylon-6,6 and PMMA) depends on the polymer (PP > Nylon-6,6 > PMMA > PS). The authors proposed a mechanism of action for these additives through the formation of a potassium silicate glass during the combustion. The pertinent phase diagrams do not show potassium silicate formation until 725°C. However, if sodium salts are present this temperature drops to 400°C - 500°C. Other work on inorganic glass forming fire retardants examined an analogous borate/carbonate system; B_2O_3/MCO_3. These formulations were found to form an inorganic glassy foam as a surface barrier which insulated and slowed the escape of volatile decomposition products[14].

2.3. Polymer Morphology Modification

The last representative of new trend in flame retardation is a modification of polymer physical structure (morphology) by means of polymer-polymer blends. We have found an extreme combustion behavior for a system polypropylene-polypropylene-co-polyethylene[15] which can be explained in terms of oxidative degradation of polymer materials. We studied the features of auto-oxidation and combustibility of blends of isotactic polypropylene (PP) and ethylene-propylene copolymers (PP-co-PE) as the thin films (50-80 µm). ASTM D2863 (ISO 4589, Part 2) was used to determine the oxygen index (LOI) for polymer films 140 by 52 mm.

The correlation between a polymer thermal oxidative degradation and its combustion under-diffusion flames condition may represent an interesting specific application. In general, the solid phase polymer reaction can play the very important role in reduction of polymer combustibility. If we can decrease the reaction ability of a polymer relatively to an oxygen, the critical conditions of the diffusion flame stability would change. Such a polymer will have a different fire behavior. The LOI method should be a sensitive one to these transformations.

The kinetics of oxygen consumption for PP/PP-co-PE compositions is shown in Figure 2. It is clearly seen that the composition of PP/PP-co-PE (62:38) has the highest induction period of auto-oxidation. In this circumstances it has been proposed a theoretical model of a preliminary oxidation localisation in interphase zone of a sample[16]. Apparently, the reaction ability of the compositions depends on the chemical structure of the interphase zone. It was shown that the increase of PP concentration in the PP/PP-co-PE composition from 38 to 62 % leads to the lowering of reaction ability of samples. The process of auto-oxidation begins from the most active ingredient of a polymer composition, PP or PP-co-PE. Interphase zone decelerates the polymer oxidation. Otherwise, we simply have a different polymeric system with the different kinetic parameters of an oxidation and different mechanism of the solid phase reactions.

The combustibility tests confirms this hypothesis (Table 2). The sample with the minimal reaction activity, PP/PP-co-PE (62:38), has the highest values of LOI (21) and the char yield (3.4 %).

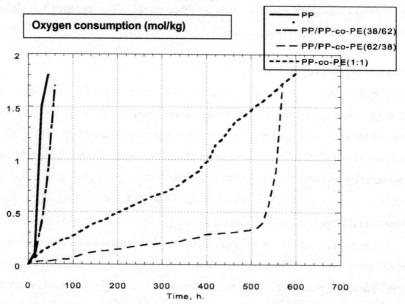

Figure 2. *Kinetic curves of the oxygen consumption vs. time for PP/PP-co-PE compositions.*

Table 2. *LOI and char yield for PP/PP-co-PE compositions.*

Composition	LOI, vol. %	Char yield, wt. %
PP	17	0
PP/PP-co-PE - 37,5/62,5	19,5	1,5
PP/PP-co-PE - 61,5/38,5	21	3,4
PP-co-PE	18,5	0,5

A correlation between LOI and a char yield for studied polymer compositions is found. This data indicate to the solid phase nature of the flame retardancy in PP/PP-co-PE compositions. The morphology modifications in the polymeric system lead to the desirable changes in combustibility.

References

1. E. Clausen, E. S. Lahaniatis, M. Behadir and D. Bienek, *Frezenius Z. Anal. Chem.*, 1987, **327**, 297.
2. D. Bienek, M. Behadir and F. Korte, *Heterocycles*, 1989, **28(2)**, 719; M. N. Pinkerton, R. J. Kociba, R.V. Petrella, D. L. McAllister, M. L. Willis, J. C. Fulfs, H. Thoma and O. Hutzinger, *Chemosphere*, 1989, **18(1-6)**, 1243; R. Dumler, D. Lenoir, H. Thoma and O. Hutzinger, *Chemosphere*, 1990, **20(10-12)**, 11867.
3. U. G. Ahlborg, G. C. Becking, L. S. Birnbaum, A. Brouwer, H. Derks, M. Freeley, G. Golor, A. Hanberg, J.C. Larsen, A. Liem, S.H. Safe, C. Schlatter, R. Waern, M. Younes, *Chemosphere*, 1994, **28**(6), 1049.
4. H. Beck, A. Dross, W. Mathar, *Environ. Health Perspect.*, **102** (Suppl 1), 173-185.
5. 5th Draft Status Report, *"OCDE Workshop on the Risk Reduction of Brominated Flame Retardant"*, Neufchatel, Switzerland, 26 May 1992 – 22-25 February 1993, OCDE – Direction de l'Environnement, April 1993.)
6. R. E. Alcock an K. C. Jones, *Environmental Science & Technology*, 1996, **30**(11), 3133.
7. U.S. Environmental Protection Agency, EPA/600, V. 1,2,3, Washington, DC, 1994.
8. G. E. Zaikov and S. M. Lomakin, *ACS Symposium Series, Fire and Materials*, Washington DC, 1995, **559**, 186.
9. G. E. Zaikov and S.M. Lomakin, *Polymer Degrad. and Stab.*, 1996, **54** (2-3), 223.
10. R. M. Hassan, *Polymer International*, 1993, **30**, 5

11. R. M. Hassan, S. A. El-Gaiar, A. M. El-Summan, *Polymer International*, 1993, **32**, 39.

12. W. J. Kroenke, *Journal Mat. Sci.*, **1986**, 21, 1123.

13. J. W. Gilman, S. J. Ritchie, T. Kashiwagi, S. M. Lomakin, *Fire and Materials*, 1997, **21**, 23-32.

14. R. E. Myers and E. Licursi, *J. Fire Sci.*, 1985, **3**, 415.

15. S. M. Lomakin, L. S., Shibryaeva, G. E. Zaikov, *Chemical Physics, (in Russian)*, 1997, in press.

16. L. S. Shibryaeva, A. A. Veretennikova, A. A. Popov, T. A. Gugueva, *Visokomolec. Soed. (In Russian)*, 1997, in press.

IDENTIFICATION OF CYANIDE IN GASEOUS COMBUSTION PRODUCTS EVOLVING FROM INTUMESCENT STYRENE - BUTADIENE COPOLYMER FORMULATIONS.

Y. Claire, E. Gaudin, C. Rossi, A. Périchaud

Laboratoire de Chimie Macromoléculaire, Université de Provence, 3,
Place V. Hugo, F-13331 Marseille Cedex 3, France.

J. Kaloustian

Laboratoire de Chimie Analytique, Faculté de Pharmacie, Université de la Méditerranée,
27, Boulevard J. Moulin, F-13385 Marseille, France.

L. El Watik

Unité de Chimie de l'Environnement, Faculté des Sciences et Techniques,
Université Moulay Ismaîl, BP 509, Boutalamine Errachidia, Morocco.

H. Zineddine

Laboratoire de Chimie-physique Appliquée , Faculté des Sciences,
Université Moulay Ismaîl, Zitoune Route d'El Hajeb, Méknès, Morocco.

1. INTRODUCTION

The most heavily used fire retardants for polymers have been the halogenated agents. Today, these products can be criticised on account of their high toxicity during the fire (release of halogenated acids). The replacement of these additives by intumescent systems seems to be a good choice. These systems are generally composed of a polyacid, a polyalcohol and a nitrogenated compound. In our case, we used the mixture: ammonium polyphosphate (APP), pentaerythritol (PER) and melamine (MEL), which was applied to the fireproof a styrene - butadiene copolymer[1].

However, the cyanide can be obtained in the gaseous combustion products at of high temperature. We have used a polarographic method with surimposed potential of constant amplitude for the detection and determination of such cyanide during the combustion of the copolymer styrene-butadiene intumescent mixture.

2. EXPERIMENTAL PROCEDURE.

2.1.Fire retardancy of the polymeric mixtures.

Samples used : HIPS (Lacqrène 7240 (as supplied by ATOCHEM), copolymer styrene – butadiene); APP (Hostaflam AP 422, as supplied by Hoechst), PER and MEL (as supplied by Hoechst).

Combustion studies were carried out in the ATS (code 100 4050) Limiting Oxygen Index (LOI) apparatus using 35 vol. % oxygen concentration. Tubing from the top of the chimney leads the evolved gases to the hydroxide solution which traps any evolved cyanide for subsequent analysis, see Figure 1; This gas exhaust system also has a glass-wool section to trap soot particles produced during the combustion. The amount of soot produced were determined by the increase in weight of the pre-weighed glass-wool.

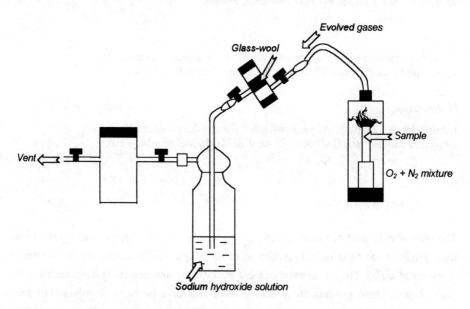

Figure 1. *LOI apparatus modified to pass evolved combustion products through a soot collector (glass-wool) and into sodium hydroxide solution.*

Each sample, 120 mm in length, 6 mm wide and 3 mm thick, was weighed before and after the experiment to determine the amount of actually burnt. About a 50 mm length of sample was burnt. Evolved cyanide absorbed in the 50 ml of 0.5 mol. dm^{-3} sodium hydroxide solution was determined polarographically.

The soot samples were tested for absorbed cyanide by placing them in 50 ml of sodium hydroxide solution. Subsequent analysis failed to detect any trace of cyanide. It was, therefore, concluded that all the evolved cyanide was in the gases passed into the sodium hydroxide trap.

2.2. Polarographic determinations.

The Tacussel apparatus is comprised of an impulsional polarograph PRG5, a EPL2 recorder, a TV11 GD electrical system and a CPRA polarographic Kit. The alkaline solutions collected from the LOI apparatus as previously described, are deoxygenated for 15 min. by sweeping with nitrogen sweeping.

Our method involves measuring the mercury limiting oxidation current in the presence of cyanide ion in the alkaline medium. This oxidation is carried out in aqueous sodium hydroxide at 0,5 mol. cm^{-1} in order to reduce the shift of the oxidation curve to negative potentials. Under these conditions the possible presence of halogen has no influence on the determination of cyanide ion, because of the higher stability of the cyanide complex of mercury in comparison with the halogenated complex which might be formed on the electrode only in a potential domain much higher then that corresponding the cyanide one. Figure 2 illustrates the classical polarograms obtained with continuous current for different cyanide concentrations. We observe the oxidation of the mercury catalysed by the cyanide through a two stages process which can be explained as follows:

- first for the most negative potentials, we see a anodic polarographic wave with the height limited by the diffusion of the cyanide ions, with a characteristic half-wave potential of -400mV/SCE. This wave corresponds to the formation of a mercury complex according to the equation (1) :

$$Hg + 4\ CN^- \rightarrow Hg\,(CN)_4^{2-} + 2\,e^- \qquad (1)$$

The dissociation constant of this mercuric complex is at 4.10^{-12}. This complex is more stable than the mercurous complex which is 100 time higher.

- during the recording of the polarogram curve, and in the case where the mercury electrode is at the highest potential, much more oxidation of the mercury is observed, allowing the formation of mercurous ions by association (2):

$$Hg^{2+} + Hg = Hg_2^{2+} \qquad \text{with } K = \ Hg_2^{2+}/Hg^{2+} = 166 \qquad (2)$$

Under these conditions, the formation of the mercurous complex leads to the appearance of a second mercury oxidation wave with a height also limited by the cyanide ion diffusion. From the dissociation constant values of the two complex and the standard tensions given in literature[2]:

$$e^0_{Hg/Hg^{2+}} = 0,85 \text{ V/NHE and } e^0_{Hg_2^{2+}/Hg^{2+}} = 0,92 \text{ V/NHE}$$

and from the association constant, the half-wave potential difference must agree with theoretical value of 130mV, which is what we observed in Figure 2.

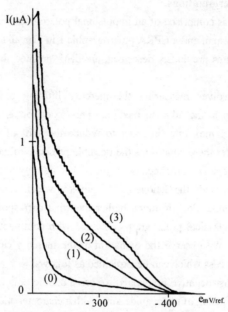

Figure 2. *Classical DC polarograph.*
(at first: (0) aqueous sodium hydroxide solution (0.5 mol. dm^{-3}; 20 dm^3); after cyanide ion addition: (1) $\Delta c = 5.10^{-6}$ mol. dm^{-3}; (2): $\Delta c = 1.10^{-5}$ mol. dm^{-3}; (3): $\Delta c = 1.5.10^{-5}$ mol. dm^{-3})

Table 1. *Example of peak current variation as a function of increased cyanide concentration (ΔC) data plotted in Figure 4.*

ΔC (mol. dm^{-3})	**Ipeak** (nA)	Correlation coefficient	C_0 (mol. dm^{-3})
0	25.5		
2.44 10^{-6}	35.0		
4.76 10^{-6}	44.0	0.9998	6.89 10^{-6}
6.97 10^{-6}	52.0		$\pm 3.1 \, 10^{-7}$
9.09 10^{-6}	59.9		
11.11 10^{-6}	67.1		

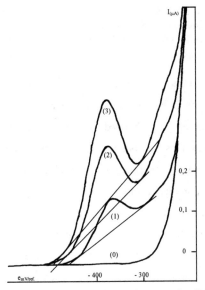

Figure 3. *Differential pulse Polarography (DPP).*
(at first: (0) aqueous sodium hydroxide solution (0.5 mol. dm⁻³; 20 dm³); after cyanide ion addition: (1) Δc = 5.10⁻⁶ mol. dm⁻³; (2): Δc = 1.10⁻⁵ mol. dm⁻³; (3): Δc = 1.5.10⁻⁵ mol. dm⁻³)

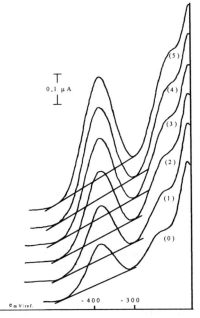

Figure 4. *Differential pulse Polarography (DPP).*

Experimental of Figure4 - *a cyanide solution is prepared by passing combustion gases from the sample into aqueous sodium hydroxide solution (0.5 mol. dm⁻³); at first: (0) before standard addition of cyanide; after cyanide ion additions: curves 1 to 5 correspond to successive and equal additions(500 μl) of standard KCN solution (1.10⁻⁴ mol. dm⁻³).*

In order to minimise the effect of capacitive current, we have replaced the classical polarographic method (Figure 2) by the method of constant amplitude pulsed polarography (Figure 3) . These two figures were obtained for the same solution and so can be directly compared. Under these conditions we obtained, for the formation of the mercury complex, a well defined peak with a height proportional to the cyanide ion concentration in the solution, in agreement with the polarographic theory.

An application of this method is described below for the determination of the cyanide ion obtained from the combustion of a sample containing PS (70 wt. %), APP (2,5 wt. %), PER (12,5 wt. %) and MEL (15 wt. %). The curve 0 in the Figure 4 corresponds to the sample with C_0 molar concentration, the curves 1 to 5 are obtained for increasing levels of additions of cyanide ions.

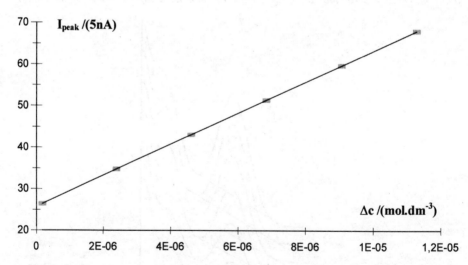

Figure 5 .*Plot of maximum of ion current against increasing cyanide concentration; data from Table 1 (DPP).*

The figure 5 plots the variation of the peak current due to the formation of a mercuric/cyanide complex with increasing of the concentration of cyanide additions. The straight lines represent the variation of the peak current with change in concentration ΔC: ($I_{peak} = 25{,}81 + 3{,}74.10^6.\Delta C$), the experimental points (table 1) fall in a straight line. The intercept on the y-axis corresponds to the unknown concentration C_0 of cyanide originally absorbed in the sodium hydroxide solution. In this particular case, $C_0 = (6{,}89 (+/-) 0{,}31).10^{-6}$ mol. dm^{-3}.

This result emphasises the feasibility of this method for determination of cyanide concentration. It is sensitive, right, selective, and perfectly reproducible. Moreover, this method can be applied without deoxygenating. Provided concentrations of the cyanide ion are greatest than or equal to 1.10^{-5} mol. dm^{-3}.

3. ANALYSIS OF THE RESULTS.

For the 70% polystyrene content, Table 2 presents all the results determinated by the polarographic method. These are plotted in Figure 6: the x-axis represents the percentage of released cyanide with respect to the original weight M_T of the sample actually burnt, the y-axis represents the percentage of APP, PER or MEL, versus M_T. This Figure illustrates the levels of cyanide evolved as a function of varying composition of the additives in the mixture.

Table 2. *Variation of the percentage of the evolved cyanide gas (referred to the initial mass of the sample actually burnt as a function of the intumescent mixture composition (containing PS 70 wt. %).*

PS (70 wt. %)			
MEL (wt. %)	**PER (wt. %)**	**APP (wt. %)**	**CN⁻ (mol. dm^{-3})**
0.0	5.0	25.0	$3.6\ 10^{-5}$
0.0	0.0	30.0	$1.0\ 10^{-4}$
0.0	25.0	5.0	$2.6\ 10^{-4}$
0.0	15.0	15.0	$4.7\ 10^{-4}$
15.0	2.5	12.5	$1.1\ 10^{-3}$
15.0	12.5	2.5	$1.4\ 10^{-3}$
30.0	0.0	0.0	$1.6\ 10^{-3}$

Thus without melamine, the data in Table 2 show an increase in the formation of cyanide, which although small, is directly related to an increasing of APP from 25 to 30 wt. %. With the APP at 5 wt. %, an increase in PER (25 wt. %) again results in increasing in cyanide. The largest amount of cyanide obtained in the absence of was for a ratio PER/APP equal to 1.

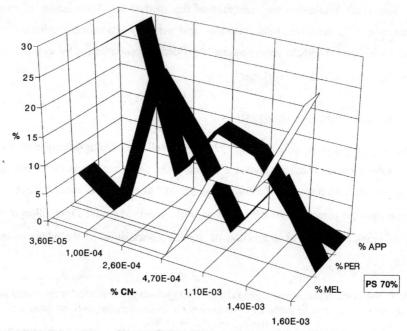

Figure 6. *Variation of the cyanide percentage in the evolved combustion gas (referred to the initial mass of the sample actually burnt as a function of the intumescent mixture composition (containing PS 70 wt. %).*

From this, it can be seen that that the generation of cyanide depends on the amount of fireproofing additive and the stoichiometric ratio PER/APP. These results show clearly the important role played by the PER in the combustion mechanism. During the esterification reaction, most of the APP reacts with the excess of PER. The greater the amount of ester formed, the greater is the amount of cyanide generated. The nitrogenated ions are responsible of the cyanide emission, even in absence of PER, which, if present in high quantity will react with the polyphosphate in accordance with the "stoichiometric" equation.

In the presence of MEL (30 wt. %) and in absence of APP and PER, cyanide is produced in greater quantities than all the other mixtures studied. The combustion of PS-MEL shows that it is not a good fire retardant mixture but is a good producer of cyanide. For a same quantity of MEL (15 wt. %), we obtained the most cyanide with a low percentage of APP (2.5 wt. %) and 12.5 wt. %of PER. This increase is due in a large part to the structure of the melamine molecule (3 groups NH_2 and 3 nitrogen atoms) and the importance of the role played by the PER in the combustion reaction: the more the nitrogenated ions (provided from the polyphosphate and melamine), the greater the amount of cyanide. This increase is directly related to the presence of the PER.

We have reported in the table 3 and illustrated by the figure 7, the increasing amounts of cyanide for the different PS compositions studied.

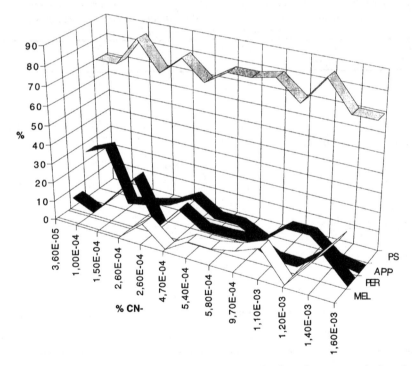

Figure 7. *Variation of the cyanide percentage in the evolved combustion gas (referred to the initial mass of the sample actually burnt) as a function of the intumescent mixture composition and different PS percentages.*

Cyanide ions form when the polymer content is high and the initial amount of intumescent additive is low. The formation of these ions may be in that case explained by the reaction of the evolved gas, in the bright zones of the flame, in the endothermic zones (pyrolysis without oxygen) or in the low temperature ignition zones.

If we compare the previously obtained values of the LOI [1], it can be seen that the maximum is obtained for the ratio PER/APP = 1 and 30 wt. % intumescent mixture. This may be explained by the much longer ignition time (82s), reduced soot emission (90 mg) but with a carbonaceous coating (char) which is thicker and more compact. However these exist a slight risk in the use this composition. The cyanide concentration is higher than the others in the intumescent mixtures with the same PS content.

Table 2. *Variation of the percentage of the evolved cyanide gas (referred to the initial mass of the sample actually burnt as a function of the intumescent mixture composition and different PS percentages.*

CN⁻ mol. %	MEL wt. %	PER wt. %	APP wt. %	PS wt. %	LOI vol. %
$3.6\ 10^{-5}$	0	5	25	70	22.7
$1.0\ 10^{-4}$	0	0	30	70	19.2
$1.5\ 10^{-4}$	0	12.5	2.5	85	21.2
$2.6\ 10^{-4}$	0	25	5	70	20.2
$2.6\ 10^{-4}$	10	1.7	8.3	80	20.1
$4.7\ 10^{-4}$	0	15	15	70	28.0
$5.4\ 10^{-4}$	7.5	7.5	7.5	77.5	21.2
$5.8\ 10^{-4}$	7.5	7.5	7.5	77.5	21.2
$9.7\ 10^{-4}$	10	8.3	1.7	80	19.6
$1.1\ 10^{-3}$	15	2.5	12.5	70	21.6
$1.2\ 10^{-3}$	0	2.5	12.5	85	22.4
$1.4\ 10^{-3}$	15	12.5	2.5	70	20.2
$1.6\ 10^{-3}$	30	0	0	70	20.6

The cyanide formation [3-8] can be explained by active carbonaceous radical formation in either the luminous (1200°C) or oxygen free pyrolysis (1000°C) zones. We can also advance the hypothesis of the presence, in small amounts, of nitrogenated radicals originating from the polyphosphate or melamine which might react to form cyanhidric acid. The distribution of such active radicals is dependant of the flame temperature[5] and the ratio of PER and APP.

4. CONCLUSION

The cyanide species generated from the combustion of intumescent fire retardant polystyrene, in variable ratio of PER/APP/MEL were measured by a polarographic method. The variation of the intumescent content in the fireproof polymer and chiefly of the MEL, causes the variation of cyanide content in the gaseous combustion products. Likewise, the closer the PER percentage is to the stoichiometric conditions, the greater is its effect on the amount of cyanide evolved. We propose that the cyanide is formed from nitrogenated free radicals produce from the APP, the MEL and carbonaceous free radicals liberated from the polymeric skeleton.

References

1. L. El Watik, P. Antonetti, Y. Claire, H. Zineddine, M. Sergent, J. Kaloustian, C. Rossi and A. Périchaud, *J. Therm. Anal.*, under press.
2. M. Wendell and Latimer. '*The Oxidation States of the Elements and their Potentials Inaqueous Solutions*', Printice-Hall Inc., Englewood Cliffs, New York, 1964, p 179.
3. G. Camino, L. Costa, M. P. Luda., *Fire Retardant Polymers*, (1993), 74, 71.
4. N. I. Sax, "*Dangerous Properties of Industrial Materials*", 6th Edition, Van Nostrand, Reinhold Company, New York, 1984.
5. G. Camino, L. Costa, L. Trossarelli, *Polym. Deg. Stab.*, (1984), 7, 221.
6. J. E. Stephenson ,in "*Proceedings Fire Retardants*", Wiley Interscience, New York, 1970.
7. G. Bertelli, G. Braca, O. Cicchetti, A. Pagliari, in "*New Flame Retardants in the International Marketplace, 4ème BBC Conference on Flame Retardancy*", Stanford, Connecticut, 18-20 May 1993.
8. D. Scharf, "*Environmentally Friendly Fire Systems Conference*", Cleveland, Ohio, 22-23 Sept. 1992.

Acknowledgement

The Authors thank Dr. Dennis Price (Science Research institute, University of Salford, UK) for helpful discussions and improvement of this paper.

Subject Index